Studies in Logic
Volume 26

Philosophical Aspects of Symbolic Reasoning in Early Modern Mathematics

Studies in Logic Series Editor
Dov Gabbay   dov.gabbay@kcl.ac.uk

Philosophical Aspects of Symbolic Reasoning in Early Modern Mathematics

Edited by

Albrecht Heeffer

and

Maarten Van Dyck

ISBN 978-1-84890-017-2

College Publications
Scientific Director: Dov Gabbay
Managing Director: Jane Spurr
Department of Computer Science
King's College London, Strand, London WC2R 2LS, UK

http://www.collegepublications.co.uk

Original cover design by orchid creative   www.orchidcreative.co.uk
Printed by Lightning Source, Milton Keynes, UK

Foreword

This book presents a selection of peer-reviewed papers which were presented on a conference organized in Ghent, Belgium, from 27 till 29 August, 2009. The conference was given the title *Philosophical Aspects of Symbolic Reasoning in Early modern Science and Mathematics* (PASR). For this book we selected papers which deal with the consequences for mathematics in particular, hence the omission of 'science' in the title of this book. Another selection, dealing with the understanding of nature and a broader range of topics, will appear in the journal *Foundations of Science*.

The conference was sponsored by the Research Foundation Flanders (FWO) and Ghent University, which indirectly made this book possible. We also have to thank the other members of the programme committee Marco Panza, Chikara Sasaki, and Erik Weber and our keynote speakers Jens Høyrup, Doug Jesseph, Eberhard Knobloch, Marco Panza, Mathias Schemmel and Michel Serfati. Five of their papers are included in this volume. Most of the papers benefited from valuable and sometimes substantive comments by our referees which must remain anonymous. Special thanks to Michael Barany who assisted in the editorial process.

Preface

The novel use of symbolism in early modern science and mathematics poses both philosophical and historical questions. The historical questions evidently are *when* and *how* symbolism was introduced into mathematics. Often François Viète is considered to be the father of symbolic algebra. But how we should then understand the centuries of algebraic practice before Viète? The abbaco tradition applied algebra to the solution of merchant problems on exchange, bartering, partnership, allegation of metals, etc., since the beginning of the fourteenth century. Some sort of symbolism was emerged within that tradition and was fully in place during the sixteenth century. Is there a fundamental difference in mathematical practice before and after Viète?

The philosophical questions relate to the nature of such symbolism and its impact on mathematical reasoning and early-modern understanding of knowledge. Is the use of short-hand notations and abbreviations the same as symbolism? Or we should understand symbolism as involving a more intricate model of reasoning, different from geometrical or arithmetical reasoning? So, what precisely do symbolic representations contribute to mathematical reasoning?

Against this background, it is striking that at the beginning of the seventeenth century, the idea took ground that there might be a universal symbolic language which facilitates the representation of all reasoning in a clear and distinct way. In what way does the idea of a *mathesis universalis* or a *characteristica universalis* depend on the symbolization of mathematics? To what extent was the project of devising such new language ever achieved?

Of course, not all our questions could be answered over the course of a three days conference, let alone on the limited number of pages of the current volume. However, a representative state of the art is here provided on three main themes:

- **The development of algebraic symbolism.** Our first three contributions cover a consecutive period of historical events from the beginning of the introduction of Arabic algebra into Europe (thirteenth century), through the abbaco period (1300-1500), the sixteenth century, all the way up to Leibniz. Taking up a diplomatic stance towards the precise meaning of "symbolism", Jens Høyrup meticulously traces the development of notations for different mathematical objects, from formal fractions over the powers of the unknown to the confrontation sign, or what later would become the equation. He attributes a distinctive role to Maghreb practices on early European notations (e.g. the fraction bar). Despite a continuous process in abbaco manuscripts towards more intricate symbolism, Høyrup concludes that this whole development was one which was neither understood, nor intended by the participants in the process.

Albrecht Heeffer picks up where Høyrup concluded his analysis – with the German cossic tradition – and continues with the innovations by Cardano and the French humanists. He shows how one particular representational difficulty – an unambiguous symbolism for multiple unknowns – shaped the very concept of an equation. The symbolic representation of conditions involving multiple unknowns facilitated the process of substitution and operations on equations. According to Heeffer, it is precisely because operations on equations and between equations became possible that the equation became a mathematical object and hence the corresponding concept developed. Challenging the generally accepted view of Viète as the father of symbolic algebra, he argues that the development of algebraic symbolism was a gradual process involving many minor achievements by several actors.

Starting by formulating six functional criteria for symbolic representations, Michel Serfati discusses the contribution by Viète and Descartes against the background of earlier achievements by Cardano and Stifel. He elaborates on two of these patterns: the dialectic of indeterminacy and the representation of compound concepts. The first contributed to the concept of an indeterminate, the second to one of the most essential operations in symbolic mathematics: substitution. Where the development of symbolism in the abbaco period was an unconscious process for the participants according to Høyrup and in the sixteenth century a technical struggle of representation for Heeffer, for Serfati it became no less than a symbolic revolution in the seventeenth century: "one of the major components of the scientific revolution".

- **The interplay between diagrams and symbolism**. Diagrams and early symbolism both added non-discursive elements to mathematical texts. Both functioned as additional sources of epistemic justification to the argumentative and rhetorical structure of the text. Four contributions deal with the interactions between these two. Michael Barany's paper deals with

English translations of Euclid during the sixteenth century. But 'translating' here has a double meaning. Not only was Euclid made accessible to a
broad public where it previously was 'locked up in straunge tongues'", these
translators also provided a new context which was dramatically different
from the original Euclid. Barany focuses on the variety of diagrammatic
approaches by five authors who published between 1551 and 1571: Robert
Recorde, Henry Billingsley, John Dee and Leonard and Thomas Digges. He
shows how different representational strategies and pedagogical views led
to equally different notions of what constitutes a diagram. The notion of a
point in a diagram is one such example.

Maria Rosa Massa Esteve takes up the work *Cursus mathematicus* by the
enigmatic author Pierre Hérigone, demonstrating how Euclid's *Elements*
became rendered into a purely symbolic language. Hérigone's ambitions
clearly show how symbolism had changed mathematics in the seventeenth
century: "I have devised a new method, brief and clear, of making demonstrations, without the use of any language". He devised his own set of notations, including a terse format for referring to propositions, lemmas and
corollaries, with the intention of not only representing objects of mathematics but the very process of axiomatic-deductive reasoning. Although
he did not find any followers in his idiosyncratic system, his whole enter-
prize is exemplary for the further development of mathematics during the
seventeenth century.

While most contributors to this book take the explicit (Heeffer and Serfati)
or less explicit position that the development of symbolism was responsible for the transformation of mathematics during the seventeenth century,
Marco Panza challenges this view. Starting from a classical construction
problem, proposition VI.30 of the *Elements*, he argues that a conceptual
transformation occurred, independent from developments in symbolic representations. This transformation took place already in the Arabic works
of al-Khwārizmī and Thābit ibn Qurra who conceived the same problem
as a configuration of pure quantities. According to Panza it was this shift
in conception that functioned as a necessary condition for the application
of the literal formalism of early modern algebra in a purely syntactic way.
Where Euclid's solution to the proposition is entirely diagrammatic, literal formalism exploits the purely quantitative aspect of such construction
problems.

A fourth contribution to the relation of algebra and geometry, or the interplay between diagrams and symbolism, is the discussion of Bombelli's
algebra linearia by Roy Wagner. Where previous chapters deal mostly
with the symbolic interpretations of geometrical problems, Wagner analyzes Bombelli's geometrical representation of algebra, which became a

crucial issue for the justification of algebraic procedures. One central problem in this practice is homogeneity: to keep the constructed geometrical objects invariant with respect to the unit of measurement. Wagner shows how Bombelli's strategy of regimenting representations enabled him to go well beyond the limits of Cardano's approach. Bombelli's *algebra linearia* explored the troublesome relations between algebra and geometry during the sixteenth century. It was a decisive step towards rigorous practices to impose well-regimented relations between the two, on which post-Cartesian analytic geometry would depend.

- **Mathesis universalis** and **charateristica universalis**. The final part of the book deals with developments and refections on symbolism in the later half of the seventeenth century. After the initial achievement of symbolic algebra – for which Descartes' *Geometry* stands as a milestone – methodological discussions arose on the all-encompassing role of a symbolic language for all 'scientific' reasoning, the notion known as *Mathesis universalis*. Doug Jesseph distinguishes two camps, which he calls algebraic and geometric foundationalists. The first group, consisting of Descartes, Wallis and others, considered algebra as the foundation of all mathematics. They were met with skepticism by the geometric foundationalists, such as Hobbes, who scorned them for representing only a "scab of symbols", ignoring the real contents of mathematics, such as quantity, measure and proportion. Such a discussion is now absent in the philosophy of mathematics. For Jesseph this is a nice illustration of how foundational issues get relocated to other contexts. The opposition was replaced by one on the different views on the new calculus at the end of the seventeenth century.

A *charateristica universalis* in which all problems are represented in a symbolic language and resolved by calculation, is Leibniz's version of algebraic foundationalism. Eberhard Knobloch describes the toolbox that Leibniz created to fulfill that aim: the *ars characteristica* or the art of inventing suitable characters and signs, the *ars combinatoria* or the art of combination, and the *ars inveniendi* for inventing new theorems and methods. He shows that it is not without reason that Cajori called him "the master-builder of mathematical notations". With well chosen examples, Knobloch demonstrates how Leibniz builds layers of symbolic representations to tackle advanced problems in differential equations, power sums and elimination theory.

Ghent, Belgium   *Albrecht Heeffer*
15 September, 2010   *Maarten Van Dyck*

List of Contributors

- **Jens Høyrup** is Professor emeritus at Roskilde University, Denmark, Section for Philosophy and Science Studies. Much of his research has dealt with the cultural and conceptual history of pre-Modern mathematics, including in particular Babylonian and practitioners' mathematics. In recent years he has worked on the abbacus tradition of thirteenth-fifteenth-century Italy. His book *Jacopo da Firenze's Tractatus Algorismi and Early Italian Abbacus Culture* (Science Networks, 34. Basel etc.: Birkhäuser, 2007) appeared exactly 700 years after Jacopo's treatise. e-mail: `jensh@ruc.dk`
- **Albrecht Heeffer** is post-doctoral fellow of the FWO (Research Foundation Flanders), affiliated with the Center of History of Science at Ghent University in Belgium. He publishes on the history of mathematics and the philosophy of mathematical practice with a special interest in cross-cultural influences. Albrecht has been a visiting fellow at Kobe University in Japan in 2008, The Center of Research in Mathematics Eduction at Khon Kaen University in Thailand in 2009 and at the Sydney Center for the Foundation of Science in Australia in 2011 where he will prepare book on the mathematical and experimental practices of *Récréations Mathématiques* (1624). e-mail: `albrecht.heeffer@ugent.be`
- **Michel Serfati** is Honorary Professor holding a Higher Chair of Mathematics in Paris. He has been for many years the head of the seminar on Epistemology and History of Mathematical Ideas held at the Institut Henri Poincaré, University de Paris VII. He has organized many conferences on the history and philosophy of mathematics and is the author and editor of works in both disciplines. His most recent books are *La Révolution symbolique. La constitution de l'écriture symbolique mathématique* (2005), De la Méthode (2003), *Mathématiciens français du XVIIème siècle. Pascal. Descartes. Fermat* (co-edition 2008). He is preparing a book on Descartes' mathematical work. He holds doctorates in mathematics and philosophy. In

mathematics, his research concerns the algebraic supports of multi-valued logics (i.e Post algebras). In philosophy, his work focuses on the philosophy of mathematical symbolic notation as a part of the philosophy of language. His research also deals with various aspects of the history of mathematics in the 17th century, as a specialist in Descartes and Leibniz. He also worked in the history and philosophy of contemporary mathematics (Category Theory, rising of spectral methods, Emil Post's early work in Logic). e-mail: `serfati@math.jussieu.fr`

- **Michael Barany** is a first-year PhD student in the Program in History of Science at Princeton University, where he studies the material and social history of mathematical rigor. Focusing on forms and manifestations of mathematical witnessing, intuition, and self-evidence, his research topics include Early Modern translations of Euclid, Victorian views on the prehistory of numbers and measures, and the development of the mathematical theory of distributions over the last half-century. e-mail: `mbarany@princeton.edu`, web site: http://www.princeton.edu/mbarany.

- **M. Rosa Massa-Esteve** (Palamós (Girona), 1954) is professor aggregate of Mathematics and History of Science and Technology at the Polytechnical University of Catalonia (UPC). She has published numerous articles on sixteenth and seventeenth century mathematics in Catalonia, in Spain and in international journals. She is teaching history of algebra at the Doctoral Programme of the Autonomous University of Barcelona. She has organized many symposiums and workshops on several aspects of the history of mathematics. Currently, she is preparing a book on Mengoli's mathematical work. Her most recent publication (with Amadeu Delshams) is "Euler's beta integral in Pietro Mengoli's works", *Archive for History of Exact Sciences*, (2009) 63, p. 325–356. e-mail: `m.rosa.massa@upc.edu`.

- **Marco Panza** is research director at the CNRS (IHPST, Paris). He is a historian and philosopher of mathematics and has especially worked on early-modern mathematics and analytical mechanics. His books include *Newton et les origins de lanalyse: 1664-1666*, Blanchard, Paris, 2005. He is co-founder of the recently established Association for the Philosophy of Mathematical Practice.

- **Roy Wagner** received a Ph.D. from Tel Aviv University's maths department in 1997, and ten years later a Ph.D. the same university's Cohn Institute for the History and Philosophy of Science and Ideas. He publishes papers in mathematics, history and philosophy of mathematics, and cultural studies. Roy Wagner held visiting positions in Paris VI, Cambridge University, Boston University, the Max Planck Institute for the History of Science and the Hebrew University's Edelstein Center. He is currently a senior lecturer at the school of computer science in the Academic College of Tel-Aviv-Jaffa and a Buber fellow at the Hebrew university in Jerusalem.

- **Douglas M. Jesseph** is Professor of Philosophy at the University of South Florida. He is the author of several books and numerous articles on seventeenth and eighteenth century philosophy, mathematics, and methodology, including *Berkeley's Philosophy of Mathematics* and *Squaring the Circle: The War between Hobbes and Wallis.* He is currently editing three volumes of Hobbes's mathematical works for the *Clarendon Edition of the Works of Thomas Hobbes.* e-mail: `dmjesseph@mac.com`
- **Eberhard Knobloch**, born in 1943, professor emeritus of history of science and technology at the Technische Universität Berlin and academy professor at the Berlin-Brandenburgische Akademie der Wissenschaften (BBAW, former Prussian Academy of Sciences). President of the International Academy of the History of Science (Paris) and Past president of the European Society for the History of Science; research interests: history and philosophy of mathematical sciences and their applications. Project leader of the research group 'Alexander von Humboldt', of series 4 (Political writings) and 8 (Scientific, technical, medical writings) of the academy edition of Leibniz's complete writings and letters at the BBAW. e-mail: `eberhard.knobloch@mailbox.tu-berlin.de`

Contents

**3   Symbolic revolution, scientific revolution: mathematical
     and philosophical aspects** 103
Michel Serfati

Part II   The interplay between diagrams and symbolism

4   Translating Euclid's diagrams into English, 1551–1571 125
Michael J. Barany

**5   The symbolic treatment of Euclid's *Elements* in
     Hérigone's *Cursus mathematicus* (1634, 1637, 1642)** 165
Maria Rosa Massa Esteve

Part III *Mathesis universalis* **and** *charateristica universalis*

List of Figures

Part I
The development of algebraic symbolism

Chapter 1
Hesitating progress – the slow development toward algebraic symbolization in abbacus-and related manuscripts, c. 1300 to c. 1550

Jens Høyrup

Ian Mueller in memoriam

Abstract From the early fourteenth century onward, some Italian Abbacus manuscripts begin to use particular abbreviations for algebraic operations and objects and, to be distinguished from that, examples of symbolic operation. The algebraic abbreviations and symbolic operations we find in German *Rechenmeister* writings can further be seen to have antecedents in Italian manuscripts. This might suggest a continuous trend or perhaps even an inherent logic in the process. Without negating the possibility of such a trend or logic, the paper will show that it becomes invisible in a close-up picture, and that it was thus not understood – nor intended – by the participants in the process.

Key words: Abbacus school, Algebra, Symbolism

1.1 Before Italy

Ultimately, Italian abbacus algebra[1] descended from Arabic algebra – this is obvious from its terminology and techniques. I shall return very briefly to some of the details of this genealogy – not so much in order to tell what

---

[1] The "abbacus school" was a school training merchant youth and a number of other boys, 11-12 years of age, in practical mathematics. It flourished in Italy, between Genoa-Milan-Venice to the north and Umbria to the south, from c. 1260 to c. 1550. It taught calculation with Hindu numerals, the rule of three, partnership, barter, alligation, simple and composite interest, and simple false position. Outside this curriculum, many of the abbacus books (teachers' handbooks and notes, etc.) deal with the double false position, and from the fourteenth century onward also with algebra.

happened as to point out how things *did not* happen; this is indeed the best we can do for the moment.

First, however, let us have a look at Arabic algebra itself under the perspective of "symbolism".[2]

The earliest surviving Arabic treatise on the topic was written by al-Khwārizmī somewhere around the year 820.[3] It is clear from the introduction that al-Khwārizmī did not invent the technique: the caliph al-Maʿūn, so he tells, had asked him to write a compendious introduction to it, so it must have existed and been so conspicuous that the caliph knew about it; but it may have existed as a technique, not in treatise form. If we are to believe al-Khwārizmī's claim that he choose to write about what was subtle and what was noble in the art (and why not believe him?), al-Khwārizmī's treatise is likely not to contain everything belonging to it but to leave out elementary matters.

It is not certain that al-Khwārizmī's treatise was the first of its kind, but of the rival to this title (written by the otherwise little known ibn Turk) only a fragment survives (ed. Sayılı, 1962). In any case it is clear that one of the two roughly contemporary treatises has influenced the other, and for our purpose we may take al-Khwārizmī's work to represent the beginning of written Arabic algebra well.

Al-Khwārizmī's algebra (proper) is basically a rhetorical algebra. As al-Khwārizmī starts by saying (ed. Hughes, 1986, p. 233), the numbers that are necessary in *al-jabr wa'l-muqābalah* are *roots*, *census* and simple numbers. *Census* (eventually *censo* in Italian) translates Arabic *māl*, a "possession" or "amount of money", the *root* (*radix/jidhr*, eventually *radice*) is its square root. As al-Khwārizmī explains, the root is something which is to be multiplied by itself, and the *census* that which results when the root is multiplied by itself; while the fundamental second-degree problems (on which presently) are likely to have *originated* as riddles concerned with a real amount of money and its square root (similar to what one finds, for instance, in Indian problem collections),[4] we see that the root is on its way to take over the role as basic unknown quantity (but only on its way), whereas "dirham" serves in

---

[2] I shall leave open the question of what constitutes an algebraic "symbolism", and adopt a fairly tolerant stance. Instead of delimiting by definition I shall describe the actual character and use of notations.

[3] The treatise is known from several Arabic manuscripts, which have now appeared in a critical edition (Rashed, 2007), and from several Latin translations, of which the one due to Gherardo of Cremona (ed. Hughes, 1986) is not only superior to the other translations as a witness of the original but also a better witness of the original Arabic text than the extant Arabic manuscripts as far as it goes (it omits the geometry and the chapter on legacies, as well as the introduction) – both regarding the grammatical format (Høyrup, 1998) and as far as the contents is concerned (Rashed, 2007, p. 89).

[4] Correspondingly, the "number term" is originally an amount of *dirham* (in Latin *dragmata*), no pure number.

al-Khwārizmī's exposition simply as the denomination for the number term, similarly to Diophantos's *monás*. In the first steps of a problem solution, the basic unknown may be posited as a *res* or *šayʾ*, "a thing" (*cosa* in Italian); but in second-degree problems it eventually becomes a *root*, as we shall see.

As an example of this we may look at the following problem (ed. Hughes, 1986, p. 250):[5]

> I have divided ten into two parts. Next I multiplied one of them by the other, and twenty-one resulted. Then you now know that one of the two sections of ten is a thing.[6] Therefore multiply that with ten with a thing removed, and you say: Ten with a thing removed times a thing are ten things, with a *census* removed, which are made equal to twenty-one. Therefore restore ten things by a census, and add a *census* to twenty-one; and say: ten things are made equal to twenty-one and a *census*. Therefore halve the roots, and they will be five, which you multiply with itself, and twenty-five results. From this you then take away twenty-one, and four remains. Whose root you take, which is two, and you subtract it from the half of the things. There thus remains three, which is one of the parts.

This falls into two sections. The first is a rhetorical-algebraic reduction which more or less explains itself.[7] There is not a single symbol here, not even a Hindu-Arabic numeral. The second section, marked in sanserif, is an unexplained algorithm, and indeed a reference to one of six such algorithms for the solution of reduced and normalized first- and second-degree equations which have been presented earlier on.

Al-Khwārizmī is perfectly able to multiply two binomials just in the way he multiplies a monomial and a binomial here; slightly later (ed. Hughes, 1986, p. 249) he states that "ten with a thing removed" multiplied by itself yields "hundred and a *census* with twenty things removed". He would thus have no difficulty in finding that a "root diminished by five" multiplied by itself gives a "*census* and twenty-five, diminished by ten roots". But he cannot go the other way, the rhetorical style and the way the powers of the unknown are labeled makes the dissolution of a trinomial into a product of two binomials too opaque either for al-Khwārizmī himself or for his "model reader". In consequence, when after presenting the algorithms al-Khwārizmī wants to give proofs for these, his proofs are geometric, not algebraic – geometric proofs not of his own making (as are his geometric illustrations of how to deal with binomials), but that is of no importance here.

It is not uncommon that rhetorical algebra like that of al-Khwārizmī is translated into letter symbols, the *thing* becoming x and the *census* becoming

---

[5] My translation, as everywhere in the following when no translator into English is identified.

[6] This position was already made in the previous problem about a "divided ten".

[7] However, those who are already somewhat familiar with the technique may take note of a detail: we are to restore ten things with a *census*, *and then* add a *census* to 21. "Restoring" (*al-jabr*) is thus not the addition to both sides of the equation (as normally assumed, in agreement with later usage) but a reparation of the deficiency on that side where something

x^2. The above problem and its solution thereby becomes

$$\begin{cases} 10 = x + (10 - x) \\ x(10 - x) = 21 \end{cases}$$

$$10x - x^2 = 21$$

$$10x = 21 + x^2$$

$$x = \frac{10}{2} - \sqrt{\left(\frac{10}{2}\right)^2 - 21}$$

To the extent that this allows us to follow the steps in a medium to which we are as accustomed as the medieval algebraic calculators were to the use of words, it may be regarded as adequate. But only to this extent: the letter symbolism makes it so much easier to understand the dissolution of trinomials into products that the need for geometric proofs becomes incomprehensible – which has to do with the theme of our meeting.

Geometric proofs recur in many later Arabic expositions of algebra – not only in Abū Kāmil but also in al-Karajī's *Fakhrī* (Woepcke, 1853, pp. 65–71), even though al-Karajī's insight in the arithmetic of polynomials[8] would certainly have allowed him to offer purely algebraic proofs (his *Al-Badīʿ* explicitly shows how to find the square root of a polynomial (ed. Hebeisen, 2008, p. 117–137)). What is more: he brings not only the type of proof that goes back to al-Khwārizmī but also the type based directly on *Elements* II (as introduced by Thābit ibn Qurrah, ed. (Luckey, 1941)).

Some Arabic writers on algebra give no geometric proofs – for instance, ibn Badr and ibn al-Bannāʾ. That, however, is because they give no proofs at all; algebraic proofs for the solution of the basic equations are absent from the entire Arabic tradition.[9]

This complete absence is interesting by showing that we should expect no direct connection between the existence of an algebraic symbolism and the creation of the kind of reasoning it seems with hindsight to make possible. It has indeed been known to historians of mathematics since Franz Woepcke's work

---

is lacking; this is followed by a corresponding addition to the other side.

[8] Carried by a purely rhetorical exposition, only supplemented by use of the particle *illā* ("less") – still a word, but used contrary to the rules of grammar in the phrase *wa illā*, "and less" – to mark a subtractive contribution. As pointed out by Abdeljaouad (2002, p. 38), this implies that *illā* has become an attribute (namely subtractivity) of the number.

[9] An interesting variant is found in ibn al-Hāʾim's *šarḥ al-Urjūzah al-Yasmīnya*, "Commentary to al-Yāsamin's *Urjuza*" (ed., trans. Abdeljaouad, 2004, pp. 18f). Ibn al-Hāʾim explains that the specialists have a tradition for giving geometric proofs, by lines (viz, as Thābit) or by areas (viz, as al-Khwārizmī), which however presuppose familiarity with Euclid. He therefore gives an arithmetical argument, fashioned after *Elements* II.4. For use of this theorem he is likely to have had precursors, since Fibonacci also seems to model his first *geometric* proof after this proposition (ed. Boncompagni, 1857, p. 408) (his second proof is "by lines").

فإذا قيل لك اضرب ثمانية أشياء إلا أربعة من العدد في ستة أموال إلا ثلاثة

أشياء فأنزل ذلك هكذا $|8^{ش}$ الا 4 في 6 ^{م} الا $3^{ش}|$.

ثم اضرب الثمانية في الستة يخرج لك ثمانية وأربعون (171) كعبا ، لأن أس
المضروبين $6^{م}$، و$8^{ش}$ ثلاثة ، احفظها أولا ، ثم اضرب الثمانية في الثلاثة يخرج لك
أربعة وعشرون مالا ، وهو ناقص ، لأنه من ضرب زائد في ناقص ، احفظه بحرف
الاستثناء ، ثم اضرب الأربعة في الستة يخرج لك اربعة وعشرون مالا ناقصا أيضا ،
ضعه مع نظيره ، ثم اضرب أيضا الاربعة في الثلاثة يخرج لك اثنا عشر شيئا زائدا
لأنه من ضرب [و33/] ناقص في مثله ، اجعله مع المحفوظ الأول ، فيكون الخارج
اثني عشر شيئا وثمانية واربعين كعبا إلا ثمانية واربعين مالا هكذا : $|12^{ش}$ $48^{ك}$ الا|

$|^{م}48|$

Fig. 1.1: Al-Qalaṣādī's explanation of how to multiply "8 things less 4" by
"6 census less 3 things" in Souissi's edition (1988, p. Ar. 96) – symbolic
notations in frames (added here).

in (1854) that elements of algebraic symbolism were present in the Maghreb,
at least in the mid-fifteenth century (they are found in al-Qalaṣādī's *Kašf*,[10]
but also referred to by ibn Khaldūn). Woepcke points to symbols for powers of the unknown and to signs for subtraction, square root and equality;
symbols for the powers[11] are written above their coefficient, and the root

---

[10] The use of the symbols can thus be seen in Mohamed Souissi's edition of he Arabic text
(1988). His translation renders the same expressions in post-Cartesian symbols; edition as
well as translation change the format of the text (unless this change of format has already
taken place in the manuscript he uses, which is not to be excluded). Woepcke's translation
(1859) renders the formulae more faithfully (using **K** for the cube, **Q** for the square and **C**
for the unknown itself), and also renders the original format better (putting the symbolic
notations outside the text). Figures 1.1 and 1.2 confront Woepcke's translation with Souissi's
Arabic text.

[11] There are individual signs for the *thing*, the *census* and the *cube*. Higher powers are represented by products of these (the fifth power thus with the signs for *census* and *cube*, one
written above or in continuation of the other, corresponding to the verbal name *māl kaᶜb*.
However, the arithmetization of the sequence of "powers" (i.e., exponents) was present. Ibn
al-Bannāʾ must have known it, since he says (he was a purist) that it is not "allowed" to
speak of the power of the *māl* (as 2), viz because it is an entity of its own; ibn Qunfudh
(1339–1407), in the commentary from which we know this prohibition, states that other
writers on algebra did not agree, and speaks himself of the power of the *number* as "nothing", that is, 0 (Djebbar, 2005, pp. 95*f*). The individual names for the powers should thus

Donc si l'on vous dit : multipliez huit choses moins quatre en nombre par six carrés moins trois choses, posez cela ainsi :

C.

4   moins   8

C   Q

3   moins   6

Ensuite multipliez le huit par le six. Vous aurez pour résultat quarante huit cubes , parce que le fond des deux facteurs est trois. Réservez cela. Après cela multipliez de nouveau le huit par les trois choses. Vous aurez pour résultat vingt quatre carrés, ce qui est négatif, parce que cela (provient) de la multiplication du positif par le négatif. Réservez cela (en le plaçant) après la particule de l'exception. Puis multipliez le quatre par le six. Vous aurez pour résultat vingt quatre carrés. Mais cela est de nouveau négatif. Placez-le avec son analogue (***). Ensuite multipliez encore le quatre par le trois. Vous aurez pour résultat douze choses positives, parce que cela (provient) de la multiplication du négatif par le négatif. Réservez cela avec le premier (produit) réservé. Le résultat sera douze choses et quarante huit cubes moins quarante huit carrés, ainsi :

Q   K   C

48   moins   48   12

Fig. 1.2: The same in Woepckes translation (1859, p. 427)

sign above the radicand. He shows that these symbols (derived from the initial letters of the corresponding words, prolonged so as to be able to cover composite expressions, that is, to delimit algebraic parentheses[12]) are used

---

not have been a serious impediment to the development of algebraic proofs, had the intention been there to develop them.

[12] Three points should perhaps be made here. One concerns terminology. "Parenthesis" does not designate the bracket but the expression that is marked off, *for example* by a pair of brackets; but pauses may also mark off a parenthesis in the flow of spoken words, and a couple of dashes may do so in written prose. What characterizes an *algebraic parenthesis* is that it marks off a single entity which can be submitted to operations as a whole, and therefore has to be calculated first in the case of calculations. When division is indicated by a fraction line, this line delimits the numerator as well as the denominator as parentheses if they happen to be composite expressions (for instance, polynomials). Similarly, the modern root sign marks off the radicand as a parenthesis.

The remaining points are substantial, one of them general. The possibility of "embedding" parentheses is fundamental for the unrestricted development of mathematical thought, as I discuss in (Høyrup, 2000). An algebraic language without full ability to form parentheses and manipulate them is bound to remain "close to earth".

The last point, also substantial, is specific and concerns the Maghreb notation. It did not use the parenthesis function to the full. The fraction line and the root sign might mark off polynomials as parentheses; the signs for powers of the unknown, on the other hand,

to write polynomials and equations, and even to operate on the equations. Making the observation (p. 355) that

la condition indispensable pour donner à des signes conventionnels quelconques le caractère d'une notation, c'est qu'ils soient toujours employeés quand il y a lieu, et toujours de la même manière

he shows that one manuscript at his disposal fulfils this condition (another one not, probably because of "la negligence d'un copiste ou d'une succession de copistes").

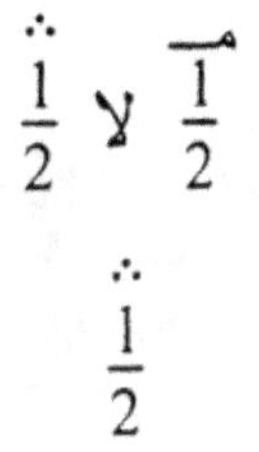

$$\frac{\overset{..}{1}}{2} \quad \mathsf{y} \quad \frac{\overline{1}}{2}$$

$$\frac{\overset{..}{1}}{2}$$

Fig. 1.3: Ibn al-Yāsamin's scheme for multiplying $\frac{1}{2}$ māl less $\frac{1}{2}$ šai› by $\frac{1}{2}$ šai›

Ibn Khaldūn's description made Woepcke suspect that the notation goes back to the twelfth century, as has now been confirmed by two isolated passages in ibn al-Yāsamin's *Talqīḥ al-afkār* reproduced by Mahdi Abdeljaouad (2002, p. 11) after Touhami Zemmouli's master thesis and corresponding exactly to what al-Qalaṣādi was going to do – one of them is shown in Figure 1.3.

Though manuscripts differ in this respect (as observed by Woepcke), the symbolic calculations appears to have been often made separate from the running text (as shown in Woepcke's translation of al-Qalaṣādi), usually preceded by the expression "its image is". They illustrate and duplicate the expressions used by words. They may also stand as marginal commentaries, as in the "Jerba manuscript" (written in Istanbul in 1747) of ibn al-Hā›im's *šarḥ al-Urjūzah al-Yasmīnya*, "Commentary to al-Yāsamin's *Urjuza*" (originally written in 1387 – manuscripts preceding the one from Jerba are without these marginalia) (ed. Abdeljaouad, 2004), of which Figure 1.4 shows a page. According to ibn Munᶜim (†1228) and al-Qalaṣādī, these marginal calculations may correspond to what was to be written in a *takht* (a dustboard, in particular used for calculation with Hindu numerals) or a *lawha* (a clayboard used for

---

might at most mark off a composite numerical expression – see (Abdeljaouad, 2002, pp. 25–34) for a much more detailed exposition. This should not surprise us: even Descartes

Fig. 1.4: A page from the "Jerba manuscript" of ibn al-Hāʾim's *Šarḥ al-Urjūzah al-Yasmīnīya* (ed. Abdeljaouad, 2004, p. Ar. 45)

---

eschewed general use of the parenthesis – for instance, expressions like $(y-3)^2$, as pointed out by Michel Serfati (1998, p. 259).

temporary writing) – see (Lamrabet, 1994, p. 203) and (Abdeljaouad, 2002, pp. 14, 19f). The use of such a device would explain that the examples of symbolic notation we find in manuscripts normally do not contain intermediate calculations, nor erasures (Abdeljaouad, 2002, p. 20).

We are accustomed to consider the notation for fractions as something quite separate from algebraic symbolism. In twelfth-century Maghreb, the two probably belonged together,[13] and from al-Ḥaṣṣār's *Kitāb al-bayān wa'l-tadhkār* onward Maghreb mathematicians used the various fraction notations with which we are familiar from Fibonacci's *Liber abbaci* (and other works of his): simple fractions written with the fraction line, ascending continued fractions ($\frac{e\,c\,a}{f\,d\,b}$ meaning $\frac{a}{b} + \frac{c}{bd} + \frac{e}{bdf}$), and additively and multiplicatively compounded fractions – see (Lamrabet, 1994, pp. 180f) and (Djebbar, 1992, pp. 231–234).

1.2 Latin algebra: *Liber mahamaleth*, *Liber abbaci*, translations of al-Khwārizmī – and Jordanus

The earliest documents in our possession from "Christian Europe" which speak of algebra are the *Liber mahamaleth* and, with a proviso, Robert of Chester's translations of al-Khwārizmī's *Algebra* (c. 1145); slightly later is Gherardo da Cremona's translation of al-Khwārizmī's treatise. All of these are from the twelfth century. From 1228 we have the algebra chapter in Fibonacci's *Liber abbaci* (the first edition from 1202 was probably rather similar, but we do not know *how* similar). In his *De numeris datis*, Jordanus de Nemore presented an *alternative to algebra*, showing how its familiar results could be based in (rather) strictly deductive manner on his *Elements of Arithmetic*, but he avoided to speak about algebra (hinting only for connoisseurs at the algebraic sub-text by using many of the familiar numerical examples) – see the analysis in (Høyrup, 1988, pp. 332–336). Finally, around 1300 a revised version of al-Khwārizmī's *Algebra* of interest for our topic was produced (ed. (Kaunzner, 1986), cf. (Kaunzner, 1985)).

The *Liber mahamaleth* and the *Liber abbaci* share certain characteristics, and may therefore be dealt with first.

All extant manuscripts of the *Liber mahamaleth*[14] have lost an introductory systematic presentation of algebra, which however is regularly referred to.[15]

---

[13] Cf. the hypothesis of Mahdi Abdeljaouad (2002, pp. 16–18), that "l'algèbre symbolique est un chapitre de l'arithmétique indienne maghrébine".

[14] I have consulted (Sesiano, 1988) and a photocopy of the manuscript Paris, Bibliothèque Nazionale, ms. latin 7377A.

[15] Thus fol. 154^v, "sicut docuimus in algebra"; fol. 161^r, "sicut ostensum est in algebra".

There are also references to Abū Kāmil,[16] and a number of problem solutions make use of algebra. Fractions are written in the Maghreb way, with Hindu numerals and fraction line;[17] there are also copious marginal calculations in rectangular frames probably rendering computation on a *lawha*. However, one finds no more traces of algebraic symbolism than in al-Khwārizmī's and Abū Kāmil's algebraic writings.

Fibonacci uses Maghreb fraction notations to the full in the *Liber abbaci* (ed. Boncompagni, 1857), writing composite fractions from right to left and mixed numbers with the fraction to the left – all in agreement with Arabic custom. Further, he often illustrates non-algebraic calculations in rectangular marginal frames suggesting a *lawha*. That systematic presentation of the algebraic technique which has been lost from the *Liber mahamaleth* is present in the *Liber abbaci*; there is no explicit reference to Abū Kāmil, but there are unmistakeable borrowings (which could of course be indirect, mediated by one or more of the many lost treatises). When the "thing" technique is used in the solution of commercial or recreational first-degree problems,[18] it is referred to as *regula recta*, not as algebra. But in one respect their algebras are similar: they are totally devoid of any hint of algebraic symbolism.[19] Inasfar as the *Liber mahamaleth* is concerned, this could hardly be otherwise – it antedates the probable creation of the Maghreb algebraic notation.

Equally devoid of any trace of symbolism is Gherardo's translation of al-Khwārizmī, which is indeed very faithful to the original – to the extent indeed that no Hindu numerals nor fraction lines occur, everything is completely verbal.

Robert does use Hindu numerals heavily in his translation (as we know it), but apart from that his translation is also fully verbal. It has often been believed, on the faith of Karpinski's edition (1915, p. 126) that his translation describes an algebraic formalism. It is true that the manuscripts contain a final list of *Regule 6 capitulis algabre correspondentes* making use of symbols for *census*, *thing* and *dragma* (the "unit" for the number term, we remember);

---

[16] Thus fol. 203^r, "modum agendi secundum algebra, non tamen secundum Auoqamel"; cf. (Sesiano, 1988, pp. 73f, 95f). We may observe that the spelling "Auoqamel" reflects an Iberian pronunciation.

[17] However, ascending continued fractions are written in a mixed system and not in Maghreb notation – e.g., "$\frac{4}{5}$ et $\frac{2}{5}$ unius sue $\frac{e}{5}$" (fol. 167^{rl}. − 9) for $\frac{4}{5} + \frac{2}{5} \cdot \frac{1}{5}$ ($\frac{e}{5}$ means "quinte").

[18] The *Liber mahamaleth* contains several pseudo-commercial problems involving the square root of an amount of money, leading to second-degree problems – see (Sesiano, 1988, pp. 80, 83). The *Liber abbaci* contains nothing of the kind, and no second-degree problems outside the final chapter 15.

[19] Florian Cajori (1928, I, p. 90) has observed a single appearance of ℞ in the *Pratica geometrie* (ed. Boncompagni, 1862, p. 209). Given how systematically Fibonacci uses his notations for composite fractions we may be sure that this isolated abbreviation is a copyist's slip of the pen (the manuscript is from the fourteenth century, where this abbreviation began to spread). Marginal reader's notes in a manuscript of the *Flos* are no better evidence of

they are classified as an appendix by Barnabas Hughes (1989, p. 67), but even he appears (p. 26) to accept them as genuine. However, the symbols are those known from the southern Germanic area of the later fifteenth century,[20] and all three manuscripts were indeed written in this area during that very period (Hughes, 1989, p. 11–13). The appendix has clearly crept in some three centuries after Robert made his translation.

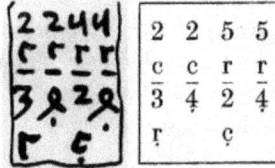

Fig. 1.5: From Oxford, Bodleian Library, Lyell 52, fol. 45^r (Kaunzner, 1986, pp. 64f)

Far more interesting from the point of view of symbolism is the anonymous ak-Khwārizmī redaction from around 1300. It contains a short section *Qualiter figurentur census, radices et dragma*, "How *census*, roots and *dragmas* are represented" (ed. Kaunzner, 1986, pp. 63f).[21] Here, *census* is written as *c*, roots as *r*, and *dragmata* (the unit for number) as *d* or not written at all. If a term is subtractive, a dot is put under it. These symbols are written below the coefficient, not above, as in the Maghreb notation. In Figure 1.5 we see (redrawn from photo and following Wolfgang Kaunzner's transcription) "2 *census* less 3 *roots*", "2 *census* less 4 *dragmata*", "5 *roots* less 2 *census*, and "5 *roots* less 4 *dragmata*". Outside this section, the notation is not used, which speaks against its being an invention of the author of the redaction; it rather looks as if he reports something he knows from elsewhere, and which, as he says, facilitates the teaching of algebraic computation. He refers not only

---

what Fibonacci did himself.

[20] One of them is an abbreviation of the spelling *zenso/zensus*, the spelling of many manuscripts from northern Italy (below, note 86). The spelling *zensus* as well as the abbreviation were taken over in Germany (as the north-Italian spelling *cossa* was taken over as *coss*); the spelling was unknown in twelfth-century Spain, and the corresponding abbreviation could therefore never have been invented in Spain in 1145.

[21] This redaction is often supposed to be identical with a translation made by Guglielmo de Lunis. However, all references to this translation (except a false ascription of a manuscript of the Gherardo translation) borrow from it a list of Arabic terms with vernacular explanation which is absent from the present Latin treatise. It is a safe conclusion that Guglielmo translated *into Italian*; that his translation is lost ; and that the present redaction is to be considered anonymous.

to additive-subtractive operations but also to multiplication, stating however only the product of *thing* by *thing* and of *thing* by *number*. He can indeed do nothing more, he has not yet explained the multiplication of binomials. The notation is certainly not identical with what we find in the Maghreb texts; the similarity to what we find in ibn al-Yāsamin and al-Qalaṣādī is sufficiently great, however, to suggest some kind of inspiration – very possibly indirect. However that may be: apart from an Italian translation from c. 1400 (Vatican, Urb. lat. 291), where *c* is replaced by *s* (for *senso*) and *r* by *c* (for *cose*), no influence in later writings can be traced. A brief description of a notation which is not used for anything was obviously not understood to be of great importance (whether the redactor believed it to be can also be doubted, given that he does not insist by using it in the rest of the treatise).

Jordanus de Nemore's *De numeris datis* precedes this redaction of al-Khwārizmī by a small century or so.[22] It is commonly cited as an early instance of symbolic algebra, and as a matter of fact it employs letters as general representatives of numbers. At the same time it is claimed to be very clumsy – which might suggest that the interpretation as symbolic algebra could be mistaken. We may look at an example:[23]

> If a given number is divided into two and if the product of one with the other is given, each of them will also be given by necessity.
>
> Let the given number *abc* be divided into *ab* and *c*, and let the product of *ab* with *c* be given as *d*, and let similarly the product of *abc* with itself be *e*. Then the quadruple of *d* is taken, which is *f*. When this is withdrawn from *e*, *g* remains, and this will be the square on the difference between *ab* and *c*. Therefore the root of *g* is extracted, and it will be *b,* the difference between *ab* and *c*. And since *b* will be given, *c* and *ab* will also be given.

As we see, Jordanus does not operate on his symbols, every calculation leads to the introduction of a *new* letter. What Jordanus has invented here is a symbolic representation of an *algorithm*, not clumsy symbolic algebra.

The same letter symbolism is used in Jordanus's *De elementis arithmetice artis*, which is presupposed by the *De numeris datis* and hence earlier. In the

---

[22] As well known, the only certain date *ante quem* for Jordanus is that all his known works appear in Richard de Fournival's *Biblionomina* (ed. de Vleeschauwer, 1965), which was certainly written some time before Richard's death in 1260 (Rouse, 1973, p. 257). However, one manuscript of Jordanus's *Demonstratio de algorismo* (Oxford, Bodleian Library, Savile 21) seems to be written by Robert Grosseteste in 1215–16, and in any case at that moment (Hunt, 1955, p. 134). This is the revised version of Jordanus's treatise on algorism. In consequence, Jordanus must have been beyond his first juvenile period by then. It seems likely (but of course is not certain) that the arithmetical works (the *Elements* and the *Data* of arithmetic) are closer in time to the beginning of his career that works on statics and on the geometry of the astrolabe, and that they should therefore antedate 1230.

[23] Translated from (Hughes, 1981, p. 58) (Hughes' own English translation is free and therefore unfit for the present purpose). Juxtaposition of letters is meant as aggregation, that is, addition (in agreement with the Euclidean understanding of number and addition).

algorithm treatises, letters are used to represent unspecified digits (Eneström, 1907, p. 146); in the two demonstrations that are quoted by Eneström (pp. 140f), the revised version can be seen also to use the mature notation, while it is absent from the early version. The assumption is close at hand that Jordanus developed the notation from the representation of digits by letters in his earliest work; it is hard to imagine that it can have been inspired in any way by the Maghreb notations. This representation of digits *might* have given rise to an algebraic symbolism – but as we see, that was not what Jordanus aimed at. Actually – as mentioned above – he did not characterize his *De numeris datis* as algebra even though he shows that he knows it to be at least a (theoretically better founded) *alternative to algebra*.

There are few echoes of this alternative in the following centuries. When taking up algebra in the mid-fourteenth century in his *Quadripartitum numerorum* ((ed. l'Huillier, 1990), cf. (l'Huillier, 1980)), Jean de Murs borrows from the *Liber abbaci*, not from Jordanus. Somewhere around 1450, Peurbach refers in a poem to "what algebra calculates, what Jordanus demonstrates" (ed. Größing, 1983, p. 210), and in his Padua lecture from 1464 (ed. Schmeidler, 1972, p. 46) Regiomontanus refers in parallel to Jordanus's "three most beautiful books about given numbers" and to "the thirteen most subtle books of Diophantos, in which the flower of the whole of arithmetic is hidden, namely the art of the thing and the *census*, which today is called algebra by an Arabic name". Regiomontanus thus seems to have been aware of the connection to algebra, and he also planned to print Jordanus's work (but suddenly died before any of his printing plans were realized).[24]

Two German algebraists from the sixteenth century knew, and used, Jordanus's quasi-algebra: Adam Ries and Johann Scheubel. The codex known as Adam Ries' *Coß* (ed. Kaunzner and Wußing, 1992) includes a fragment of an originally complete redaction of the *De numeris datis*, containing the statements of the propositions in Latin and in German translation, and for each statement an alternative solution of a numerical example by cossic technique; Jordanus's general proofs as well as his letter symbols have disappeared (Kaunzner and Wußing, 1992, II, pp. 92–100). From Scheubel's hand, a complete manuscript has survived. It has the same character – as Barnabas Hughes says in his description (1972, pp. 222f), "Scheubel's revision and elucidation [...] has all the characteristics of an original work save one: he used the statements of the propositions enunciated by Jordanus". Both thus did to Jordanus exactly what Jordanus had done to Arabic algebra: they took over his problems and showed how their own technique (basically that of Arabic algebra) allowed them to deal with them in what *they* saw as a more satisfactory man-

---

[24] As we shall see, these prestigious representatives of Ancient and university culture had no impact on Regiomontanus's own algebraic practice.

ner. Jordanus's treatise must thus have had a certain prestige, even though his technique appealed to nobody.[25]

I only know of two works where Jordanus's letter formalism turns up after his own times, both from France. One is Lefèvre d'étaples' edition of Jordanus's *De elementis arithmetice artis* (Lefèvre d'étaples, 1514) (first edition 1494). The other is Claude Gaspar Bachet's *Problemes plaisans et delectables, que se font par les nombres* (1624) (first edition 1612), where (for the first and only time?) Jordanus's technique is used actively and creatively by a later mathematician.[26]

1.3 Abbacus writings before algebra

The earliest extant abbacus treatises are roughly contemporary with the al-Khwārizmī-redaction (at least the originals – what we have are later copies). They contain no algebra, but their use of the notations for fractions is of some interest.

Traditionally, a *Livero dell'abbecho* (ed. Arrighi, 1989) conserved in the codex Florence, Ricc. 2404, has been supposed to be the earliest extant abbacus book, "internal evidence" suggesting a date in the years 1288–90. Since closer analysis reveals this internal evidence to be copied from elsewhere, all we can say on this foundation is that the treatise postdates 1290 (Høyrup, 2005, p. 47 n. 57) – but not by many decades, see imminently.

The treatise claims in its incipit to be "according to the opinion" of Fibonacci. Actually, it consists of two strata – see the analysis in (Høyrup, 2005). One corresponds to the basic abbacus school curriculum, and has nothing to do with Fibonacci; the other contains advanced matters, translated from the *Liber abbaci* but demonstrably often with scarce understanding.

The Fibonacci-stratum copies his numbers, not only his mixed numbers with the fraction written to the left ($\frac{2}{7}10$ where we would write $10\frac{2}{7}$) but also his ascending continued fractions (written, we remember, in Maghreb notation, and indeed from right to left, as done by al-Ḥaṣṣār, cf. above). However, the compiler does not understand the notation, at one place (ed.

---

[25] Vague evidence for prestige can also be read from the catalogue the books belonging to a third Vienna astronomer (Andreas Stiborius, c. 1500). Three neighbouring items in the list are *dedomenorum euclidis. Iordanus de datis. Demonstrationes cosse* (Clagett, 1978, p. 347). Whether it was Stiborius (in the ordering of his books) or Georg Tannstetter (who made the list) who understood *De numeris datis* as belonging midway between Euclid's Data and algebra remains a guess.

[26] In order to discover that one has to go to the seventeenth-century editions. Labosne's "edition" (1959) is a paraphrase in modern algebraic symbolism. Ries and Stifel were not the last of their kind.

Arrighi, 1989, p. 112), for instance, he changes

$$\frac{33 \quad 6 \quad 42 \quad 46}{53 \quad 53 \quad 53 \quad 53}$$

standing in the *Liber abbaci* (ed. Boncompagni, 1857, p. 273) for

$$46 + \cfrac{42 + \cfrac{6 + \cfrac{33}{53}}{53}}{53}$$

into $\frac{3364246}{53535353}$. It is obvious, moreover, that he has not got the faintest idea about algebra: he mostly omits Fibonacci's alternative solutions by means of *regula recta*; on one occasion where he does not (fol. 83^r, ed. Arrighi 1989: 89) he skips the initial position and afterwards translates *res* as an ordinary, not an algebraic *cosa*.[27]

The basic stratum contains ordinary fractions written with a fraction line but none of the composite fractions. Very strange is its way to speak of concrete mixed numbers. On the first few pages they look quite regular – e.g. "d. $6\frac{27}{28}$ de denaio", meaning "*denari* 6, $\frac{27}{28}$ of a *denaro*". Then, suddenly (with some slips that show the compiler to copy from material written in the normal way) the system changes, and we find expressions like "d. $\frac{2}{7}4$ de denaio", "*denari* $\frac{2}{7}4$ of a *denaro*" – obviously a misshaped compromise between Fibonacci's way to write mixed numbers with the way of the source material, which hence can *not* have been produced by Fibonacci (all his extant works write simple and composite fractions as well as mixed numbers in the same way as the *Liber abbaci*). All in all, the *Livero dell'abbecho* is thus evidence, firstly, that the Maghreb notations adopted by Fibonacci had not gained foothold in the early Italian abbacus environment (which it would by necessity have, had Fibonacci's works been the inspiration) ; secondly, that the aspiration of the compiler to dress himself in the robes of the famous culture hero was not accompanied by understanding of these notations (nor of other advanced matters presented by Fibonacci).

The other early abbacus book is the *Columbia Algorism* (New York, Columbia University, MS X511 AL3, ed. (Vogel, 1977)). The manuscript was written in the fourteenth century, but a new reading of a coin list which it contains dates this list to the years 1278–1284 (Travaini, 2003, pp. 88–92). Since the shapes of numerals are mostly those of the thirteenth century (with occasional slips, where the scribe uses those of his own epoch) (Vogel, 1977, p. 12), a dating close to the coin list seems plausible – for which reason we

---

[27] This total ignorance of everything algebraic allows us to conclude that the treatise cannot be written many decades after 1290.

must suppose the Columbia Algorism to be (a fairly scrupulous copy of) the oldest extant abbacus book.

There is no trace of familiarity with algebra, neither a systematic exposition nor an occasional algebraic *cosa*. *A fortiori*, there is no algebraic symbolism whatsoever, not even rudiments. Another one of the Maghreb innovations is present, however (Vogel, 1977, p. 13). Ascending continued fractions turn up several times, sometimes in Maghreb notation, but once reversed and thus to be read from left to right ($\frac{1}{4}\,\frac{1}{2}$ standing for $\frac{3}{8}$). Nothing else suggests any link to Fibonacci. Moreover, the notation is used in a way never found in the *Liber abbaci*, the first "denominator" being sometimes the metrological denomination – thus $\frac{1}{gran}\,\frac{1}{2}$ being used for $1\frac{1}{2}$ *gran* (or rather, as it would be written elsewhere in the manuscript, for 1 *gran* $\frac{1}{2}$). Next, the *Columbia Algorism* differs from all other Italian treatises (including those written in Provence by Italians) in its formulation of the rule of three – but in a way which approaches it to Ibero-Provençal writings of abbacus type – see (Høyrup, 2008, pp. 5*f*). Finally, at least one problem in the *Columbia Algorism* is strikingly similar to a problem found in a Castilian manuscript written in 1393 (copied from an earlier original) while not appearing elsewhere in sources I have inspected – see (Høyrup, 2005, p. 42 n. 32). In conclusion it seems reasonable to assume that the *Columbia Algorism* has learned the Maghreb notation for ascending continued fractions *not* from Fibonacci but from the Iberian area.

1.4 The beginning of abbacus algebra

The earliest abbacus *algebra* we know of was written in Montpellier in 1307 by one Jacopo da Firenze (or Jacobus de Florentia; otherwise unknown as a person). It is contained in one of three manuscripts claiming to represent his *Tractatus algorismi* (Vatican, Vat. lat. 4826; the others are Florence, Riccardiana 2236, and Milan, Trivulziana 90).[28] As it follows from in-depth analysis of the texts (Høyrup, 2007a, pp. 5–25 and *passim*), the Florence and Milan manuscripts represent a revised and abridged version of the original, while the Vatican manuscript is a meticulous copy of a meticulous copy of the shared archetype for all three manuscripts (extra intermediate steps not being excluded, but they must have been equally meticulous if they exist);

---

[28] The Vatican manuscript can be dated by watermarks to c. 1450, the Milan manuscript in the same way to c. 1410. The Florence manuscript is undated but slightly more removed from the precursor it shares with the Milan manuscript (which of course does not automatically make it younger but disqualifies it as a better source for the original).

this shared archetype could be Jacopo's original, but also a copy written well before 1328.[29]

Jacopo may have been aware of presenting something new. Whereas the rest of the treatise (and the rest of the vocabulary in the algebra chapter) employs the standard abbreviations of the epoch and genre, the algebraic technical vocabulary is never abbreviated.[30] Even *meno*, abbreviated ⑪ in the coin list, is written in full in the algebra section. Everything here is rhetorical, there is not the slightest hint of any symbolism. We may probably take this as evidence that Jacopo was aware of writing about a topic the reader would not know about in advance (the book is stated also to be intended for independent study), and thus perhaps that his algebra is not only the earliest extant Italian algebra but also the first that was written. As we shall see, however, several manuscripts certainly written later also avoid the abbreviation of algebraic core terms – even around 1400, authors of general abbacus treatises may have suspected their readers to possess no preliminary knowledge of algebra.

Not only symbolism but also the Maghreb notations for composite fractions are absent from the treatise, even though they turned up in the Columbia Algorism. None the less, Jacopo's algebra must be presumed to have its direct roots in the Ibero-Provençal area, with further ancestry in al-Andalus and the Maghreb; there is absolutely no trace of inspiration from Fibonacci nor of direct influence of Arabic classics like al-Khwārizmī or Abū Kāmil (nor any Arabisms suggesting direct impact of other Arabic writings or settings). Jacopo offers no geometric proofs but only rules, and the very mixture of commercial and algebraic mathematics is characteristic of the Maghreb–al-Andalus tradition (as also reflected in the *Liber mahamaleth*). A particular multiplicative writing for Roman numerals (for example $\frac{m}{cccc}$, used as explanation of the Hindu-Arabic number 400000) *could* also be inspired by the Maghreb algebraic notation (it may also have been an independent invention, Middle Kingdom Egyptian scribes and Diophantos sometimes put the "de-

---

[29] Comparing only lists of the equation types dealt with in various abbacus algebras and believing in a steady progress of their number within each family, Warren Van Egmond claims (2008, p. 313) that the algebra of the Vatican manuscript "falls entirely within the much later and securely dated Benedetto tradition and was undoubtedly added to a manuscript containing some sections from Jacopo's earlier work" (actually, it contains fewer types than the manuscript from c. 1390 which Van Egmond takes as the starting point for this tradition). If he had looked at the words used in the manuscripts he refers to he would have discovered that the Vatican algebra agrees verbatim with a section of an algebra manuscript from c. 1365, which however fills out a calculational lacuna left open in the Vatican manuscript and therefore represents a more developed form of the text (and combines it with other material – details in (Høyrup, 2007a, pp. 163*f*)). Van Egmond's dating can be safely dismissed.

[30] There is one instance of R (fol. 44^r, ed. (Høyrup, 2007a, p. 326); as the single appearance of R in Fibonacci's *Pratica geometrie* (see note 19), this is likely to be a copyist's *lapsus calami*.

nomination" above the "coefficient" in a similar way, and there is no reason to believe that these notations were connected to the Maghreb invention).

In 1328, also in Montpellier, a certain Paolo Gherardi (as Jacopo, unknown apart from the name) wrote a *Libro di ragioni*, known from a later copy (Florence, Bibl. Naz. Centr., Magl. XI, 87, ed. (Arrighi, 1987, p. 13–107)). Its final section is another presentation of algebra.[31] Part of this presentation is so close to Jacopo's algebra that it must descend either from that text (by reduction) or from a close source; but whereas Jacopo only deals (correctly) with 20 (of the possible 22) quadratic, cubic and quartic basic equations ("cases") that can be solved by reduction to quadratic equations or by simple root extraction,[32] Gherardi (omitting all quartics) introduces false rules for the solution of several cubics that cannot be solved in these ways (with examples that are "solved" by means of the false rules). Comparison with later sources show that they are unlikely to be of his own invention. A couple of the cases he shares with Jacopo also differ from the latter in their choice of examples, one of them agreeing at the same time with what can be found in a slightly later Provençal treatise (see imminently).

Gherardi's algebra is almost as rhetorical as Jacopo's, but not fully. Firstly, the abbreviation R is used copiously though not systematically. This *may* be due to the copyist – the effort of Jacopo's and Fibonacci's copyists to conserve the features of the original was no general rule; but it could also correspond to Gherardi's own text. More important is the reference to a diagram in one example (100 is first divided by some number, next by five more, and the sum of the two quotients is given); this diagram is actually missing in the copy, but so clearly described in the text that it can be seen to correspond to the diagram found in a parallel text:[33]

$$100 \diagdown\!\!\!\!\diagup 1 \text{ cosa}$$
$$100 \diagup\!\!\!\!\diagdown 1 \text{ cosa piu } 5$$

The operations performed on the diagram ("cross-multiplication" and the other operations needed to add fractions) are described in a way that implies underlying operations with the "formal fractions" $\frac{100}{1\,cosa}$ and $\frac{100}{1\,cosa\,piu\,5}$. No abbreviations being used, we may speak of what goes on as a beginning of symbolic *syntax* without symbolic *vocabulary*.

Such formal fractions, we may observe, constitute an element of "symbolic algebra" that does not presuppose that "*cosa*" itself be replaced by a sym-

---

[31] Beyond Arrighi's complete edition of the treatise (1987, pp. 97–107), there is an edition of the algebra text with translation and mathematical commentary in (Van Egmond, 1978).
[32] The lacking equations are the two mixed biquadratics that correspond to al-Khwārizmī's (and Jacopo's) fifth and sixth case. Only the six simple cases (linear and quadratic) are provided with examples – ten in total, half of which are dressed as commercial problems. For the others, only rules are offered.
[33] Florence, Ricc. 2252, see (Van Egmond, 1978, p. 169).

bol – but certainly an isolated element only. It must be acknowledged, on the other hand, that this isolated element already made possible calculations that were impossible within a purely rhetorical framework. Jacopo, as already al-Khwārizmī, could get rid of one division by a binomial via multiplication. However, problems of the type where Gherardi and later abbacus algebra use two formal fractions were either solved geometrically by al-Khwārizmī, Abū Kāmil and Fibonacci, as I discuss in a forthcoming paper,[34] or they were replaced before being expressed algebraically *without explanation* by a different problem, namely the one resulting from multiplication by the denominators (al-Khwārizmī, ed. (Hughes, 1986, p. 51)).

A third abbacus book written in Provence (this one in Avignon) is the *Trattato di tutta l'arte dell'abbacho*. As shown by Jean Cassinet (2001), it must be dated to 1334. Cassinet also shows that the traditional ascription to Paolo dell'Abbaco is unfounded.[35] Exactly how much should be counted to the treatise is not clear. The codex Florence, Bibl. Naz. Centr., fond. princ. II.IX.57 (the author's own draft according to (Van Egmond, 1980, p. 140)) contains a part that is not found in the other copies[36] but which is informative about algebra and algebraic notation; however, since this extra part is in the same hand as the main treatise (Van Egmond, 1980, p. 140), it is unimportant whether it went into what the author eventually decided to put into the final version.

There is no systematic presentation of algebra nor listing of rules in this part,[37] only a number of problems solved by a rhetorical *censo-cosa* technique.[38] The author uses no abbreviations for *cosa*, *censo* and *radice* – but at one point (fol. 159^r) an astonishing notation turns up: $\frac{10}{cose}$, meaning "10 *cose*". The idea is the same as we encountered in the *Columbia Algorism* when it writes $\frac{1}{gran\ 2}$ meaning "1 *gran* $\frac{1}{2}$": that what is written below the line is a denomination; indeed, many manuscripts write "il $\frac{1}{3}$" in the sense of "the

---

[34] "'Proportions' in the Liber abbaci", to appear in the proceedings of the meeting "Proportions: Arts – Architecture – Musique – Mathématiques – Sciences", Centre d'études Supérieures de la Renaissance, Tours, 30 juin au 4 juillet 2008.

Al-Khwārizmī (ed. Hughes, 1986, p. 255) does not make the geometric argument explicit, but a division by 1 betrays his use of the same diagram as Abū Kāmil (ed. Sesiano, 1993, p. 370).

[35] Arguments speaking against the ascription are given in (Høyrup, 2008, p. 11 n. 29).

[36] I have compared with Rome, Acc. Naz. dei Lincei, Cors. 1875, from c. 1340. For other manuscripts, see (Cassinet, 2001) and (Van Egmond, 1980, passim).

[37] The codex contains a list of four rules (fol. 171^v), three of which are followed by examples, written on paper from the same years (according to the watermark) but in a different hand than the recto of the sheet and thus apparently added by a user of the manuscript. It contains one of the examples which Gherardi does not share with Jacopo, confirming that his extra examples came from what circulated in the Provençal area. It contains no algebraic abbreviations nor anything else suggesting symbolism.

[38] Jean Cassinet (2001, pp. 124–127) gives an almost complete list.

third" (as ordinal number as well as fraction) – that is, the notation for the fraction was understood as an *image of the spoken form*, not of the division procedure (cf. also the writing of *quinte* as $\frac{e}{5}$ in the *Liber mahamaleth*, see note 17).

The compiler of the *Trattato di tutta l'arte* was certainly not the first to use this algebraic notation – who introduces a new notation does not restrict himself to using it a single time in a passage well hidden in an odd corner of a text. He just happens to be our earliest witness of a notation which for long was in the way of the development of one that could serve symbolic calculation.

This compiler was, indeed, not only not the first but also not the last to use this writing of monomials as quasi-fractions. It is used profusely in Dardi of Pisa's *Aliabraa Argibra* from 1344,[39] better known for being the first Italian-vernacular treatise dedicated exclusively to algebra and for its presentation of rules for solving no less than 194+4 algebraic cases, 194 of which are solved according to generally valid rules (with two slips, explained by Van Egmond (1983, p. 417)), while the rules for the last four cases are pointed out by Dardi to hold only under particular (unspecified) circumstances.[40]

Dardi uses algebraic abbreviations systematically. *Radice* is always R, *meno* ("less") is $\tilde{m}$, *cosa* is c, *censo* is ς, *numero/numeri* are $n\tilde{u}o/n\tilde{u}i$. *Cubo* is unabridged, *censo de censo* (the fourth power) appears not as $\varsigma\varsigma$ but in the expanded linguistic form ς *de* ς, which we may take as an indication that Dardi merely thinks in terms of abbreviation and nothing more. Roots of composite entities are written by a partially rhetorical expression, for instance (fol. 9^v) "R *de zonto* $\frac{1}{4}$ *cō* R *de* 12" (meaning $\sqrt{\frac{1}{4} + \sqrt{12}}$; *zonto* corresponds to Tuscan *gionto*, "joined").

As just mentioned, Dardi also employs the quasi-fraction notation for monomials, and does so quite systematically in the rules and the examples (but only here).[41] When coefficients are mixed numbers Dardi also uses the

---

[39] See (Van Egmond, 1983). The three principal manuscripts are Vatican, Chigi M.VIII.170 written in Venetian in c. 1395; Siena, Biblioteca Comunale I.VII.17 from c. 1470 (ed. Franci, 2001); and a manuscript from Mantua written in 1429 and actually held by Arizona State University Temple, which I am grateful to know from Van Egmond's personal transcription. In some of the details, the Arizona manuscript appears to be superior to the others, but at the level of overall structure the Chigi manuscript is demonstrably better – see (Høyrup, 2007a, pp. 169*f*). Considerations of consistency suggests it to be better also in its use of abbreviations and other quasi-symbolism, for which reason I shall build my presentation on this manuscript (cross-checking with the transcription of the Arizona-manuscript – differences on this account are minimal); for references I shall use the original foliation.

A fourth manuscript from c. 1495 (Florence, Bibl. Med.-Laur., Ash. 1199, partial ed. (Libri, 1838, III, pp. 349–356)) appears to be very close to the Siena manuscript.

A critical edition of the treatise should be forthcoming from Van Egmond's hand.

[40] Dardi reaches this impressive number of resolvable cases by making ample use of radicals.

[41] This notation appears only to be present in the Chigi and Arizona manuscripts; Franci

formalism systematically in a way which suggests ascending continued fractions, writing for instance $2\frac{1}{2}c$ not quite as $\frac{2\,1}{c\,2}$ but as $\frac{2\,1}{c\,2}$ (which however *could* also mean simply "2 *censi* and $\frac{1}{2}$". Often, a number term is written as a quasi-fraction, for example as $\frac{325}{n}$. How far this notation is from any operative symbolism is revealed by the way multiples of the *censo de censo* are sometimes written – namely for example as $\frac{81}{\wp}$ *de* ς (fol. 46^{v}).

None the less, symbolic operations are not absent from Dardi's treatise. They turn up when he teaches the multiplication of binomials (either algebraic or containing numbers and square roots) – for instance, for $(3-\sqrt{5})\cdot(3-\sqrt{5})$,

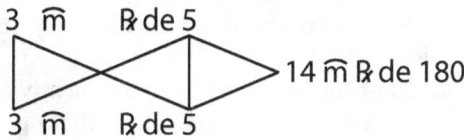

Noteworthy is also Dardi's use of a similar scheme

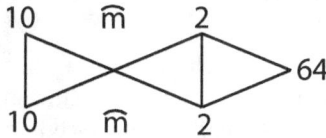

as support for his proof of the sign rule "less times less makes plus" on fol. 5^{v}:

> Now I want to demonstrate by number how less times less makes plus, so that every times you have in a construction to multiply less times less you see with certainty that it makes plus, of which I shall give you an obvious example. 8 times 8 makes 64, and this 8 is 2 less than 10, and to multiply by the other 8, which is still 2 less than 10, it should similarly make 64. This is the proof. Multiply 10 by 10, it makes 100, and 10 times 2 less makes 20 less, and the other 10 times 2 less makes 40 less, which 40 less detract from 100, and there remains 60. Now it is left for the completion of the multiplication to multiply 2 less times 2 less, it amounts to 4 plus, which 4 plus join above 60, it amounts to 64. And if 2 less times two less had been 4 less, this 4 less should have been detracted from 60, and 56 would remain, and thus it would appear that 10 less 2 times 10 less two had been 56, which is not true. And so also if 2 less times 2 less had been nothing, then the multiplication of 10 less 2 times 10 less 2 would come to be 60, which is still false. Hence less times less by necessity comes to be plus.

---

does not mention it in her edition of the much later Siena manuscript, and composite

Such schemes were no more Dardi's invention than the quasi-fraction notation (even though he may well have been more systematic in the use of both than his precursors). The clearest evidence for this is offered by an anonymous *Trattato dell'alcibra amuchabile* from c. 1365 (ed. Simi, 1994), contained in the codex Florence, Ricc. 2263. This is the treatise referred to in note 29, part of which agrees verbatim with Jacopo's algebra. It also has Gherardi's false rules. However, here the agreement is not verbatim, showing Gherardi not to be the immediate source (a compiler who follows one source verbatim will not use another one freely) – cf. (Høyrup, 2007a, p. 163).

The treatise consists of several parts. The first presents the arithmetic of monomials and binomials, the second contains rules and examples for 24 algebraic cases (mostly shared with Jacopo or Gherardi), the third a collection of 40 algebraic problems. All are purely rhetorical in formulation, except for using R in the schemes of the first part (see imminently). However, the first and third part contain the same kinds of non-verbal operations as we have encountered in Gherardi and Dardi, and throws more light on the former.

In part 3, there are indeed a number of additions of formal fractions, for example (in problem #13) $\frac{100}{1\,cosa} + \frac{100}{1\,cosa+5}$. This is shown as

$$\frac{100 \qquad\qquad 100}{per\ una\ cosa \quad per\ una\ cosa\ e\ 5}$$

and explained with reference to the parallel $\frac{24}{4} + \frac{24}{6}$ (cross-multiplication of denominators with numerators followed by addition, multiplication of the denominators, etc.). Gherardi's small scheme (see just after note 33) must build on the same insights (whether shared by Gherardi or not).

Part 1 explains the multiplication of binomials with schemes similar to those used by Dardi – for example

$$
\begin{array}{cccccc}
5 & e & piu & R & di & 20 \\
\text{via} \\
5 & e & meno & R & di & 20
\end{array}
$$

As we see, the scheme is very similar to those of Dardi but more rudimentary. It also differs from Dardi in its use of the ungrammatical expressions *e più* and *e meno*, where Dardi uses the grammatical *e* for addition and the abbreviation $\hat{m}$ for subtraction.[42] There is thus no reason to suppose it should

---

expressions where their presence might be revealed show no trace of them. They are also absent from Guglielmo Libri's extract of the Florence manuscript.

[42] The expression *e meno n*, as we remember, corresponds to what was done by al-Karajī, see note 8. The appearance of the parallel expression *e più n* shows that the attribute "subtractivity" was seen to ask for the existence of a corresponding attribute "additivity" – another instance of "symbolic syntax" without "symbolic vocabulary" (or, in a different terminology but with the same meaning, the incipient shaping of the language of algebra

be borrowed from Dardi's earlier treatise – influence from which is on the whole totally absent. Schemes of this kind must hence have been around in the environment or in the source area for early abbacus algebra before 1340, just as the calculation with formal fractions must have been around before 1328, and the quasi-fractions for monomials before 1334.[43] On the whole, this tells us how far the development of algebraic symbolic operations had gone in abbacus algebra in the early fourteenth century – and that all that was taken over from the Maghreb symbolism was the calculation with formal fractions; a very dubious use of the ascending continued fractions; and possibly the idea of presenting *radice*, *cosa* and *censo* by single-letter abbreviations (implemented consistently by Dardi but not broadly, and not necessarily a borrowing).

1.5 The decades around 1400

The Venetian manuscript Vatican, Vat. lat. 10488 (*Alchune ragione*), written in 1424, connects the early phase of abbacus algebra with its own times. The manuscript is written by several hands, but clearly as a single project (hands may change in the middle of a page; we should perhaps think of an abbacus master and his assistants). From fol. 29^v to fol. 32^r it contains a short introduction to algebra, taken from a text written in 1339 by Giovanni di Davizzo, a member of a well-known Florentine abbacist family, see (Ulivi, 2002, pp. 39, 197, 200). At first come sign rules and rules for the multiplication of algebraic powers, next a strange section with rules for the division of algebraic powers where "roots" take the place of negative powers;[44] then a short section about the arithmetic of roots (including binomials containing roots)[45] somehow but indirectly pointing back to al-Karajī; and finally 20 rules for algebraic cases without examples, of which one is false and the rest parallel to those of Jacopo (not borrowed from him but sharing the same source tradition). Everywhere, *radice* is ℞, but "less", *cosa* and *censo* all appear unabbreviated (*censo* mostly as *zenso*, which cannot have been the Florentine Giovanni's spelling).

---

as an *artificial* language).

In the proof that "less times less makes plus" (see above), Dardi speaks of subtractive numbere, e.g., as "2 *meno*"/"2 less", etc., whereas additive numbers are not characterized explicitly as such.

[43] This latter presence leads naturally to the question whether the notation in the al-Khwārizmī–redaction from c. 1300 should belong to the same family. This cannot be completely excluded, but the absence of a fraction line from the notation of the redaction speaks against it. It remains more plausible that the latter notation is inspired from the Maghreb, or an independent invention.

[44] An edition, English translation and analysis of this initial part of the introduction can be found in (Høyrup, 2007b, pp. 479–484).

[45] Translation in (Høyrup, 2009, pp. 56f).

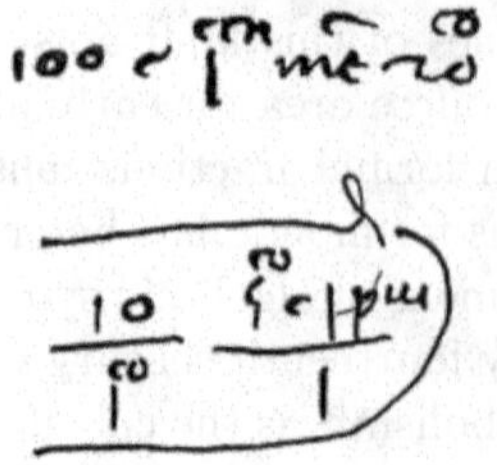

Fig. 1.6: The "equations" from VAT 10488 fol. 37^v (top) and fol. 39^v (bottom)

This introduction comes in the middle of a long section containing number problems mostly solved by means of algebra (many of them about numbers in continued proportion).[46] Here, abbreviations abound. *Radice* is always R, *meno* is often ⓜ, $\hat{m}$, or $\widehat{me}$ (different shapes may occur in the same line). More interesting, however, is the frequent use of *co*, □, (occasionally *ce*) and *no* written above the coefficient, precisely as in the Maghreb notation (and quite likely inspired by it). However, these notations are not used systematically, and only used once for formal calculation, namely in a marginal "equation" without equation sign[47] on fol. 39^v – see Figure 1.6, bottom.[48] In another place (fol. 37^r, Figure 1.6 top) the running text formulates a genuine equation, but this is merely an abbreviation for *100 è 1 censo meno 20 cose*. It serves within the rhetorical argument without being operated upon.

Later in the text comes another extensive collection of problems solved by means of algebra (some of them number problems, others dressed as business problems), and inside it another collection of rules for algebraic cases (17 in total, only 2 overlapping the first collection). In its use of abbreviations, this second cluster of problems and rules is quite similar to the first cluster, the only exception being a problem (fols. 95r–96v) where the use of coefficients

---

[46] Even these are borrowed en bloc, as revealed by a commentary within the running text on fol. 36^r, where the compiler tells how a certain problem should be made *al parere mio*, "in my opinion". The several hands of the manuscripts are thus not professional scribes copying without following the argument.

[47] Two formal fractions are indicated to be equal; the hand seems to be the same as that of the main text and of marginal notes adding words that were omitted during copying.

[48] The treatment of the problem is quite interesting. The problem asks for a number which, when divided into 10 yields 5 times the same number and 1 more. Instead of writing "$\frac{\frac{10}{co}}{1} = \frac{co}{5}$ e 1 piu" it expresses the right-hand side as a fraction $\frac{\frac{co}{5} \text{ e } 1 \text{ } piu}{1}$, thus opening the way to the usual cross-multiplication.

As in several cases below, I have had to redraw the extract from the manuscript in order to get clear contours, my scanned microfilm being too much grey in grey.

with superscript power is so dense (without being fully systematic) that it may possibly have facilitated understanding of the argument by making most of the multiples of *cosa* and *censo* stand out visually.

In the whole manuscript, addition is normally indicated by a simple *e*, "and". I have located three occurrences of *più*,[49] none of them abbreviated. The expressions *e più* and *e meno* appear to be wholly absent.

It is fairly obvious that this casual use of what could be a symbolism was not invented by the compilers of the manuscript, and certainly not something they were experimenting with. They used for convenience something which was familiar, without probing its possibilities. If anybody else in the abbacus environment had used the notation as a symbolism and not merely as a set of abbreviations (and the single case of an equation between formal fractions suggests that this may well have been the case), then the compilers of the present manuscript have not really discovered – or they reveal, which would be more significant, that the contents of abbacus algebra did not call for and justify the effort needed to implement a symbolism to which its practitioners were not accustomed.[50] They might *almost* as well have used Dardi's quasi-fractions – only in the equation between formal fractions would the left-hand side have collided with it by meaning simply "10 *cose*".

Though not using the notation as a symbolism, the compilers of Vat. lat. 10488 at least show that they knew it. However, this should not make us believe that every abbacus writer on algebra from the same period was familiar with the notation, or at least not that everybody adopted it. As an example we may look at two closely related manuscripts coming from Bologna, one (Palermo, Biblioteca Comunale 2 Qq E 13, *Libro merchatantesche*) written in 1398, the other (Vatican, Vat. lat. 4825, Tomaso de Jachomo Lione, *Libro da razioni*) in 1429.[51] They both contain a list of 27 algebraic cases with examples followed by a brief section about the arithmetic of roots (definition, multiplication, division, addition and subtraction). The former has a very fanciful abbreviation for *meno*, namely ℞, which corresponds, however, to the way *che* and various other non-algebraic words are abbreviated, and is thus merely a personal style of the scribe; the other writes *meno* in full, and none of the two manuscripts have any other abbreviation whatsoever of algebraic terms – not even ℞ for *radice* which they are unlikely not to have known, which suggests but does not prove that the other abbreviations were also avoided consciously.

---

[49] In a marginal scheme and the running text of a problem about combined works (fol. 90^r), and once in an algebra problem (fol. 94^r). There may be more instances, but they will be rare.

[50] The latter proviso is needed. For us, accustomed as we are to symbolic algebra, it is often much easier to follow a complex abbacus texts if we make symbolic notes on a sheet of paper.

[51] More precisely, 7 March 1429 – which with year change at Easter means 1430 according to our calendar, the date given in (Van Egmond, 1980, p. 223).

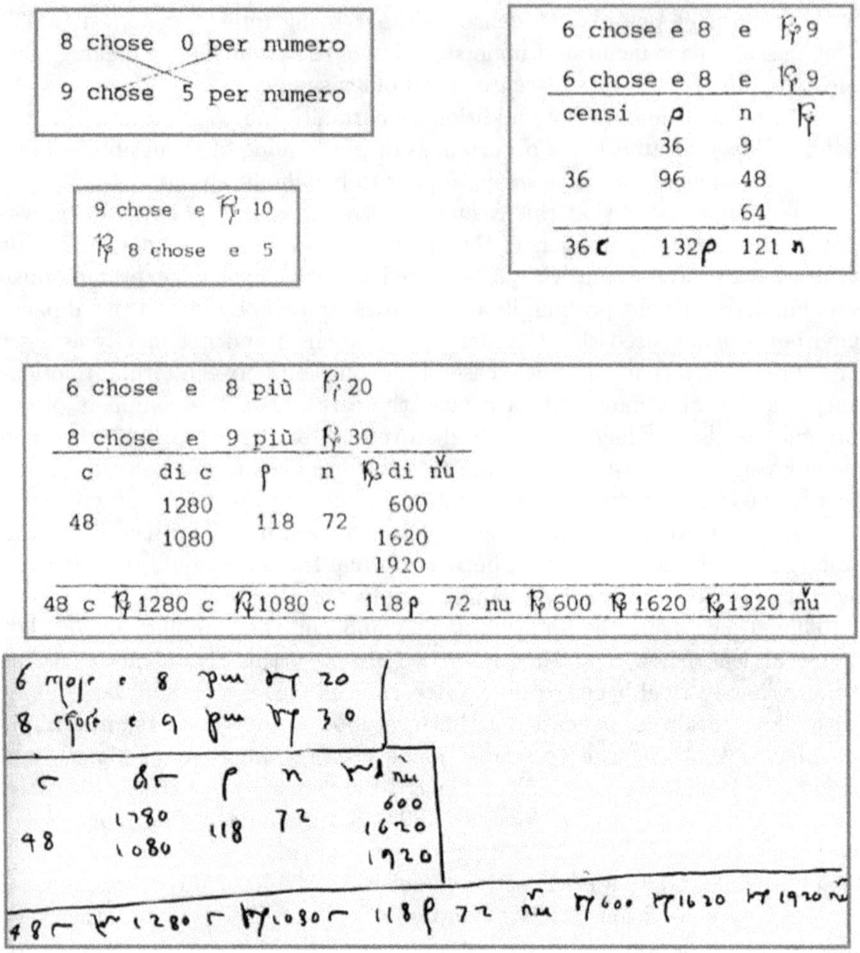

8 chose   0 per numero

9 chose   5 per numero

---

6 chose e 8 e ℞ 9

6 chose e 8 e ℞ 9

censi	ρ	n	℞
	36	9	
36	96	48	
		64	
36 c	132ρ	121 n	

---

9 chose   e ℞ 10

℞ 8 chose   e 5

---

6 chose   e   8 più   ℞ 20

8 chose   e   9 più   ℞ 30

c	di c	ρ	n	℞ di nu
48	1280	118	72	600
	1080			1620
				1920

48 c ℞1280 c ℞1080 c 118ρ 72 nu ℞600 ℞1620 ℞1920 nu

Fig. 1.7: Schemes for the multiplication of polynomials, from (Franci and Pancanti, 1988, pp. 812), and from the manuscript, fol. 146ᵛ

Maybe we should not be surprised not to find any daring development in these two manuscripts. In general, they offer no evidence of deep mathematical insight. In this perspective, the manuscript Florence, Bibl. Naz. Centr., fondo princ. II.V.152 (*Tratato sopra l'arte della arismetricha*) is more illuminating. Its algebraic section was edited by Franci and Pancanti (1988).[52] It

---

[52] I have controlled on a scan of a microfilm, but since it is almost illegible my principal basis for discussing the treatise is this edition.

was written in Florence in c. 1390, and offers both a clear discussion of the sequence of algebraic powers as a geometric progression and sophisticated use of polynomial algebra in the transformation of equation types – see (Høyrup, 2008, pp. 30–34).

In the running text, there are no abbreviations nor anything else which foreshadows symbolism. However, inserted to the left we find a number of schemes explained by the text and showing multiplication of polynomials with two or three terms (numbers, roots and/or algebraic powers), of which Figure 1.7 shows some examples – four as rendered by Franci and Pancanti, the last of these also as appearing in the manuscript (redrawn for clarity).

Those involving only binomials are easily seen to be related to what we find in the *Trattato dell'alcibra amuchabile* and in Dardi's *Aliabraa Argibra* – but also to schemes used in non-algebraic sections of other treatises, for instance the Palermo-treatise discussed above, see Figure 1.8, which should warn us against seeing any direct connection.

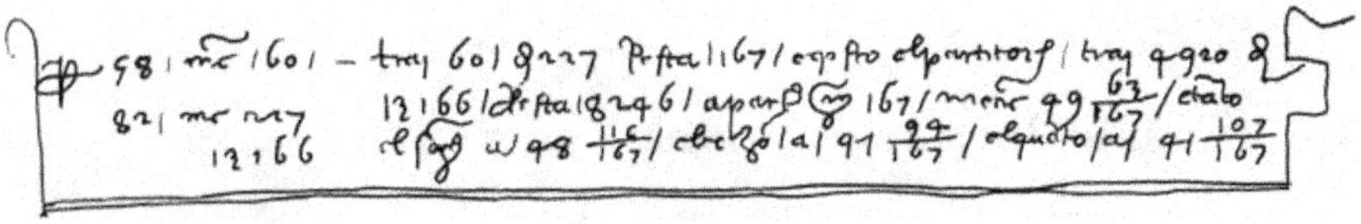

Fig. 1.8: Non-algebraic scheme from Palermo, Biblioteca Comunale 2 Qq E 13, fol. 38^v

The schemes for the multiplication of three-term polynomials are different. They emulate the scheme for multiplying multi-digit numbers, and the text itself justly refers to multiplication *a chasella* (ed. Franci and Pancanti, 1988, p. 9). The *a casella* version of the algorithm uses vertical columns, while the scheme for multiplying polynomials used in the Jerba manuscript (ed. Abdeljaouad, 2002, p. 47) follows the older algorithm *a scacchiera* with slanted columns; none the less inspiration from the Maghreb is plausible, in particular because another odd feature of the manuscript suggests a pipeline to the Arabic world. In a wage problem, an unknown amount of money is posited to be a *censo*, whereas Biagio *il vecchio* (ed. Pieraccini, 1983, p. 89f) posits it to be a *cosa* in the same problem in a treatise written at least 50 years earlier. But the present author does not understand that a *censo* can be an amount of money, and therefore feels obliged to find its square root – only to square it again to find the amount of money asked for. He thus uses the terminology without understanding it, and therefore cannot have shaped the solution himself; nor can the source be anything of what we have discussed so far.

Schemes of this kind (and other schemes for calculating with polynomials) turn up not only in later abbacus writings (for instance, in Raffaello Canacci, see below) but also in Stifel's *Arithmetica integra* (1544, fols. $238^r - 239^r$), in Jacques Peletier's *L'Algèbre* (1554, pp. 15–22) and in Petrus Ramus's *Algebra* (1560, fol. *Aiii^r*).

Returning to the schemes of the present treatise we observe that the *cosa* is represented (within the calculations, not in the statement lines) by a symbol looking like ρ, and the *censo* by *c*. *Radice* is R in statement as well as calculation. The writing of *meno* is not quite systematic – whether it is written in full, abbreviated *me* or as ⑰ (rendered "m." by Franci and Pancanti) seems mostly to depend on the space available in the line. Addition may be *e* or *più* (*più* being mostly but not always nor exclusively used before R); when space is insufficient, and only then, *più* may be abbreviated *p*.[53] All in all, the writer can be seen to have taken advantage of this incipient symbolism but not to have felt any need to use it systematically – it stays on the watershed, between facultative abbreviation and symbolic notation.

1.6 The mid-15th-century abbacus encyclopediæ

Around 1460, three extensive "abbacus encyclopediae" were written in Florence. Most famous among these is, and was, Benedetto da Firenze's *Trattato de praticha d'arismetrica* – it is the only one of them which is known from several manuscripts.[54]

Earliest of these is Siena, Biblioteca Comunale degli Intronati, L.IV.21, which I have used together with the editions of some of its books.[55] According to the colophon (fol. 1^r) it was "conpilato da B. a uno suo charo amicho negl'anni di Christo MCCCCLXIII". It consists of 495 folios, 106 of which deal with algebra.

The algebra part consists of the following books:

- XIII: Benedetto's own introduction to the field, starting with a 23-lines' excerpt from Guglielmo de Lunis's lost translation of al-Khwārizmī (cf. note 21). Then follows a presentation of the six fundamental cases with geomet-

---

[53] The phrases *e più* and *e meno* occur each around half a dozen times, but apparently in a processual meaning, "and (then) added" respectively "and (then) subtracted". Nothing suggest a use of *più* and *meno* as attributes of numbers, even though the author does operate with negative (not merely subtractive) numbers in his transformation of cubic equations – see (Høyrup, 2008, p. 33).

[54] On Benedetto and his historical setting, see the exhaustive study in (Ulivi, 2002).

[55] (Salomone, 1982); (Pieraccini, 1983); (Pancanti, 1982); (Arrighi, 1967). All of these editions were made from the same Siena manuscript, which is also described in detail with extensive extracts in (Arrighi, 2004/1965).

ric proofs, built on al-Khwārizmī; a second chapter on the multiplication
and division of algebraic powers (*nomi*, "names") and the multiplication of
binomials; and a third chapter containing rules and examples for 36 cases
(none of them false);

- XIV: a problem collection going back to Biagio *il vecchio* († c. 1340 according to Benedetto);
- XV: containing a translation of the algebra chapter from the *Liber abbaci*,
  provided with "some clarifications, specification of the rules in relation
  to the cases presented in book XIII, and the completion of calculations,
  which the ancient master had often neglected, indicating only the result"
  (Franci and Toti Rigatelli, 1983, p. 309); a problem collection going back
  to Giovanni di Bartolo (fl. 1390–1430, a disciple of Antonio de' Mazzinghi);
  and Antonio de' Mazzinghi's *Fioretti* from 1373 or earlier (Ulivi, 1998, p.
  122).

The basic problem in using this manuscript is to which extent we can rely on
Benedetto as a faithful witness of the notations and possible symbolism of the
earlier authors he cites. A secondary problem is whether we should ascribe
to Benedetto himself or to a later user a number of marginal quasi-symbolic
calculations.

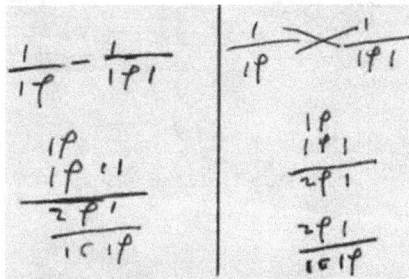

Fig. 1.9: A marginal calculation accompanying the same problem from Antonio's *Fioretti* in Siena L.IV.21, fol. 456^r and Ottobon. lat. 3307, fol. 338^v

Regarding the first problem we may observe that there are no abbreviations
or any other hints of incipient symbolism in the chapters borrowed from Fibonacci and al-Khwārizmī. This suggests that Benedetto is a fairly faithful
witness, at least as far as the presence or absence of such things is concerned.
On the other hand it is striking that the symbols he uses are the same throughout;[56] this could mean that he employed his own notation when rendering the

---

[56] One partial exception to this rule is pointed out below, note 59.

notations of others, but could also be explained by the fact that all the abbacists he cites from Biagio onward belong to his own school tradition – as observed by Raffaella Franci and Laura Toti Rigatelli (1983, p. 307), the *Trattato* is not without "a certain parochialism".

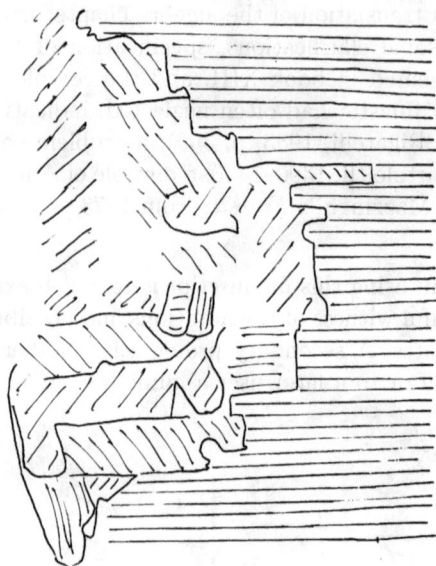

Fig. 1.10: The structure of Siena, L.IV.21, fol. 263^v. To the right, the orderly lines of the text proper. Left a variety of numerical calculations, separated by Benedetto by curved lines drawn ad hoc.

Marginal calculations along borrowed problems can obviously not be supposed a priori to be borrowed, and not even to have been written by the compiler. However, the marginal calculations in the algebraic chapters appear to be made in the same hand as marginal calculations and diagrams for which partial space is made in indentions in book XIII, chapter 2 as well as in earlier books of the treatise. Often, the irregular shape of the insertions shows these earlier calculations and diagrams to have been written before the main text, cf. fol. 263^v as shown in Figure 1.10.[57] This order of writing shows that the manuscript is Benedetto's original, and that he worked out the calculations

---

[57] This page presents a particularly striking case, and contains calculations for a very complicated problem dealing with two unknowns, a *borsa*, "[the unknown contents of] a purse", and a *quantità*, the share received by the first of those who divide its contents.

while making it – in particular because the marginal calculations are never indented in the algebra chapters copied from earlier authors.

Comparison of the marginal calculations accompanying a problem in the excerpt from Antonio's *Fioretti* and the same problem as contained in the manuscript Vatican, Ottobon. lat. 3307 from c. 1465 (on which below) show astonishing agreement, proving that these calculations were neither made by a later user nor invented by Benedetto and the compiler of the Vatican manuscript – see Figure 1.9. In principle, the calculations in the two manuscripts *could* have been added in a manuscript drawn from the *Fioretti* that had been written after Antonio's time and on which both encyclopedias build; given that the encyclopedias do not contain the same selection it seems reasonable, however, to assume that they reflect Antonio's own style – not least, as we shall see, because we are not far from what can be found in the equally Florentine *Tratato sopra l'arte della arismetricha* c. 1390, discussed around note 52.

What Benedetto does when he approaches symbolism can be summed up as follows: He uses ρ (often a shape more or less like φ) and (much less often) c and c^o for *cosa* respectively *censo* (and their plurals), but almost exclusively within formal fractions.[58] Even in formal fractions, *censo* may also be written in full. *Meno* is mostly abbreviated $\widehat{me}$ in formal fractions.[59] *Radice* may be abbreviated R in the running text, but often, and without system, it is left unabridged; within formal fractions, where there is little space for the usual abbreviation, it may become r or ra. Both when written in full and when appearing as R, it may be encircled if it is to be taken of a composite expression. In later times (e.g., in Pacioli's *Summa*, see below) this root was to be called *radice legata* or *radice universale*; the use of the circle to indicate it goes back at least to Gilio of Siena's *Questioni d'algebra* from 1384 (Franci, 1983, p. xxiii), and presumably to Antonio, since Gilio's is likely to have been taught by him or at least to have known his works well (*ibid.* pp. ivf). The concept itself, we remember, was expressed by Dardi as "R *de zonto* ... *con* ...", close in meaning to *radice legata*.

All of this suggests that the "symbolism" is only a set of facultative abbreviations, and not really an incipient symbolism. However, in a number of

---

[58] Outside such fractions, I have noticed ρ three times in the main text of the *Fioretti*, *viz* on fols. 453r, 469r and 469v (of which the first occurrence seems to be explained by an initial omission of the word chosa leaving hardly space for the abbreviation), and c^o once, on fol. 458^r. Arrighi (1967, p. 22) claims another c^o on fol. 453^r, but the manuscript writes *chosa* in the corresponding place.

[59] Additively composite symbolic expressions are mostly constructed by juxtaposition (in running text as well as marginal computations); in rhetorical exposition, e or (when a root and a number are added) an unabbreviated *più* is used. A few marginal diagrams in the section copied from Bartolo mark additive contributions to a sum by p, and all subtractive contributions by m.

marginal calculations it does serve as carrier of the reasoning. One example
was shown in Figure 1.9, another one (fol. 455^r, see Figure 1.11) performs a
multiplication which, in slightly mixed notation, looks as follows:

$$(1\rho \, \widehat{me} \, \mathrm{R}[13\tfrac{1}{2} \, \widehat{me} \, 1 \ c]) \times (1\rho \ \mathrm{p[iù]} \ \mathrm{R}[13\tfrac{1}{2} \, \widehat{me} \, 1 \ c])$$

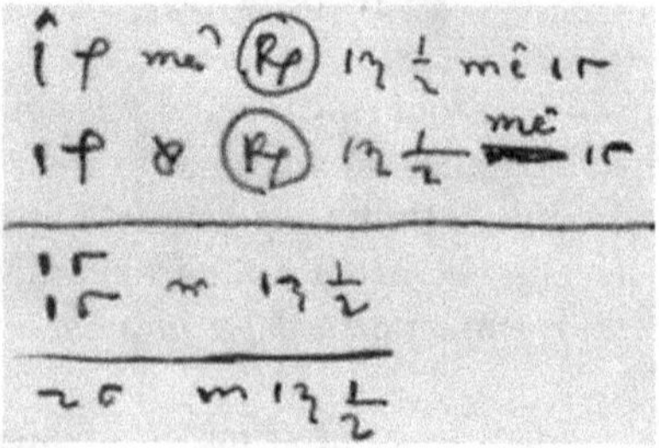

Fig. 1.11: The multiplication of $1\rho - \sqrt{13\tfrac{1}{2} - 1c}$ by $1\rho + \sqrt{13\tfrac{1}{2} - 1c}$

Formal fractions *without* abbreviation are used in the presentation of the
arithmetic of algebraic powers in Book XIII (fols. 372r–373r). At first in this
piece of text we find

> Partendo chose per censi ne viene rotto nominato da chose chome partendo 48 chose
> per 8 censi ne viene $\frac{6}{1\ chosa}$.

in translation

> Dividing *things* by *censi* results in a fraction denominated by *things*, as dividing 48
> *things* by 8 *censi* results in $\frac{6}{1\ chosa}$.

Afterwards we find denominators "1 *censo*", "1 *cubo*", "1 *cubo di censo*", etc.
When addition of such expressions and the division by a binomial are taught,
we also find denominators like "3 *cubi* and 2 *cose*".[60]

Long before we come to the algebra, namely on fols. 259v–260v, there is an
interesting appearance of formal fractions in problems of combined works,
involving not a *cosa* or a *censo* but a *quantità* – such as $\frac{8}{1\ quantita}$ and

---

[60] This whole section looks as if it was inspired by al-Karajī or the tradition he inaugurated;
but more or less independent invention is not to be excluded: once the notation for fractions
is combined with interest in the arithmetic of algebraic monomials and binomials things
should go by themselves.

$\frac{1\ quantita\ meno\ 8}{1\ chosa}$.[61] These fractions are written without any abbreviation.[62] Together with the explanation of the division of algebraic powers they demonstrate (as we already saw it in the *Trattato dell'alcibra amuchabile*) that the use of and the argumentation based on formal fractions do not depend on the presence of standard abbreviations for the unknown (even though calculations involving products of unknown quantities become heavy without standard abbreviations).

The manuscript Vatican, Ottobon. lat. 3307, was already mentioned above.[63] Like Benedetto's *Trattato*, it was written in Florence; it dates from c. 1465, and is also encyclopedic in character but somewhat less extensive than Benedetto's treatise, of which it is probably independent in substance.[64] It presents itself (fol. 1^r) as *Libro di praticha d'arismetrica, cioè fioretti tracti di più libri facti da Lionardo pisano* – which is to be taken *cum grano salis*, Fibonacci is certainly not the main source.

Judged as a mathematician (and as a Humanist digging in his historical tradition), the present compiler does not reach Benedetto's shoulders. However, from our present point of view he is very similar, and the manuscript even presents us with a couple of innovations (which are certainly not of the compiler's own invention).

Even in this text, margin calculations are often indented into the text in a way that shows them to have been written first, indicating that it is the compiler's autograph.[65] Already in an intricate problem about combined works (not the same as Benedetto's, but closely related) use is made of formal fractions involving an unknown (unabbreviated) *quantità*. Now, even the square of the *quantità* turns up, as *quantità di quantità*.

---

[61] Benedetto would probably see these solutions not as applications of algebra but of the *regula recta* – which he speaks of as *modo retto/repto/recto* in the *Tractato d'abbaco*, ed. (Arrighi, 1974, pp. 153, 168, 181), everywhere using *quantità* for the unknown.

[62] However, in the slightly later problem about a *borsa* and a *quantità* mentioned in note 57, these are abbreviated in the marginal computations – perhaps not only in order to save space (already a valid consideration given how full the page is) but also because it makes it easier to schematize the calculations.

[63] Description with extracts in (Arrighi, 2004/1968).

[64] The *idea* of producing an encyclopedic presentation of abbacus mathematics may of course have been inspired by Benedetto's *Trattato* from 1463 – unless the inspiration goes the other way, the dating "c. 1465" is based on watermarks (Van Egmond, 1980, p. 213) and is therefore only approximate. If the present compiler had emulated Benedetto, one might perhaps expect that he would have indicated it in a heading, as does Benedetto when bringing a whole sequence of problems borrowed from Antonio. In consequence, I tend to suspect that the Ottoboniano manuscript precedes Benedetto's *Trattato*.

[65] This happens seven times from fol. 48^v to fol. 54^v. On fols. 176^v and 211^v there are empty indentions, but these are quite different in character, wedge-shaped and made in the beginning of problems, and thus expressions of visual artistry and not evidence that the earlier indentions were made as empty space while the text was written and then filled out afterwards by the compiler or a user.

When presenting the quotients between powers, the compiler writes the names of powers in full within the formal fractions, just as done by Benedetto. The details of the exposition show beyond doubt, however, that the compiler does not copy Benedetto but that both draw on a common background; it seems likely that the present author makes an attempt to be creative, with little success. In the present treatise, the first fractional power is introduced like this (fol. 304^v):

> Partendo dramme per chose ne viene un rocto denominato da chose, chome partendo 48 dramme per 6 chose ne viene questo rotto cioè $\frac{48\,dramme}{1\,chosa}$.

The second example makes the same numerical error. From the third example onward, it has disappeared. The fourth one looks as follows (fol. 305^r):

> Partendo chose per chubi ne viene rotto nominato da chubi, come partendo 48 chose per 6 chubi, ne viene questo rotto, cioè $\frac{8\,chose}{1\,chubo}$.

Only afterwards is the reduction of the ratio between powers (*schifare*) introduced, for instance, that $\frac{8\,chose}{1\,chubo}$ is $\frac{8\,dramme}{1\,censo}$.

Abbreviations for the powers are absent not only from this discussion but also from the presentation of the rules. When we come to the examples, however, marginal calculations with binomials expressed by means of abbreviations abound. That for *cosa* changes between ρ and φ, that for *censo* between c (written C) and σ (actually σ); in both cases the difference is simply the length of the initial stroke; since all intermediate shapes are present, a single grapheme is certainly meant for *cosa* as well as *censo*. c^o appears to be absent. In the marginal computations, *più* may appear as p, whereas *meno* may be may be m or $m\hat{e}$.[66] However, addition may also indicated by mere juxtaposition. The marginal calculations mostly have the same character as those of Benedetto, cf. Figure 1.9; in the running text abbreviations are reserved for formal fractions and otherwise as absent as from Benedetto's *Trattato*.

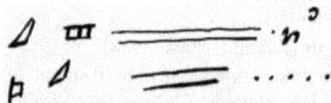

Fig. 1.12: The marginal note from Ottobon. lat. 3307 fol. 309^r

On two points the present manuscript goes slightly beyond Benedetto. Alongside a passage in the main text which introduces cases involving *cubi* and *censi di censi* (fol. 309^r), the margin contains the note shown in Figure 1.12.

---

[66] m and $m\hat{e}$ appear in the same calculation on fol. 31^v – by the way together with p.

n^o being *numero* and the superscript square being known (for instance from Vat. lat. 10488, cf. above) to be a possible representative for *censo*, it is a reasonable assumption (which we shall find fully confirmed below) that the triangle stands for the cube and the double square for *censo di censi*, the whole diagram thus being a pointer to the equation types "*cubi* and *censi di censi* equal number" and "*censi* and *cubi* equal number". We observe that equality is indicated by a double line.[67] As we shall see imminently, the compiler and several other fifteenth-century writers indicate equality by a single line. This, as well as the deviating symbols for the powers, suggests that this particular note was made by a later user of the manuscript.

The other innovation can be safely ascribed to the hand of the compiler if not (as an innovation) to his mind. It is a marginal calculation found on fol. 331^v, alongside a problem $\frac{100}{1\,\rho} + \frac{100}{1\,\rho+7} = 40$ (these formal fractions, without + and =, stand in the text). The solution follows from a transformation

$$\frac{100\rho + 100 \cdot (\rho + 7)}{(1\rho) \cdot (1\rho + 7)} = \frac{100\rho + (100\rho + 700)}{1\sigma + 7\rho} = 40$$

whence $200\rho + 700 = 40\sigma + 280\rho$. In the margin, the same solution is given schematically:

$$
\begin{array}{l}
100\rho \\
\underline{100\rho \ \ 700} \\
\underline{200\rho \ \ 700} \\
\ \ \underline{1\sigma \ \ 7\rho} \\
\hphantom{200\rho \ \ 700 \ \ \ } 40
\end{array}
$$

$$200\rho \ \ 700 \ \text{———} \ 40\sigma \qquad \langle 280\rho\rangle$$

(the omitted $\langle 280\rho\rangle$ in the last line is present within the main text). The strokes before 40 and 40σ appear to be meant as equation signs. It might be better, however, to understand them as all-purpose "confrontation signs" – in the margin of fol. 338^r, ——— means that one commercial partner has $\frac{3000}{1\rho\,5000}$, the other $\frac{4000}{1\rho\,6000}$ (see Figure 1.13).[68]

---

[67] The double line is also used for equality in a Bologna manuscript from the mid-sixteenth century reproduced in (Cajori, 1928, I, p. 129); whether Recorde's introduction of the same symbol in 1557 was independent of this little known Italian tradition is difficult to decide. In any case, the combination with the geometric symbols indicates that the present example (and thus the Italian tradition) predates Recorde by at least half a century or so.

[68] As we shall see, Raffaello Canacci also uses the line both for equality and for confrontation. Even Widmann (1489) uses the long stroke for confrontation: fols. 12r, 21r–v, 23r, 27r, 38v when confronting the numbers 9 and 7 with the schemes for casting out nines and sevens, fol. 193^v (and elsewhere) when stakes and profits in a partnership are confronted.

This is one of Antonio's problems. In Benedetto's manuscript, we find the same problem and the same diagram on fol. 456^r – with the only difference that the line is replaced by an X indicating the cross-multiplication that is to be performed – see Figure 1.9. The "confrontation line" is thus not part of the inheritance from Antonio (nor, in general, of the inheritance shared with Benedetto). Though hardly due to the present compiler, it *is* an innovation.

The reason to doubt the innovative role of our compiler is one of Regiomontanus's notes for the Bianchini correspondence from c. 1460 (ed. Curtze, 1902, p. 278). For the problem $\frac{100}{1\rho} + \frac{100}{1\rho+8}$, he uses exactly the same scheme, including the "confrontation line":

$$\frac{100}{1\rho} \qquad\qquad \frac{100}{1\rho+8}$$

$$100\rho \text{ et } 800$$
$$\underline{100\rho}$$
$$\overline{200\rho \text{ et } 800}$$
$$1\rho \text{ et } 8\ \sigma \quad\text{———}\quad 40$$
$$40\ \sigma \text{ et } 320\rho \quad\text{———}\quad 200\rho \text{ et } 800$$
$$40\ \sigma \text{ et } 120\rho \quad\text{———}\quad 800$$
$$1\ \sigma \text{ et } 3\rho \quad\text{———}\quad 20$$

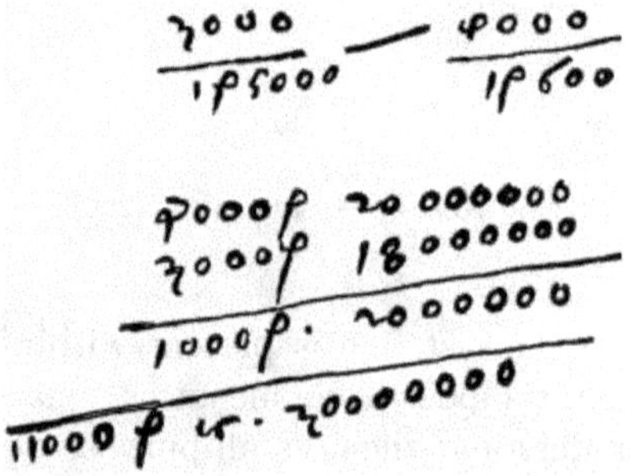

Fig. 1.13: The confrontation sign of Ottobon. lat. 3307 fol. 338^r

(Regiomontanus extends the initial stroke of ρ even more than our compiler, to Υ; his variant of σ, *census*, is $\mathcal{C}$, possibly a different extension of c)[69].

A third Florentine encyclopedic abbacus treatise is Florence, Bibl. Naz. Centr., Palat. 573.[70] Van Egmond (1980, p. 124) dates it to c. 1460 on the

---

[69] Curtze does not show these shapes in his edition, but see (Cajori, 1928, I, p. 95).

[70] Described with sometimes extensive extracts from the beginnings of all chapters in

basis of dates contained in problems, but since the compiler refers (fol. 1^r) to Benedetto's *Trattato* (from 1463) as having been made "already some time ago" (*già è più tenpo*), a date around 1470 seems more plausible. This is confirmed by the watermarks referred to by Van Egmond – even this manuscript can be seen from marginal calculations made before the writing of the main text to be the compiler's original, whose date must therefore fit the watermarks.

As regards algebraic notations and incipient symbolism, this treatise teaches us nothing new. It does not copy Benedetto (in the passages I checked) but does not go beyond him in any respect; it uses the same abbreviations for algebraic powers, in marginal calculations and (sparingly) in formal fractions within the main text – including the encircled *radice* and ℞. In the chapter copying Fibonacci's algebra it has no marginal calculations (only indications of forgotten words), which confirms that the compilers of the three encyclopedic treatises copied the marginal calculations and did not add on their own when copying – at least not when copying venerated predecessors mentioned by name.

1.7 Late fifteenth-century Italy

The three encyclopediae confirm that no systematic effort to develop notations or to extend the range of symbolic calculation characterizes the mid-century Italian abbacus environment – not even among those masters who, like Benedetto and the compiler of Palat. 573, reveal scholarly and Humanist ambitions by including such matters as the Boethian names for ratios in their treatises and by basing their introduction of algebra on its oldest author (al-Khwārizmī).[71] The experiments and innovations of the fourteenth century – mostly, so it seems, vague reflections of Maghreb practices – had not been developed further.[72] In that respect, their attitude is not too far from that of mid-fifteenth–century mainstream Humanism.

(Arrighi, 2004/1967).

[71] Benedetto (ed. Salomone, 1982, p. 20) gives this argument explicitly; the compiler of Palat. 573 speaks of his wish that "the work of Maumetto the Arab which has been almost lost be renovated" (Arrighi, 2004/1967, p. 191).

[72] It is true that we have not seen the quotients between powers expressed as formal fractions in earlier manuscripts; however, the way they turn up independently in all three encyclopædiæ shows that they were already part of the heritage – *perhaps* from Antonio. The interest in such quotients is already documented in Giovanni di Davizzo in 1339, who however makes the unlucky choice to identify negative powers with roots – see (Høyrup, 2007c, pp. 478–484) (and cf. above, before note 44).

Fig. 1.14: The two presentations of the algebraic powers in Bibl. Estense, ital. 578

Towards the end of the century we have evidence of more conscious exploration of the potentialities of symbolic notations. A first manuscript to be mentioned here is Modena, Bibl. Estense, ital. 578 from c. 1485 (according to the orthography written in northern Italy – e.g., *zonzi* and *mazore* where Tuscan normal orthography would have *giongi* and *magiore*).[73] It contains (fols. 5r–20r) an algebra, starting with a presentation of symbols for the powers with a double explanation, first with symbols and corresponding "degrees", *gradi* (fol. 5^r), next by symbols and signification (fol. 5^v) – see Figure 1.14.

As we see, the symbol for the *cosa* is the habitual *c*. For the *censo*, *z* is used, in agreement with the usual northern orthography *zenso* – however, in a writing which is quite different from the *z* used in full writing of *zenso* (**z** respectively **ʒ**, see also Figure 1.15); the *cubo* is *Q*, the fourth power is *z di z*. The fifth power is *c di zz*, obviously meant as a multiplicative composition (as the traditional *cubo di censo*), the sixth instead *z di Q*, that is, composed by embedding. The seventh degree is *c di z di Q*, mixing the two principles, the eight again made with embedding as *z di zz*. So is the ninth, *QQ*.

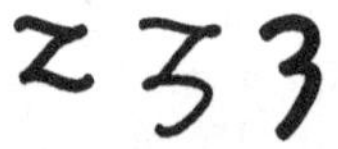

Fig. 1.15: Three graphemes from Bibl. Estense, ital. 578. Left, *z* abbreviating *zenso* in the initial overview; centre, *z* as written as part of the running text; right, the digit 3

Then follow the significations. *c* is "that which you find", *z* "the root of that", *Q* "the cube root of that", and *z di z* "the root of the root of that". Already now we may wonder – why "roots"? I have no answer, but discuss possible

---

[73] (Van Egmond, 1986) is an edition of the manuscript. It has some discussion of its symbolism but does not go into details with the written shapes, for which reason I base my

hints in (Høyrup, 2008, p. 31), in connection with the *Tratato sopra l'arte della arismetricha* (see just before note 52), from where these "root-names" are known for the first time.[74] It is reasonable to assume a connection – this *Tratato* has the same mixture of multiplicative and embedding-based formation of the names for powers, though calling the fifth degree *cubo di censo*, and the sixth (like here) *censo di cubo*.[75]

The root names go on with "root of this" for the fifth power – which is probably meant as "5th root of this", since the seventh power is "the 7th root of this". The names for the sixth, eighth and ninth degree are made by embedding.

After explaining algebraic operations and the arithmetic of monomials and binomials the manuscript offers a list of algebraic cases followed by examples illustrating them. Here the same symbols are used within the text (there are no marginal calculations) – with one exception, instead of z a sign is used which is a transformed version of Dardi's ς – ς, with variations that sometimes make it look like a z provided with an initial and a final curlicue.[76]

The problems are grouped in *capitoli* asking for the same procedure in spite of involving different powers – chapter 14, for instance, combines "*zz* and *z di zz* equal to n^o" and "*c di zz* and *QQ* equal to *c*". The orderly presentation of the powers in a scheme and the concept of numerical *gradi*, "degrees", (our exponents) has facilitated this further ordering. This is clear from the presentation – in chapter 14, "When you find three names of which one is 4 degrees more than the other ...". Beyond this, the abbreviations seem to serve as nothing but abbreviations, though used consistently.

---

discussion on the manuscript.

[74] Van Egmond (1986, 20) "explains" them $Z = R$, $x^2 = n \rightarrow x\sqrt{n}$ etc., which however, while being an impeccable piece of mathematics, is completely at odds with the words of the text.

[75] This difference may tell us something about the spontaneous psychology of embedding: it seems to be easier to embed within a single than within a repeated multiplication – that is, to grasp *censo* of P as $(P)^2$ than to understand *cubo* of R as $(R)^3$.

[76] There are a few slips. In the initial list, a full *zenso* is once written *çenso* (written with ς), and ς itself appears once; within the list of cases and the examples a few instances of zenso abbreviated z (written $\mathfrak{z}$, not $\mathbf{z}$) occur. Van Egmond (1986, p. 23) reads these as "3", and takes this as evidence that the manuscript was made by a copyist who did not really understand but had a tendency to replace a z used in the original by ς. However, even though the writings of z and 3 are similar, magnification shows them quite clearly to be different, and makes it clear that the copyist did not write 3 where he should have written z (see Figure 1.15). Other errors pointed out by Van Egmond demonstrate beyond doubt that the beautifully written manuscript *is* a copy. However, the almost systematic distinction between the abbreviations $\mathbf{z}$ and ς, as well as the general idea of applying stylized shapes of letters when used as symbols, is likely to reflect the ways of the original – an unskilled copyist would hardly introduce them.

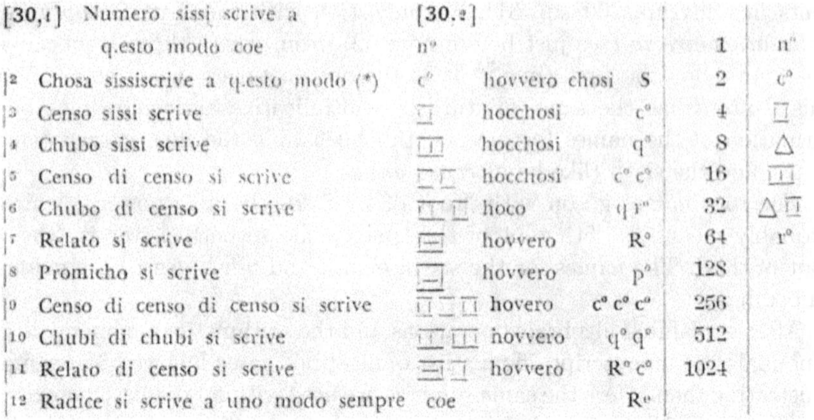

42   Jens Høyrup

[30,1]		[30.2]				
Numero sissi scrive a q.esto modo coe		n°			1	n°
|2 Chosa sississcrive a q.esto modo (*)	c°	hovvero chosi	S		2	c°
|3 Censo sissi scrive		hocchosi	c°		4	
|4 Chubo sissi scrive		hocchosi	q°		8	
|5 Censo di censo si scrive		hocchosi	c°c°		16	
|6 Chubo di censo si scrive		hoco	q r"		32	
|7 Relato si scrive		hovvero	R°		64	r°
|8 Promicho si scrive		hovvero	p°		128	
|9 Censo di censo di censo si scrive		hovero	c° c° c°		256	
|10 Chubi di chubi si scrive		hovvero	q°q°		512	
|11 Relato di censo si scrive		hovvero	R° c°		1024	
|12 Radice si scrive a uno modo sempre coe			R°			

Fig. 1.16: Canaccis scheme with the naming of powers, after (Procissi, 1954, p. 432)

Raffaello Canacci's use of schemes for the calculation with polynomials (including multiplication *a casella*) in the *Ragionamenti d'algebra*[77] from c. 1495 (ed. Procissi 1954, pp. 316–323) was mentioned above. In a couple of these he employs geometric signs for the powers, but mostly he writes *s* for *cosa* and *censo* in full. Addition may be indicated by juxtaposition, by *e*, by *più* or by *p*, subtraction by *m̂* or *me*.[78] Later he presents an ordered list, with three different systems alongside each other – see Figure 1.16. To the right we find an extension of a different "geometric" system – namely the one which was found in a (secondary) marginal note in the Ottoboniano encyclopædia. Next toward the left we find powers of 2 corresponding to the algebraic powers (an explanatory stratagem also used by Pacioli in the *Summa*); then letter abbreviations; and then finally, just to the right of the column with Canacci's full names, his own "geometric" system (not necessarily invented by him, cf. imminently, but the one he uses in the schemes) – better planned for the economy of drawing than as a support for operations or algebraic thought. According to Cajori (1928, I, pp. 112*f*) the system turns up again in Ghaligai's *Pratica d'arithmetica* from 1552 (and probably in the first edition from 1521, entitled *Summa de arithmetica*), where their use is ascribed to Ghaligai's teacher Giovanni del Sodo.

---

[77] Florence, Bibl. Naz. Centr., Palat. 567. I have not seen the manuscript but only Angiolo Procissi's diplomatic transcriptions.

[78] However, *p n* and *p n°* stand for "per numero". In schemes showing the stepwise calculation of products (pp. 313f), *m* stands for multiplication. In one scheme p. 318), a first *p* stands for *più*, a second in this way for *per*.

Canacci uses these last geometric signs immediately afterwards in a brief exposition of the rules for multiplying powers – and then no more. In a couple of marginal notes to the long collection of problems (ed. Procissi, 1983, pp. 58, 62–64) he uses the letter abbreviations (only s and c^o) – but also the line as an indication, once of equality, twice of confrontation or correspondence not involving equality. The running text, including formal fractions, writes the powers unabridged (except *numero*, which once is n^o); even *più* and *meno* are mostly written in full, but *meno* sometimes (pp. 21–23) with a brief stroke "–" – the earliest occurrence of the minus sign in Italy I know of.[79]

Three works by Luca Pacioli are of interest: the Perugia manuscript from 1478, the *Summa de arithmetica* from 1494, and his translation of Piero della Francesca's *Libellus de quinque corporibus regularibus* as printed in (Pacioli, 1509).

Since there is only one brief observation to make on the latter work, I shall start by that. According to the manuscript Vatican, Urb. lat. 632 as edited by G. Mancini (1916, pp. 499–501), Piero uses the familiar superscript square for *censo* when performing algebraic calculations, or he writes words; for *res* he uses a horizontal stroke over the coefficient, but mostly also keeps the word.[80] Pacioli (1509, fols. 3v–26r, *passim*) instead uses a sign ◇ for the *cosa* and □ for the *censo* (or, in the old unsystematic way, words). *Censo di censi* is □□ on fol. 4^r and □ *de* □ on fols. 4r and 11v. These geometric signs are absent from Pacioli's other works, and they must rather be considered a typographic experiment – given that their use is not systematic, they can hardly be understood as an instance of mathematical exploration beyond what Pacioli had done before. It is difficult to agree with Paola Manni (2001, p. 146) that they should represent "progress of mathematical symbolism" with respect to the more systematic use of letter abbreviations in the Perugia manuscript and the *Summa* (see imminently; and cf. the quotation from Woepcke after note 12). Indeed, the *Libellus* is an appendix to Pacioli's *Divina proportione*, in which Pacioli (1509, fol. 3^v) explains that various professions, among whom *le mathematici per algebra*, use specific *caratheri e abreviature* "in order to avoid prolixity in writing and also of reading".[81]

The 1478 Perugia manuscript *Suis carissimis disciplis ...* (Vatican, Vat. lat. 3129) has lost the systematic algebra chapters listed in the initial table

---

[79] As well known, "–" is already used in the *Deutsche algebra* from 1481 (ed. Vogel, 1981, p. 20). Whether this is part of the very mixed Italian heritage of this manuscript (see below, note 88 and surrounding text) or a German innovation eventually borrowed by Ghaligai is undecidable unless supplementary evidence should turn up.

[80] The same (lack of) system is found in his abbacus treatise, see (Arrighi, 1970, p. 12).

[81] That Pacioli really thinks in terms of abbreviations is confirmed by a list of examples given in the manuscript of the treatise (Milano, Biblioteca Ambrosiana, Ms. 170 Sup., written in 1498), see (Maia Bertato, 2008, 13): it mixes the abbreviations for *radice*, *più*, *meno*, *quadrato* (*cosa* and *censo* are absent) with others for, *inter alia*, *linea*, *geometria*

of contents,[82] but it does contain a large amount of algebraic calculation. Everywhere here – in the main text as well as in the margin, and in the neat original prepared in 1478 as well as in fols. 350r–360v, added at a later moment and obviously very private notes – we find the signs from Canacci's right-hand column (Figure 1.16) written superscript and to the right – on fol. 360^v extended until ⊟, *censi di censi di censi*. *Meno* is Ⓜ and *più* (both signifying addition and as a normal word) a corresponding encircled *p*. This is thus the system which Pacioli used when calculating for himself, at least at that moment.[83] He uses the equality line in the margin (but also the same line indicating confrontation/correspondence, e.g., fol. 130^r).

Most important (in the sense that it was immensely influential and the other two works not) is of course the *Summa* (Pacioli, 1494). Typographic constraints are likely to have caused Pacioli to give up his usual notation. In ordinary algebraic explanation and computation, he now uses *.co.* and *.ce.* written on the line, and *più* and *meno* have become *p̃* and *m̃* (*meno* sometimes *mē*) – both as operators and as indicators of positivity and negativity (not only additivity and subtractivity).[84] However, he also has more systematic presentations. The first, in the margin of fol. 67^v, shows how the sequence *.co.-.ce.* is to be continued, namely (third power) *cubo*, (4th) *censo de censo*, (5th) *primo relato*, (6th) *censo de cubo/cubo de senso*, (7th) *secundo relato*, (8th) *censo de censo de censo*, (9th) *cubo de cubo*, (10th) *censo de primo relato*, (11th) *terzo relato*, etc. until the 29th power. As we see, the embedding principle has taken over completely, creating problems for the naming of prime-number powers. For each power the "root name" is indicated, number being "R *prima*", *cosa* "R *2a*", *censo* "R *3a*", etc.[85] As we see, the "root number" is *not* the exponent, but the exponent augmented by 1. This diminishes the heuristic value of the concept: it still permits to see directly that "6th roots and 4th roots equal 2nd roots" must be equivalent to "5th roots and 3rd roots equal 1st roots", but it requires as much thinking as in Jacopo's days almost 200 years earlier to see that this is a biquadratic problem that must be solved in the same way as "3rd roots and 2nd roots equals 1st roots".

---

and *arithmetica*).

[82] See the meticulous description in (Derenzini, 1998), here p. 173. Since all abbreviations except the superscript symbols are expanded in the edition (Calzoni and Gavalzoni, 1996), I have used a scan of the manuscript.

[83] This restriction is probably unnecessary. At least the encircled *p* and *m* and the square are in the list offered by the 1498 manuscript, cf. note 81.

[84] E.g., on see fol. 114^r, "a partir .m̃.16.p.m̃.2. ne vene .p̃.8", and the proof that "meno via meno fa più" on fol. 11^3r, which is characterized as "absurda" and referred to the concept of a debt – if only subtractive numbers were involved, as in Dardi's corresponding proof, nothing would be absurd.

[85] Pacioli believes (or at least asserts) that these names go back to "the practice of algebra according to the Arabs, first inventors of this art". Could he have been led to this belief by the equivalence of "root" and *thing/cosa* in al-Khwārizmī's algebra?

After this list comes a list of symbols for "normal" roots: ℞ meaning *radici*; ℞℞ meaning *radici de radici*; ℞u. meaning *radici universale* or *radici legata*, that is, root of a composite expression following the root sign (encircled in Benedetto's *Trattato* and spoken of as "℞ *de zonzo*" by Dardi, we remember); and ℞ cu., cube root.

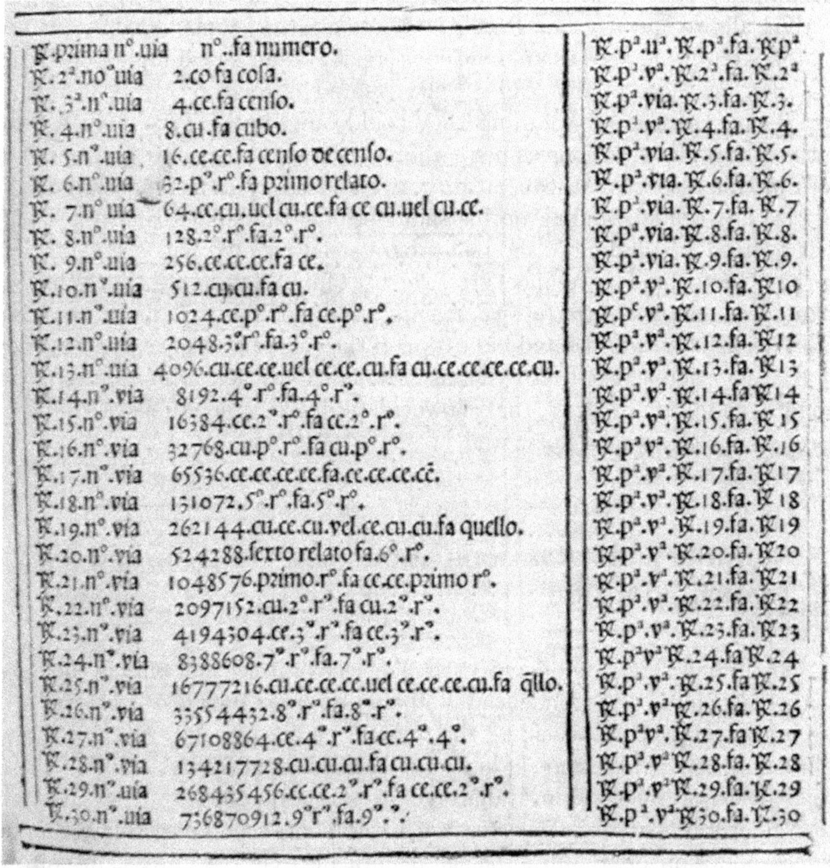

℞.prima n°.uia	n°..fa numero.	℞.pª.uª.℞.pª.fa.℞pª
℞.2ª.no°uia	2.co fa cosa.	℞.pª.vª.℞.2ª.fa.℞.2ª
℞.3ª.n°.uia	4.ce.fa censo.	℞.pª.via.℞.3.fa.℞.3.
℞.4.n°.uia	8.cu.fa cubo.	℞.pª.vª.℞.4.fa.℞.4.
℞.5.n°.uia	16.ce.ce.fa censo de censo.	℞.pª.via.℞.5.fa.℞.5.
℞.6.n°.uia	32.p°.r°.fa primo relato.	℞.pª.via.℞.6.fa.℞.6.
℞.7.n° uia	64.ce.cu.uel cu.ce.fa ce cu.uel cu.ce.	℞.pª.via.℞.7.fa.℞.7
℞.8.n°.uia	128.2°.r°.fa.2°.r°.	℞.pª.via.℞.8.fa.℞.8.
℞.9.n°.uia	256.ce.ce.ce.fa ce.	℞.pª.via.℞.9.fa.℞.9.
℞.10.n°.uia	512.cu.cu.fa cu.	℞.pª.vª.℞.10.fa.℞10
℞.11.n°.uia	1024.ce.p°.r°.fa ce.p°.r°.	℞.pª.vª.℞.11.fa.℞.11
℞.12.n°.uia	2048.3°.r°.fa.3°.r°.	℞.pª.vª.℞.12.fa.℞12
℞.13.n°.uia	4096.cu.ce.ce.uel ce.ce.cu.fa cu.ce.ce.ce.ce.cu.	℞.pª.vª.℞.13.fa.℞13
℞.14.n°.via	8192.4°.r°.fa.4°.r°.	℞.pª.vª.℞.14.fa℞14
℞.15.n°.via	16384.ce.2°.r°.fa ce.2°.r°.	℞.pª.vª.℞.15.fa.℞15
℞.16.n°.via	32768.cu.p°.r°.fa cu.p°.r°.	℞.pª vª.℞.16.fa.℞.16
℞.17.n°.via	65536.ce.ce.ce.ce.fa.ce.ce.ce.ce.	℞.pª.vª.℞.17.fa.℞17
℞.18.n°.via	131072.5°.r°.fa.5°.r°.	℞.pª.vª.℞.18.fa.℞18
℞.19.n°.via	262144.cu.ce.cu.vel.ce.cu.cu.fa quello.	℞.pª.vª.℞.19.fa.℞19
℞.20.n°.via	524288.sexto relato fa.6°.r°.	℞.pª.vª.℞.20.fa.℞20
℞.21.n°.via	1048576.primo.r°.fa ce.ce.primo r°.	℞.pª.vª.℞.21.fa.℞21
℞.22.n°.via	2097152.cu.2°.r°.fa cu.2°.r°.	℞.pª.vª.℞.22.fa.℞22
℞.23.n°.via	4194304.ce.3°.r°.fa ce.3°.r°.	℞.pª.vª.℞.23.fa.℞23
℞.24.n°.via	8388608.7°.r°.fa.7°.r°.	℞.pªvª℞.24.fa℞.24
℞.25.n°.via	16777216.cu.ce.ce.ce.uel ce.ce.ce.cu.fa q̄llo.	℞.pª.vª.℞.25.fa℞.25
℞.26.n°.via	33554432.8°.r°.fa.8°.r°.	℞.pª.vª℞.26.fa.℞.26
℞.27.n°.via	67108864.ce.4°.r°.fa ce.4°.4°.	℞.pªvª.℞.27.fa℞.27
℞.28.n°.via	134217728.cu.cu.cu.fa cu.cu.cu.	℞.pª.vª℞.28.fa.℞.28
℞.29.n°.uia	268435456.ce.ce.2°.r°.fa ce.ce.2°.r°.	℞.pª.vª℞.29.fa.℞.29
℞.30.n°.uia	536870912.9°r°.fa.9°.°.	℞.pª.vª℞30.fa.℞.30

Fig. 1.17: Paciolis scheme (1494, fol. 143ʳ) showing the powers with root names

On fol. 143ʳ follows a scheme that deals with the first 30 powers (*dignità*), and with how they are brought forth as products (*li nascimenti pratici o li 30 gradi de li caratteri algebratici*). It runs in four tangled columns and 30 rows. The first column has the numbered "root name" of the power, the second formulates in Pacioli's normal language or in abbreviations that number times this power gives the same power. The third, written inside the second, indicates the corresponding power of 2. The fourth, finally, repeats the second column, now translated into root names – see Figure 1.17.

On the next page follow further schemes, expressed in roots names, for the products of the nth root with all roots from the nth to the $(31-n)$th (meaning that all products remain within the range defined by the 30th root), $2 \leq n \leq 15$.

All in all, we may say that Pacioli explored existing symbolic notations to a greater extent (and used them more consistently) than for example Benedetto, thus offering those of his readers who wanted it matters to chew; but he hardly gave them many solutions they could build on (and as we have seen, he thought of his notations as mere abbreviations serving to avoid prolixity). Even in this respect, subsequent authors could easily have found reasons to criticize him while standing on his shoulders (as they did regularly), if only their own understanding of the real progress they offered had been sufficient for that. Tartaglia, for instance, gives the list of *dignitates* until the 29th in *La sesta parte del general trattato* (Tartaglia, 1560, fol. 2^r), with names agreeing with Pacioli's *.co.-.ce.*-list and indication of the corresponding exponents (now *segni*), alongside a text that explains how multiplication of *dignitates* corresponds to addition of *segni*; that, however, was well after Stifel's *Arithmetica integra*, which Tartaglia knew well.

1.8 Summary observations about the German and French adoption

Regiomontanus shows familiarity with algebraic practice, not only in the notes for the Bianchini-correspondence (cf. above) but also elsewhere – several articles in (Folkerts, 2006) elucidate the topic in detail. Not only the calculation before note 69 but also some of his abbreviations (and the variability of these) are evident borrowings from Italian models (Høyrup, 2007c, p. 134). It might seem a not impossible assumption that Regiomontanus was the main channel for the adoption of Italian abbacus algebra into German areas, in spite of his purely ideological ascription of the algebraic domain to Diophantos and Jordanus (above, text before note 24).

An influence cannot be excluded, even though those of Regiomontanus' algebraic notes we know about may not have circulated widely. However, those of his symbolic notations or abbreviations which are not to be identified as Italian are already present in a section of a manuscript possessed by Regiomontanus but not written by him (Folkerts, 2006, V, pp. 201f), cf. (Høyrup, 2007c, pp. 136f).[86].

That Regiomontanus was at most one of several channels can also be seen from the so-called *Deutsche Algebra* from 1481 (ed. Vogel, 1981). Its symbols[87] for *number* (*denarius*, replaces earlier *dragma*), *thing* and *census* coincide with those of the Robert-Appendix,[88] that for the *cube* with the one Regiomontanus employs for *census* – hardly evidence for inspiration from the latter. A token of Italian inspiration certainly *not* passing through Regiomontanus is occasional use of the quasi-fraction notation for powers and of 1c for *cosa* (Vogel, 1981, p. 10) – all in all, as Kurt Vogel observes, evidence that a number of sources flow together in this manuscript.

I shall not consider in detail German algebraic writings from the sixteenth century (Rudolff, Ries, Stifel, Scheubel), only sum up that with time German algebra tends to be more systematic and coherent in its use of symbolism (for notation as well as calculation) than any single Italian treatise.[89] But what the German authors do is to combine and put into system ideas that are all present in *some* Italian work. They never really go beyond the Italian inspiration *seen as a whole*, and never attain the coherence which appears to have been reached by the Maghreb algebraists of the twelfth century.[90]

I shall also be brief on what happened in French area. Scrutiny of Nicolas Chuquet's daring exploration of the possibilities of symbolism in the *Triparty* from 1484 (ed. Marre, 1880) would be a task of its own; his parenthesis (an underlining[91]) and his complete arithmetization of the notation for powers

---

[86] The *thing* symbol in the appendix to Robert of Chester's translation of al-Khwārizmī is the same as Regiomontanus's transformation of ρ ; the *census* symbol is a z provided with a final curlicue and which *could* be derived from the Ç which we find in the Modena-manuscript but is much more likely to correspond to its initial use of z in this function.

[87] Listed in (Vogel, 1981, p. 11).

[88] With ∂ as an alternative for *thing*, standing probably for *dingk*.

[89] The use of schemes for polynomial arithmetical calculation by Stifel (1544) and Scheubel (1551) was mentioned above. They also appear in Rudolff's *Coss* (1525).

[90] Quite new, as far as I know, and awkwardly related to the drive toward more systematic use of notations (but maybe more closely to the teaching of Aristotelian logic), is the idea to represent persons appearing in commercial problems by letters A, B, C, ... I have noticed it in Magister Wolack's Erfurt lecture from 1467, apparently the earliest public presentation of abbacus mathematics in German land (ed. Wappler, 1900, pp. 53f), and again in Christoff Rudolff's *Behend und hübsch Rechnung durch die kunstreichen Regeln Algebra* #128 (1525, fol. Nv^{r-v}).

[91] The only parentheses Italian symbolic notation had made use of were those marked off by the fraction line and the R *de zonzo*/*legata*/*universale*. The latter, furthermore, was ambiguous – how far does the expression go that it is meant to include? (Actually, I have not seen it go beyond two terms, which may indeed have been part of the concept.) A parenthesis as good and universal as that of Chuquet had to await Bombelli (1572), even though Pacioli (1494) uses brackets containing *textual* parentheses (e.g., on fol. 3^r). As we remember from note 12, even Descartes eschews general use of the parenthesis.

as well as roots certainly goes beyond what can be found in anything Italian until Bombelli, and (as far as the symbols for powers and roots are concerned) even beyond the Maghreb notation. However, his innovations were historical dead ends; Etienne de la Roche, while transmitting other aspects of Chuquet's mathematics in his *Larismetique* from 1520, returned to more familiar notations (Moss, 1988, pp. 120*f*). What later authors learned (or, like Buteo, refused to learn, *ibid.*, p. 123) from de la Roche could as well have been Italian.[92]

As a representative of the French mid-sixteenth century I shall choose Jacques Peletier's *L'algebre* from (1554) – interesting not least because his orthographic reform proposal (1555; 1554, final unpaged note) shows him to have reflected on notation. Peletier knows Stifel's *Arithmetica integra*, cites it often and learns from it. But he must be acquainted with the Italian abbacus tradition, and not only through Pacioli and Cardano, both of whom he cites on p. 2: he speaks of the powers as *nombres radicaus* (p. 5), and uses ℞ for the first power (this, as well as the *nombres radicaus, could* at a pinch be inspired by Pacioli) and the stylized ç (**ς**) which we know from the Modena-manuscript for the second power (following Stifel for higher powers). That certainly does not help him go beyond the combination of the most developed elements of Italian symbolism we know from the German authors – and like Stifel he does not get beyond.

1.9 Why should they?

As we have seen, Italian abbacus algebra makes use of a variety of elements that might have been (and in the main probably were) borrowed from the Maghreb, most of them already present in one or the other manuscript from the fourteenth century. But the abbacus masters do not seem to have been eager to use them consistently, to learn from each other or to surpass each other in this domain (to which extent they wanted to avoid to *teach* symbolism is difficult to know – it will not have had the same value in the competition for jobs and pupils as the ability to solve intricate questions); Benedetto and the compilers of the Ottoboniano and Palatino encyclopædiae were quite satisfied with repeating a heritage that may reach back to Antonio, and did not care about the schemes for polynomial arithmetic that had been in circulation at least since Dardi's times. Only with the Modena manuscript, with Canacci

---

[92] The question to which extent the Provençal tradition which Chuquet draws upon was independent of the Italian tradition (to some extent it certainly was) is immaterial for the present discussion; no surviving earlier or near-contemporary Provençal writings offer as much incipient symbolism as the Italian abbacus writers.

and with Pacioli's *Summa* do we find some effort to be encyclopedic (if not systematic) also in the presentation of notations.

Our meeting is about the "philosophical aspects of symbolic reasoning", and about "early modern science and mathematics". The philosophical question to raise to the material presented above is whether the abbacus masters of the fourteenth and fifteenth century, and even the algebraic writers of the early and mid-sixteenth century, had any *reason* to develop a coherent symbolic approach. The answer seems to be that they had none (cf. also note 50 and preceding text). The kind of mathematics they were engaged in (even when they applied their art to *Elements* X, as do for instance Fibonacci and Stifel) did not ask for that. They might sometimes extrapolate their technique further than their mathematical practice asked for – 29 algebraic powers is an example of that, as is of course the creation of never-used symbols for these powers. But without a genuine practice there was nothing which could force these extrapolations to merge into a consistent conceptual and operational framework. Even those abbacus authors that had scholarly ambitions – as Benedetto and his contemporary encyclopedists, Pacioli and Tartaglia – did not encounter anything within the practice of university or Humanist mathematics which asked for much more than they did. To the contrary, the aspiration to connect their mathematics to the Euclidean ideal made them re-attach geometric proofs to a tradition from which these had mostly been absent, barring thereby the insight that purely arithmetical reasoning could be made as rigorous as geometric proofs – barring it indeed to such an extent that Ries and Scheubel rejected Jordanus' arithmetical rigor and borrowed only his problems, as we have seen.

That changed in the outgoing sixteenth century. By then (if I may be allowed some concluding sweeping statements), Apollonios, Archimedes and Pappos were no longer mere names (or at most authors of difficult texts to be assimilated) but providers of problems to be worked on, and trigonometry had become an advanced topic. This was probably what created the pull on the development of symbolic reasoning and of those notations that symbolic reasoning presupposed if it was to go beyond simple formal fractions;[93] the reaction to this pull (which at first created a complex of new mathematical developments) was what ultimately transformed symbolic mathematics into

---

[93] It may perhaps be allowed to give a frivolous illustration of a sweeping statement: the problems which the 16–17 years old Huygens investigated by means of Cartesian algebra under the guidance of Frans van Schooten. Quite a few of them deal with matters from Archimedes or Apollonios (Huygens, 1908, 27–60). The problems he dealt with 4–5 years later (pp. 217–275 in the same volume) are derived from Pappos, and even they make extensive use of Descartes' technique. This is thus what a young but brilliant mathematical mind was training itself at a decade after the appearance of Descartes' *Geometrie*.

It is difficult to imagine that these problems could have been well served by *cossic* algebra, with or without the abbreviations that had been standardized in the mid-sixteenth century.

a factor that could (eventually) push the development of (some constituents of) early modern science.

Acknowledgements I started work on several of the manuscripts used above during a stay at the Max-Planck-Institut für Wissenschaftsgeschichte, Berlin, in October 2008. It is a pleasant duty to express my gratitude for the hospitality I enjoyed.

I also thank Mahdi Abdeljaouad for extensive commentaries to the pages on Maghreb algebra.

Finally, thanks are due to the the Biblioteca Estense, Modena, for permission to publish reproductions from the manuscript Bibl. Estense, ital. 587.

References

1. Abdeljaouad, Mahdi, 2002. "Le manuscrit mathématique de Jerba: Une pratique des symboles algébriques maghrébins en pleine maturité". Présenté au Septième Colloque sur l'histoire des mathématiques arabes (Marrakech, 30–31 mai et 1er juin 2002). *Mimeo.*

2. Abdeljaouad, Mahdi (ed.), 2004. Ibn al-Hā'im, *Sharh al-Urjūza al-Yasminīya.* Tunis: Publications de l'Association Tunisienne des Sciences Mathématiques, n. d. [2004].

3. Arrighi, Gino (ed.), 1967. Antonio de' Mazzinghi, *Trattato di Fioretti* nella trascelta a cura di Mᵒ Benedetto secondo la lezione del Codice L.IV.21 (sec. XV) della Biblioteca degl'Intronati di Siena. Siena: Domus Galilaeana.

4. Arrighi, Gino (ed.), 1974. Pier Maria Calandri, *Tractato d'abbacho.* Dal codice Acq. e doni 154 (sec. XV) della Biblioteca Medicea Laurenziana di Firenze. Pisa: Domus Galiaeana. [The ascription is mistaken, the treatise is by Benedetto da Firenze.]

5. Arrighi, Gino (ed.), 1987. Paolo Gherardi, *Opera mathematica: Libro di ragioni – Liber habaci.* Codici Magliabechiani Classe XI, nn. 87 e 88 (sec. XIV) della Biblioteca Nazionale di Firenze. Lucca: Pacini-Fazzi.

6. Arrighi, Gino (ed.), 1989. "Maestro Umbro (sec. XIII), *Livero de l'abbecho.* (Cod. 2404 della Biblioteca Riccardiana di Firenze)". *Bollettino della Deputazione di Storia Patria per l'Umbria* **86**, 5–140.

7. Arrighi, Gino, 2004/1965. "Il codice L.IV.21 della Biblioteca degl'Intronati di Siena e la 'bottega dell'abaco a Santa Trinita di Firenze'". pp. 129–159 *in* Gino Arrighi, *La matematica nell'Età di Mezzo. Scritti scelti,* a cura di F. Barberini, R. Franci and L. Toti Rigatelli Pisa: Edizioni ETS. First published in *Physis* **7** (1965), 369–400.

8. Arrighi, Gino, 2004/1967. "Nuovi contributi per la storia della matematica in Firenze nell'età di mezzo: Il codice Palatino 573 della Biblioteca Nazionale di Firenze", pp. 159–194 *in* Gino Arrighi, *La matematica nell'Età di Mezzo. Scritti scelti,* a cura di F. Barberini, R. Franci and L. Toti Rigatelli Pisa: Edizioni ETS. First published in *Istituto Lombardo. Accademia di scienze e lettere. Rendiconti, Classe di scienze (A)* **101** (1967), 395–437.

9. Arrighi, Gino, 2004/1968. "La matematica a Firenze nel Rinascimento: Il codice ottoboniano 3307 della Biblioteca Apostolica Vaticana", pp. 209–222 *in* Gino Arrighi, *La matematica nell'Età di Mezzo. Scritti scelti,* a cura di F. Barberini, R. Franci and L. Toti Rigatelli. Pisa: Edizioni ETS. First published in *Physis* **10** (1968), 70–82.

10. Bachet, Claude Gaspar, sieur de Meziriac, 1624. *Problemes plaisans et delectables, que se font par les nombres.* Partie recuellis de divers autheurs, partie inventez de nouveau

avec leur demonstration. Seconde Edition, reveue, corrigée, et augmentée de plusieurs propositions, et de plusieurs problèmes. Lyon: Pierre Rigaud & Associez.

11. Bombelli, Rafael, 1572. *L'Algebra*. Bologna: Giovanni Rossi, 1572 (*impr.* 1579).

12. Boncompagni, Baldassare (ed.), 1857. *Scritti* di Leonardo Pisano matematico del secolo decimoterzo. I. Il *Liber abbaci* di Leonardo Pisano. Roma: Tipografia delle Scienze Matematiche e Fisiche.

13. Boncompagni, Baldassare (ed.), 1862. *Scritti* di Leonardo Pisano matematico del secolo decimoterzo. II. *Practica geometriae* et *Opusculi*. Roma: Tipografia delle Scienze Matematiche e Fisiche.

14. Cajori, Florian, 1928. *A History of Mathematical Notations*. I. *Notations in Elementary mathematics*. II. *Notations Mainly in Higher Mathematics*. La Salle, Illinois: Open Court, 1928–29.

15. Calzoni, Giuseppe, and Gianfranco Cavazzoni (eds), 1996. Luca Pacioli, *"tractatus mathematicus ad discipulos perusinos"*. Città di Castello: Delta Grafica

16. Cassinet, Jean, 2001. "Une arithmétique toscane en 1334 en Avignon dans la citè des papes et de leurs banquiers florentins", pp. 105–128 *in Commerce et mathématiques du moyen âge à la renaissance, autour de la Méditerranée*. Actes du Colloque International du Centre International d'Histoire des Sciences Occitanes (Beaumont de Lomagne, 13–16 mai 1999). Toulouse: éditions du C.I.H.S.O.

17. Clagett, Marshall, 1978. *Archimedes in the Middle Ages*. Volume III, *The Fate of the Medieval Archimedes 1300–1565*. (Memoirs of the American Philosophical Society, 125 A+B+C). Philadelphia: The American Philosophical Society.

18. Curtze, Maximilian (ed.), 1902. *Urkunden zur Geschichte der Mathematik im Mittelalter und der Renaissance*. (Abhandlungen zur Geschichte der mathematischen Wissenschaften, vol. 12–13). Leipzig: Teubner.

19. de Vleeschauwer, H. J., 1965. "La *Biblionomia* de Richard de Fournival du Manuscript 636 de la Bibliothèque de Sorbonne. Texte en facsimilé avec la transcription de Léopold Delisle". *Mousaion* **62**.

20. Derenzini, Giovanna, 1998. "Il codice Vaticano latino 3129 di Luca Pacioli", pp. 169–191 *in* Enrico Giusti (ed), *Luca Pacioli e la matematica del Rinascimento*. Sansepolcro: Fondazione Piero della Francesca/Petruzzi.

21. Djebbar, Ahmed, 1992. "Le traitement des fractions dans la tradition mathématique arabe du Maghreb", pp. 223–254 *in* Paul Benoit, K. Chemla & J. Ritter (eds), *Histoire de fractions, fractions d'histoire*. (Science Networks, 10). Boston etc.: Birkhäuser, 1992.

22. Djebbar, Ahmed, 2005. *L'Algèbre arabe: genèse d'un art*. Paris: Vuibert.

23. Eneström, Georg, 1907. "Über eine dem Nemorarius zugeschriebene kurze Algorismusschrift". *Bibliotheca Mathematica*, 3. Folge **8** (1907–08), 135–153.

24. Folkerts, Menso, 2006. *The Development of Mathematics in Medieval Europe: The Arabs, Euclid, Regiomontanus*. (Variorum Collected Studies Series, CS811). Aldershot: Ashgate.

25. Franci, Rafaella (ed.), 1983. Mo Gilio, *Questioni d'algebra* dal Codice L.IX.28 della Biblioteca Comunale di Siena. (Quaderni del Centro Studi della Matematica Medioevale, 6). Siena: Servizio Editoriale dell'Università di Siena.

26. Franci, Raffaella (ed.), 2001. Maestro Dardi, *Aliabraa argibra*, dal manoscritto I.VII.17 della Biblioteca Comunale di Siena. (Quaderni del Centro Studi della Matematica Medioevale, 26). Siena: Università degli Studi di Siena.

27. Franci, Raffaella, and Marisa Pancanti (eds), 1988. Anonimo (sec. XIV), *Il trattato d'algibra* dal manoscritto Fond. Prin. II. V. 152 della Biblioteca Nazionale di Firenze. (Quaderni del Centro Studi della Matematica Medioevale, 18). Siena: Servizio Editoriale dell'Università di Siena.

28. Franci, Raffaella, and Laura Toti Rigatelli, 1983. "Maestro Benedetto da Firenze e la storia dell'algebra". *Historia Mathematica* **10**, 297–317.

29. Größing, Helmuth, 1983. *Humanistische Naturwissenschaft. Zur Geschichte der Wiener mathematischen Schulen des 15. und 16. Jahrhunderts.* (Saecula Spiritualia, Band 8). Baden-Baden: Valentin Koerner.

30. Hebeisen, Christophe (ed., trans.), 2008. "L'algèbre *al-Badīʿ* d'al-Karağī". 2 vols. *Travail de doctorat*, école Polytechnique Fédérale de Lausanne.

31. Høyrup, Jens, 1988. "Jordanus de Nemore, 13th Century Mathematical Innovator: an Essay on Intellectual Context, Achievement, and Failure". *Archive for History of Exact Sciences* **38**, 307–363.

32. Høyrup, Jens, 1998. "'Oxford' and 'Gherardo da Cremona': on the Relation between Two Versions of al-Khwārizmī's Algebra", pp. 159–178 *in Actes du 3me Colloque Maghrébin sur l'Histoire des Mathématiques Arabes, Tipaza (Alger, Algérie), 1–3 Décembre 1990*, vol. II. Alger: Association Algérienne d'Histoire des Mathématiques.

33. Høyrup, Jens, 2000. "Embedding: Multi-purpose Device for Understanding Mathematics and Its Development, or Empty Generalization?" Paper presented to IX Congreso de la Asociación Española de Semiótica "Humanidades, ciencia y tecnologia", Valencia, 30 noviembre – 2 de diciembre 2000. *Filosofi og Videnskabsteori på Roskilde Universitetscenter. 3. Række: Preprints og Reprints* 2000 Nr. 8. To appear in the proceedings of the conference.

34. Høyrup, Jens, 2005. "Leonardo Fibonacci and *Abbaco* Culture: a Proposal to Invert the Roles". *Revue d'Histoire des Mathématiques* **11**, 23–56.

35. Høyrup, Jens, 2007a. *Jacopo da Firenze's Tractatus Algorismi and Early Italian Abbacus Culture.* (Science Networks. Historical Studies, 34). Basel etc.: Birkhäuser.

36. Høyrup, Jens, 2007b. "Generosity: No Doubt, but at Times Excessive and Delusive". *Journal of Indian Philosophy* **35**, 469–485. DOI 10.1007/s10781-007-9028-2.

37. Høyrup, Jens, 2007c. [Review of Menso Folkerts, *The Development of Mathematics in Medieval Europe: The Arabs, Euclid, Regiomontanus.* (Variorum Collected Studies Series, CS811). Aldershot: Ashgate, 2006]. *Aestimatio* **4** (2007; publ. 17.5.2008), 122–141.

38. Høyrup, Jens, 2008. "Über den italienischen Hintergrund der Rechenmeister-Mathematik". Beitrag zur Tagung "Die Rechenmeister in der Renaissance und der frühen Neuzeit: Stand der Forschung und Perspektiven". München, Deutsches Museum, 29. Februar 2008. *Max-Planck-Institut für Wissenschaftsgeschichte, Preprint* 349 (Berlin, 2008). http://www.mpiwg-berlin.mpg.de/Preprints/P349.PDF.

39. Høyrup, Jens, 2009. "What Did the Abbacus Teachers Aim At When They (Sometimes) Ended Up Doing Mathematics? An Investigation of the Incentives and Norms of a Distinct Mathematical Practice", pp. 47–75 *in* Bart van Kerkhove (ed.), *New Perspectives on Mathematical Practices: Essays in Philosophy and History of Mathematics.* Singapore: World Scientific.

40. Hughes, Barnabas B., 1972. "Johann Scheubel's Revision of Jordanus of Nemore's *De numeris datis*: an Analysis of an Unpublished Manuscript". *Isis* **63**, 221–234.

41. Hughes, Barnabas B., O.F.M. (ed., trans.), 1981. Jordanus de Nemore, *De numeris datis.* A Critical Edition and Translation. (Publications of the Center for Medieval and Renaissance Studies, UCLA, 14). University of California Press.

42. Hughes, Barnabas, O.F.M., 1986. "Gerard of Cremona's Translation of al-Khwārizmī's *Al-Jabr*: A Critical Edition". *Mediaeval Studies* **48**, 211–263.

43. Hughes, Barnabas B. (ed.), 1989. Robert of Chester's Latin translation of al-Khwārizmī's *Al-jabr.* A New Critical Edition. (Boethius. Texte und Abhandlungen zur Geschichte der exakten Naturwissenschaften, 14). Wiesbaden: Franz Steiner.

44. Hunt, Richard William, 1955. "The Library of Robert Grosseteste", pp. 121–145 *in* Daniel A. Callus, O.P. (ed.), *Robert Grosseteste, Scholar and Bishop.* Essays in Commemoration of the Seventh Centenary of his Death. Oxford: Clarendon Press.

45. Huygens, Christiaan, 1908. *Oeuvres complètes*. XI. *Travaux mathématiques 1645–1651*. La Haye: Martinus Nijhoff.
46. Karpinski, Louis Charles (ed., trans.), 1915. *Robert of Chester's Latin Translation of the Algebra of al-Khowarizmi*. (University of Michigan Studies, Humanistic Series, vol. 11). New York. Reprint in L. C. Karpinski and J. G. Winter, *Contributions to the History of Science*. Ann Arbor: University of Michigan, 1930.
47. Kaunzner, Wolfgang, 1985. "Über eine frühe lateinische Bearbeitung der Algebra al-Khwārizmīs in MS Lyell 52 der Bodleian Library Oxford". *Archive for History of Exact Sciences* **32**, 1–16.
48. Kaunzner, Wolfgang, 1986. "Die lateinische Algebra in MS Lyell 52 der Bodleian Library, Oxford, früher MS Admont 612", pp. 47–89 *in* G. Hamann (ed.), *Aufsätze zur Geschichte der Naturwissenschaften und Geographie*. (Österreichische Akademie der Wissenschaften, Phil.-Hist. Klasse, Sitzungsberichte, Bd. 475). Wien: Österreichische Akademie der Wissenschaften.
49. Kaunzner, Wolfgang, and Hans Wußing (eds), 1992. Adam Rieß, *Coß*. Faksimile+Kommentarband. Stuttgart & Leipzig: Teubner.
50. L'Huillier, Ghislaine, 1980. "Regiomontanus et le *Quadripartitum numerorum* de Jean de Murs". *Revue d'Histoire des Sciences et de leurs applications* **33**, 193–214.
51. L'Huillier, Ghislaine (ed.), 1990. Jean de Murs, le *Quadripartitum numerorum*. (Mémoires et documents publiés par la Société de l'école des Chartes, 32). Genève & Paris: Droz.
52. Labosne, A.(ed.), 1959. Claude-Gaspar Bachet, *Problèmes plaisants et délectables qui se font par les nombres*. Cinquième édition revue, simplifiée et augmentée. Nouveau tirage augmenté d'un avant-propos par J. Itard. Paris: Blanchard.
53. Lamrabet, Driss, 1994. *Introduction à l'histoire des mathématiques maghrébines*. Rabat, published by the author.
54. Lefèvre d'Étaples, Jacques (ed.), 1514. *In hoc opere contenta. Arithmetica decem libris demonstrata. Musica libris demonstrata quatuor. Epitome in libros Arithmeticos divi Severini Boetii. Rithmimachie ludus qui et pugna numerorum appellatur*. Secundaria aeditio. Paris: Henricus Stephanus.
55. Libri, Guillaume, 1838. *Histoire des mathématiques en Italie*. 4 vols. Paris: Jules Renouard, 1838–1841. Reprint Hildesheim: Georg Olms.
56. Luckey, Paul, 1941. "Tābit b. Qurra über den geometrischen Richtigkeitsnachweis der Auflösung der quadratischen Gleichungen". *Sächsischen Akademie der Wissenschaften zu Leipzig. Mathematisch-physische Klasse. Berichte* **93**, 93–114.
57. Maia Bertato, Fábio (ed.), Luca Pacioli, *De divina proportione*. Tradução Anotada e Comentada. Versão Final da Tese (Doutorado em ilosofia – CLE/IFCH/UNICAMP). Campinas, 2008.
58. Mancini, G. (ed.), 1916. "L'opera 'De corporibus regularibus' di Pietro Franceschi detto Della Francesca usurpata da Fra Luca Pacioli". *Atti della R. Accademia dei Lincei*, anno CCCVI. Serie quinta. *Memorie della Classe di Scienze morali, storiche e filologiche*, volume XIV (Roma 1909–16), 446–580.
59. Manni, Paola, 2001. "La matematica in volgare nel Medioevo (con particolare riguardo al linguaggio algebrico)", pp. 127–152 *in* Riccardo Gualdo (ed.), *Le parole della scienza. Scritture tecniche e scientifiche in volgare (secc. XIII-XV)*. Atti del convegno, Lecce, 16–18 aprile 1999. Galatina: Congedo.
60. Marre, Aristide (ed.), 1880. "Le Triparty en la science des nombres par Maistre Nicolas Chuquet Parisien". *Bullettino di Bibliografia e di Storia delle Scienze Matematiche e Fisiche* **13**, 593–659, 693–814.
61. Moss, Barbara, 1988. "Chuquet's Mathematical Executor: Could Estienne de la Roche have Changed the History of Algebra?", pp. 117–126 in Cynthia Hay (ed.), *Mathematics from Manuscript to Print, 1300-1600*. (Oxford Scientific Publications). New York: Oxford University Press, 1988.

62. Pacioli, Luca, 1494. *Summa de Arithmetica Geometria Proportioni et Proportionalita.* Venezia: Paganino de Paganini. [All folio references are to the first part.]

63. Pacioli, Luca, 1509. *Divina proportione.* Venezia: Paganius Paganinus, 1509. [Folio references are to the third part, the translation of Piero della Francesca.]

64. Pancanti, Marisa (ed.), 1982. Giovanni di Bartolo, *Certi chasi nella trascelta a cura di Maestro Benedetto* secondo la lezione del Codice L.IV.21 (sec. XV) della Biblioteca degli Intronati di Siena. (Quaderni del Centro Studi della Matematica Medioevale, 3). Siena: Servizio Editoriale dell'Università di Siena.

65. Peletier, Jacques, 1554. *L'algebre.* Lyon: Ian de Tournes.

66. Peletier, Jacques, 1555. *Dialogue de l'Ortografe e Prononciacion Françoese, departi an deus livres.* Lyon: Ian de Tournes.

67. Pieraccini, Lucia (ed.), 1983. Mo Biagio, *Chasi exenplari alla regola dell'algibra nella trascelta a cura di Mo Benedetto* dal Codice L. VII. 2Q della Biblioteca Comunale di Siena. (Quaderni del Centro Studi della Matematica Medioevale, 5). Siena: Servizio Editoriale dell'Università di Siena.

68. Procissi, Angiolo (ed.), 1954. "I Ragionamenti d'Algebra di R. Canacci". *Bollettino Unione Matematica Italiana,* serie III, **9**, 300–326, 420–451.

69. Procissi, Angiolo (ed.), 1983. Raffaello Canacci, *Ragionamenti d'algebra. I problemi* dal Codice Pal. 567 della Biblioteca Nazionale di Firenze. (Quaderni del Centro Studi della Matematica Medioevale, 7). Siena: Servizio Editoriale dell'Università di Siena.

70. [Ramus, Petrus], 1560. *Algebra.* Paris: Andreas Wechelum.

71. Rashed, Roshdi (ed., trans.), 2007. Al-Khwārizmī, *Le Commencement de l'algèbre.* (Collections Sciences dans l'histoire). Paris: Blanchard.

72. Rouse, Richard M., 1973. "Manuscripts Belonging to Richard de Fournival". *Revue d'Histoire des Textes* **3**, 253–269.

73. Rudolff, Christoff, 1525. *Behend und hübsch Rechnung durch die kunstreichen Regeln Algebra, so gemeincklich die Coss genennt werden.* Straßburg.

74. Salomone, Lucia (ed.), 1982. Mo Benedetto da Firenze, *La reghola de algebra amuchabale* dal Codice L.IV.21 della Biblioteca Comunale de Siena. (Quaderni del Centro Studi della Matematica Medioevale, 2). Siena: Servizio Editoriale dell'Università di Siena.

75. Sayılı, Aydın, 1962. *Abdülhamid ibn Türk'ün katışık denklemlerde mantıkî zaruretler adlı yazısı ve zamanın cebri (Logical Necessities in Mixed Equations by 'Abd al Ḥamîd ibn Turk and the Algebra of his Time).* (Publications of the Turkish Historical Society, Series VII, No 41). Ankara: Türk Tarih Kurumu Basımevi.

76. Scheubel, Johann, 1551. *Algebrae compendiosa facilísque descriptio, qua depromuntur magna Arithmetices miracula.* Parisiis: Apud Gulielmum Cavellat.

77. Schmeidler, Felix (ed.), 1972. Joannis Regiomontani *Opera collectanea.* Faksimiledrucke von neun Schriften Regiomontans und einer von ihm gedruckten Schrift seines Lehrers Purbach. Zusammengestellt und mit einer Einleitung herausgegeben. (Milliaria X,2). Osnabrück: Otto Zeller.

78. Serfati, Michel, 1998. "Descartes et la constitution de l'écriture symbolique mathématique". *Revue d'Histoire des Sciences* **51**, 237–289.

79. Sesiano, Jacques, 1988. "Le Liber Mahamaleth, un traité mathématique latin composé au XIIe siècle en Espagne", pp. 69–98 *in Histoire des Mathématiques Arabes.* Premier colloque international sur l'histoire des mathématiques arabes, Alger, 1.2.3 décembre 1986. Actes. Alger: La Maison des Livres.

80. Sesiano, Jacques (ed.), 1993. "La version latine médiévale de l'Algèbre d'Abū Kāmil", pp. 315–452 *in* M. Folkerts and J. P. Hogendijk (eds), *Vestigia Mathematica.* Studies in Medieval and Early Modern Mathematics in Honour of H. L. L. Busard. Amsterdam & Atlanta: Rodopi.

81. Simi, Annalisa (ed.), 1994. Anonimo (sec. XIV), *Trattato dell'alcibra amuchabile* dal Codice Ricc. 2263 della Biblioteca Riccardiana di Firenze. (Quaderni del Centro Studi della Matematica Medioevale, 22). Siena: Servizio Editoriale dell'Università di Siena.

82. Souissi, Mohamed (ed., trans.), 1988. Qalaṣādī, *Kašf al-asrār 'an 'ilm ḥurūf al-ġubār*. Carthage: Maison Arabe du Livre.
83. Stifel, Michael, 1544. *Arithmetica integra*. Nürnberg: Petreius.
84. Tartaglia, Nicolò, 1560. *La sesta parte del general trattato de' numeri, et misure*. Venezia: Curtio Troiano.
85. Travaini, Lucia, 2003. *Monete, mercanti e matematica. Le monete medievali nei trattati di aritmetica e nei libri di mercatura*. Roma: Jouvence.
86. Ulivi, Elisabetta, 1998. "Le scuole d'abaco a Firenze (seconda metà del sec. XIII–prima metà del sec. XVI)", pp. 41–60 *in* Enrico Giusti (ed), *Luca Pacioli e la matematica del Rinascimento*. Sansepolcro: Fondazione Piero della Francesca/Petruzzi, 1998.
87. Ulivi, Elisabetta, 2002. "Benedetto da Firenze (1429–1479), un maestro d'abbaco del XV secolo. Con documenti inediti e con un'Appendice su abacisti e scuole d'abaco a Firenze nei secoli XIII–XVI". *Bollettino di Storia delle Scienze Matematiche* **22**:1, 3–243.
88. Van Egmond, Warren, 1978. "The Earliest Vernacular Treatment of Algebra: The *Libro di ragioni* of Paolo Gerardi (1328)". *Physis* **20**, 155–189.
89. Van Egmond, Warren, 1980. *Practical Mathematics in the Italian Renaissance: A Catalog of Italian Abbacus Manuscripts and Printed Books to 1600*. (Istituto e Museo di Storia della Scienza, Firenze. Monografia N. 4). Firenze: Istituto e Museo di Storia della Scienza.
90. Van Egmond, Warren, 1983. "The Algebra of Master Dardi of Pisa". *Historia Mathematica* **10**, 399–421.
91. Van Egmond, Warren (ed.), 1986. Anonimo (sec. XV), *Della radice de' numeri e metodo di trovarla (Trattatello di Algebra e Geometria)* dal Codice Ital. 578 della Biblioteca Estense di Modena. Parte prima. (Quaderni del Centro Studi della Matematica Medioevale, 15). Siena: Servizio Editoriale dell'Università di Siena.
92. Vogel, Kurt, 1977. *Ein italienisches Rechenbuch aus dem 14. Jahrhundert (Columbia X 511 AL3)*. (Veröffentlichungen des Deutschen Museums für die Geschichte der Wissenschaften und der Technik. Reihe C, Quellentexte und Übersetzungen, Nr. 33). München.
93. Vogel, Kurt (ed.), 1981. *Die erste deutsche Algebra aus dem Jahre 1481*, nach einer Handschrift aus C 80 Dresdensis herausgegeben und erläutert. (Bayerische Akademie der Wissenschaften. Mathematisch-naturwissenschaftliche Klasse. Abhandlungen. Neue Folge, Heft 160). München: Verlag der Bayerischen Akademie der Wissenschaften.
94. Wappler, E., 1900. "Zur Geschichte der Mathematik im 15. Jahrhundert". *Zeitschrift für Mathematik und Physik. Historisch-literarische Abteilung* **45**, 47–56.
95. Widmann, Johannes, 1489. *Behende und hubsche Rechenung auff allen kauffmannschafft*. Leipzig: Konrad Kacheloffen.
96. Woepcke, Franz, 1853. *Extrait du Fakhrî, traité d'algèbre* par Aboû Bekr Mohammed ben Alhaçan Alkarkhî; *précédé d'un mémoire sur l'algèbre indéterminé chez les Arabes*. Paris: L'Imprimerie Impériale.
97. Woepcke, Franz, 1854. "Recherches sur l'histoire des sciences mathématiques chez les Orientaux, d'après des traités inédits arabes et persans. Premier article. Notice sur des notations algébriques employées par les Arabes". *Journal Asiatique*, 5e série 4, 348–384.
98. Woepcke, Franz, 1859. "Traduction du traité d'arithmétique d'Aboûl Haçan Alî Ben Mohammed Alkalçâdî". *Atti dell'Accademia Pontificia de' Nuovi Lincei* **12** (1858–59), 230–275, 399–438.

Manuscripts consulted

• Florence, Bibl. Naz. Centr., Palat. 573 (*Tratato di praticha d'arismetricha*)

- Florence, Bibl. Naz. Centr., fondo princ. II.V.152 (*Tratato sopra l'arte della arismetricha*)
- Florence, Bibl. Naz. Centr., fond. princ. II.IX.57 (*Trattato di tutta l'arte dell'abbacho*)
- Florence, Ricc. 2404 (*Livero dell'abbecho*)
- Milan, Trivulziana 90 (Jacopo da Firenze, *Tractatus algorismi*, redaction)
- Modena, Bibl. Estense, ital. 578 (*L'agibra*)
- Palermo, Biblioteca Comunale 2 Qq E 13 (*Libro merchatantesche*)
- Paris, Bibliothèque Nazionale, ms. latin 7377A (*Liber mahamaleth*)
- Rome, Acc. Naz. dei Lincei, Cors. 1875 (*Trattato di tutta l'arte dell'abbacho*)
- Siena, Biblioteca Comunale degli Intronati, L.IV.21 (Benedetto da Firenze, *Trattato de praticha d'arismetrica*)
- Vatican, Chigi M.VIII.170 (Dardi, *Aliabraa argibra*)
- Vatican, Ottobon. lat. 3307 (*Libro di praticha d'arismetrica*)
- Vatican, Vat. lat. 3129 (Luca Pacioli, *Suis carissimis disciplis ...*)
- Vatican, Vat. lat. 4825 (Tomaso de Jachomo Lione, *Libro da razioni*)
- Vatican, Vat. lat. 4826 (Jacopo da Firenze, *Tractatus algorismi*)
- Vatican, Vat. lat. 10488 (*Alchune ragione*)

Chapter 2
From the second unknown to the symbolic equation

Albrecht Heeffer

Abstract The symbolic equation slowly emerged during the course of the sixteenth century as a new mathematical concept as well as a mathematical object on which new operations were made possible. Where historians have often pointed at François Viète as the father of symbolic algebra, we would like to emphasize the foundations on which Viète could base his *logistica speciosa*. The period between Cardano's *Practica Arithmeticae* of 1539 and Gosselin's *De arte magna* of 1577 has been crucial in providing the necessary building blocks for the transformation of algebra from rules for problem solving to the study of equations. In this paper we argue that the so-called "second unknown" or the *Regula quantitates* steered the development of an adequate symbolism to deal with multiple unknowns and aggregates of equations. During this process the very concept of a symbolic equation emerged separate from previous notions of what we call "co-equal polynomials".

Key words: Symbolic equation, linear algebra, Cardano, Stifel, regula quantitates.

L'histoire de la résolution des équations à plusieurs inconnues n'a pas encore donné lieu à un travail d'ensemble satisfaisant, qui donnerait d'ailleurs lieu à d'assez longues recherches. Il est intimement lié aux progrès des notations algébriques. J'ai appelé l'attention sur le problème de la resolution des equations simultanées, chaque fois que je l'ai rencontré, chez les auteurs de la fin du XVIe et du commencement du XVIIe siècle. (Bosmans, 1926, 150, footnote 16).

Centre for History of Science, Ghent University, Belgium.
Fellow of the Research Foundation Flanders (FWO Vlaanderen).

57

2.1 Introduction

This footnote, together with many similar remarks by the Belgian historian
Father Henri Bosmans (S.J.), initiated our interest in the role of the second
unknown or *regula quantitates* on the development of symbolism during the
sixteenth century.[1] Indeed, the importance of the use of multiple unknowns in
the process leading to the concept of an equation cannot be overestimated. We
have traced the use and the development of the second unknown in algebraic
problem solving from early Arabic algebra and its introduction in Europe
until its last appearance in Jesuit works on algebra during the late seventeenth
century. The first important step in abbaco algebra can be attributed to the
Florentine abbaco master Antonio de' Mazzinghi, who wrote an algebraic
treatise around 1380 (Arrighi 1967). Luca Pacioli almost literally copied the
solution method in his *Summa* of 1494, and Cardano used the second unknown
both in his *Arithmetica* and the *Ars Magna*. A second thread of influence is
to be distinguished through the *Triparty* by Chuquet and the printed works
of de la Roche and Christoff Rudolff. The *Rule of Quantity* finally culminates
in the full recognition of a system of linear equation by Buteo and Gosselin.
The importance of the use of letters to represent several unknowns goes much
further than the introduction of a useful system of notation. It contributed to
the development of the modern concept of unknown and that of a symbolic
equation. These developments formed the basis on which Viète could build
his theory of equations.

It is impossible to treat this whole development within the scope of a single
chapter. The use of the second unknown by Chuquet (1489) and de la Roche
(1520) and its spread in early sixteenth-century Europe is already treated in
Heeffer (2010a). Its reception and development on the Iberian peninsula has
recently be studied by Romero (2010). In this paper we will concentrate on
one specific aspect of the second unknown – the way it shaped the emergence
of the symbolic equation.

2.2 Methodological considerations

As argued in Heeffer (2008), the correct characterization of the Arabic concept
of an equation is the act of keeping related polynomials equal. Two of the
three translators of al-Khwārizmī's algebra, Guglielmo de Lunis and Robert
of Chester use the specific term *coaequare*. In the geometrical demonstration

---

[1] References to the second unknown are found in Bosmans (1925-6) on Stifel, Bosmans
(1906) on Gosselin, Bosmans (1907) on Peletier, Bosmans (1908a) on Nunez and Bosmans
(1926) on Girard.

of the fifth case, de Lunis proves the validity of the solution for the "equation" $x^2 + 21 = 10x$. The binomial $x^2 + 21$ is coequal with the monomial $10x$, as both are represented by the surface of a rectangle (Kaunzner, 1989, 60):

> Ponam censum tetragonum abgd, cuius radicem ab multiplicabo in 10 dragmas, quae sunt latus be, unde proveniat superficies ae; ex quo igitur 10 radices censui, una cum dragmis 21, coequantur.

Once two polynomials are connected because it is found that their arithmetical value is equal, or, in the case of the geometrical demonstration, because they have the same area, the continuation of the derivation requires them to be kept equal. Every operation that is performed on one of them should be followed by a corresponding operation to keep the coequal polynomial arithmetical equivalent. Instead of operating on equations, Arabic algebra and the abbaco tradition operate on the coequal polynomials, always keeping in mind their relation and arithmetical equivalence. Such a notion is intimately related with the *al-jabr* operation in early Arabic algebra. As is now generally acknowledged (Oaks and Alkhateeb, 2007; Heeffer 2008; Hoyrup 2010, note 7), the restoration operation should not be interpreted as adding a term to both sides of an equation, but as the repair of a deficiency in a polynomial. Once this polynomial is restored – and as a second step – the coequal polynomial should have the same term added.

At some point in the history of algebra, coequal polynomials will transform into symbolic equations. This transformation was facilitated by many small innovations and gradual changes in permissable operations. An analysis of this process therefore poses certain methodological difficulties. A concept as elusive as the symbolic equation, which before the sixteenth century did not exist in its current sense, and which gradually transformed into its present meaning, evades a full understanding if we only use our current symbolic language. To tackle the problem we present the original sources in a rather uncommon format, by tables. The purpose is to split up the historical text in segments which we consider as significant reasoning steps from our current perspective. Each of these steps is numbered. Next, a symbolic representation is given which conveys how the reasoning step would look like in symbolic algebra, not necessarily being a faithful translation of the original source. Finally, a meta-description is added to explain the reasoning and to verify its validity. So, we have two levels of description: the original text in the original language and notations, and a meta-level description which explains how the reasoning would be in symbolic algebra. Only by drawing the distinction, we will be able to discern and understand important conceptual transformations. Our central argument is that once the original text is directly translatable into the meta-description we are dealing with the modern concept of a symbolic equation.

2.3 The second unknown

Before discussing the examples, it is appropriate to emphasize the difference
between the rhetorical unknown and unknowns used in modern transliterations. Firstly, the method of using a second unknown is an exception in algebraic practice before 1560. In general, algebraic problem solving before the
seventeenth century uses a single unknown. This unknown is easily identified
in Latin text by its name *res* (or sometimes *radix*), *cosa* in Italian and *coss* or
ding in German. The unknown should be interpreted as a single hypothetical
value used within the analytic method. Modern interpretations such as an
indeterminate value or a variable, referring to eighteenth century notions of
function and continuity, do not fit the historical context. In solving problems
by means of algebra, abbacus masters often use the term 'quantity' or 'share'
or 'value' apart from the *cosa*. The rhetoric of abacus algebra requires that the
quantities given in the problem text are formulated in terms of the hypothetical unknown. The problem solving process typically starts with "suppose that
the first value sought is one cosa". These values or unknown quantities cannot be considered algebraic unknowns by themselves. The solution depends
on the expression of all unknown quantities in terms of the *cosa*. Once a value
has been determined for the *cosa*, the unknown quantities can then easily be
determined.

However, several authors, even in recent publications, confuse the unknown
quantities of a problem, with algebraic unknowns. As a result, they consider
the rhetorical unknown as an auxiliary one. For example, in his commentary on
Leonardo of Pisa's *Flos*, Ettore Picutti (1983) consistently uses the unknowns
x, y, z for the sought quantities and regards the *cosa* in the linear problems
solved by Leonardo to be an auxiliary unknown. The "method of auxiliary
variable" as a characterization by Barnabas Hughes (2001) for a problem-
solving method by ben-Ezra also follows that interpretation. We believe this
to be a misrepresentation of the original text and problem-solving method.

The more sophisticated problems sometimes require a division into sub-
problems or subsequent reasoning steps. These derived problems are also formulated using an unknown but one which is different from the unknown in the
main problem. For example, in the anonymous manuscript 2263 of the Biblioteca Riccardiana in Florence (c. 1365; Simi, 1994), the author solves the
classic problem of finding three numbers in geometric proportion given their
sum and the sum of their squares. He first uses the middle term as unknown,
arriving at the value of 3. Then the problem of finding the two extremes is
treated as a new problem, for which he selects the lower extreme as unknown.
We will not consider such cases as the use of two unknowns, but the use of
a single one at two subsequent occasions. We have given some examples of
what should not be comprehended as a second unknown, but let us turn to a

positive definition. The best characterization of the use of several unknowns
is operational. We will consider a problem solved by several unknowns if all
of the following conditions apply in algebraic problem solving:

1. The reasoning process should involve more than one rhetorical unknown
   which is named or symbolized consistently throughout the text. One of the
   unknowns is usually the traditional cosa. The other can be named *quantità*,
   but can also be a name of an abstract entity representing a share or value
   of the problem.
2. The named entities should be used as unknowns in the sense that they are
   operated upon algebraically by arithmetical operators, by squaring or root
   extraction. If no operation is performed on the entity, it has no operational
   function as unknown in solving the problem
3. The determination of the value of the unknowns should lead to the solution
   or partial solution of the problem. In some cases the value of the second
   unknown is not determined but its elimination contributes to the solution
   of the problem. This will also be considered as an instance of multiple
   unknowns.
4. The entities should be used together at some point of the reasoning process
   and connected by operators or by a substitution step. If the unknowns are
   not connected in this way the problem is considered to be solved by a single
   unknown.

In all the examples discussed below, these four conditions apply.

2.4 Constructing the equation: Cardano and Stifel

2.4.1 Cardano introducing operation on equations

As far as we know from extant abbaco manuscripts Antonio de' Mazzinghi
was the first to use the second unknown (Arrighi, 1967). Surprisingly, this was
not for the solution of a linear problem but for a series of problems on three
numbers in continuous proportion (or geometric progression, further GP). The
same problems and the method of the second unknown are discussed by Pacioli
in his *Summa*, without acknowledging de' Mazzinghi (Heeffer, 2010b). Before
turning to Cardano's use of the second unknown, it is instructive to review
his commentary on the way Pacioli treats these – and hence, Mazzinghi's –
problems. In the *Questionibus Arithmeticis*, the problem is listed as number
28 (Cardano, 1539, f. DDiii^v). Not convinced of the usefullness of the second
unknown, he shows little consideration for this novel solution as it uses too
many unnecessary steps ("Frater autem Lucas posuit ean et soluit cum maga

difficultate et pluribus operationibus superfluis"). He presents the problem
(2.1) with $a = 25$ instead of 36, as used by Pacioli.

$$\frac{x}{y} = \frac{y}{z}$$
$$x + y + z = a \qquad\qquad (2.1)$$
$$\frac{a}{x} + \frac{a}{y} + \frac{a}{z} = x + y + z = xyz$$

The solution is rather typical for Cardano's approach to problem solving.
The path of the least effort is the reduction of the problem to a form in which
theoretical principles apply. Using his previously formulated rule,[2]

$$\frac{a}{x} + \frac{a}{y} + \frac{a}{z} = x + y + z, y = \sqrt{a}$$

he immediately finds 5 for the mean term. As the product of the three, $xyz =
y^3 = 125$, is also equal to the sum of the three, the sum of the two extremes
is 120. Applying his rule for dividing a number a into two parts in continuous
progression[3] with b as mean proportional

$$\frac{a}{2} \pm \sqrt{\left(\frac{a}{2}\right)^2 - b^2},$$

he immediately arrives at

$$\left(60 + \sqrt{3575}, 5, 60 - \sqrt{3575}\right)$$

ita soluta est.

This approach is interesting from a rhetorical point of view. Abbaco treatises are primarily intended to show off the skills of the master, often involving
the excessive use of irrationals while an example with integral values would
have illustrated the demonstration with the same persuasion. These treatises are, with the exception of some preliminaries, limited to problem solving
only. With Pacioli, some recurring themes are extracted from his sources and
treated in separate sections. Cardano extends this evolution to a full body of
theory, titled *De proprietatibus numerorum mirisicis*, including 136 articles
(Cardano 1539, Chapter 42). The problem is easily solved because it is an
application of two principles expounded in this chapter.

---

[2] Cardano 1539, Chapter 42, art. 91, f. *Iii^v*: "Omnium trium quantitatum continuae proportionalium ex quarum divisione alicuius numeri proventus congregati ipsarum aggregato
aequari debeat, media illius numeri radix erit nam est eaedem necessarioeveniunt quantum
aggregatum est idem ex supposito".

[3] Cardano 1539, Chapter 42, art. 116, f. *Ivi^r*: "Si sint duo numeri utpote 24 et 10 et velis
dividere 24 in duas partes in quarum medio cadat 10 in continua proportionalitate, quadra
dimidium maioris quod est 12 sit 144. Detrahe quadratum minoris quod est 100 remanet
44, cuius R addita ad 12 et diminuta faciet duos numeros inet quos 10 cadit in medio in
contuna proportionalitate, et erunt 12 p R 44 et 10 et 12 m R 44."

Using such solution method, he completely ignores Pacioli's use of two unknowns for this problem. However Cardano adopts two unknowns for the solution of linear problems in the *Arithmetica Practicae* of 1539. Six years later he even dedicates two chapters of the *Ars Magna* (Cardano, 1545) to the use of the second unknown. The last problem he solved with two unknowns is again a division problem with numbers in continuous proportion.

Cardano used the second unknown first in chapter 51 in a linear problem (Opera Omnia, IV, 73-4). He does not use the name *regula quantitates* but *operandi per quantitatem surda*, showing the terminology of Pacioli. He uses *cosa* and *quantita* for the unknowns but will later shift to *positio* and *quantitates* in the *Ars Magna*.[4]

Let us look at problem 91 from the *Questionibus*, as this fragment embodies a conceptual breakthrough towards a symbolic algebra. The problem is a complex version of the classic problem of doubling other's money to make equal shares (Tropfke 1980, 647-8; Singmaster 2004, 7.H.4). In Cardano's problem, three men have different sums of money. The first has to give 10 plus one third of the rest to the second. The second has to give 7 plus one fourth of the rest to the third. The third had 5 to start with. The result should be so that the total is divided into the proportion 3 : 2 : 1 (Cardano 1539, Chap. 66, article 91, ff. GGviii^v – HHi^v):

> Tres ludebant irati rapverunt peccunias suas & alienas cum autem pro amicum quievissent primus dedit secundo 10 p 1/3 residui. Secundus dedit tertio 7 p residui & tertio iam remanserant 5 nummi & primus habuit 1/2 secundus 1/3 tertius 1/6 quaeritur summa omnium, & quantum habuit quilibet.

The meta-description in symbolic form is as follows:

$$a - 10 - \tfrac{1}{3}(a-10) = \tfrac{1}{2}(a+b+c)$$
$$b + 10 + \tfrac{1}{3}(a-10) - 7 - \tfrac{1}{4}\left(b + 10 + \tfrac{1}{3}(a-10) - 7\right) = \tfrac{1}{3}(a+b+c)$$
$$c + \tfrac{1}{4}\left(b + 10 + \tfrac{1}{3}(a-10) - 7\right) = \tfrac{1}{6}(a+b+c)$$
$$c = 5$$

Cardano uses the first unknown for a and the second for b ("Pone quod primus habuerit 1 co. secundus 1 quan."). He solves the problem, in the standard way, by constructing the polynomial expressions, corresponding with the procedure of exchanging the shares. Doing so he arrives at two expressions. The first one is

$$x = 21\frac{4}{7} + 3\frac{6}{7}y$$

("igitur detrae 1/8 co. ex 5/12 co remanent 7/24 co. et hoc aequivalet 6 7/24 p. 1 1/8 quan. quare 7 co. aequivalent 151 p. 27 quan. quare 1 co. aequalet

---

[4] The same problem is solved slightly different in the *Ars Magna* and is discussed below.

21 4/7 p. 3 6/7 quan."). This expression for x would allow us to arrive at a value for the second unknown. Instead, Cardano derives a second expression in x

$$x = 101\frac{4}{5} + 1\frac{4}{5}y$$

("et q*ui*a 5/12 co. aequivalent etiam 42 5/12 p. quan. igitur 5 co. aequivalebunt 509 p. 9 quan. quare 1 co. aequivalent 101 4/5 p. 1 4/5 quan."). As these two expression are equal he constructs an equation in the second unknown:

$$21\frac{4}{7} + 3\frac{6}{7}y = 101\frac{4}{5} + 1\frac{4}{5}y$$

("igitur cu*m* etia*m* aequivaleat 21 4/7 p. 3 6/7 quan. erunt 21 4/7 p. 3 6/7 quan. aequalia 101 4/5 p. 1 4/5 quan. "). The text continues with: "Therefore, subtracting the second unknowns from each other and the numbers from each other this leads to a value of 39 for the second unknown. And this is the share of the second one." ("igit*ur* ta*n*de*m* detrahe*n*do quan. ex quan. et numeru*m* ex numero fiet valor qua*n*titatis 39 et t*antu*m habuit secu*n*dus"). However, the added illustration shows us something very interesting (see Figure 2.1).

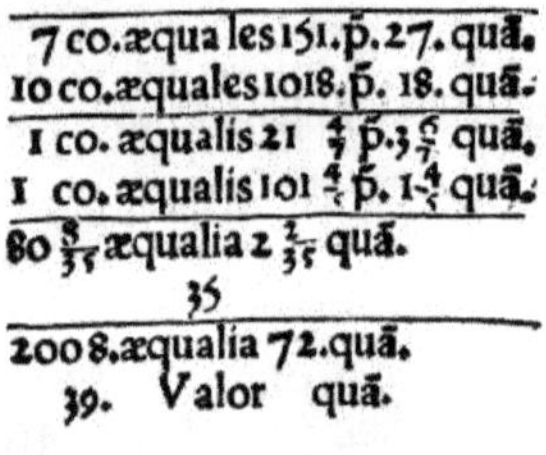
7 co. æquales 151. p̃. 27. quã.
10 co. æquales 1018. p̃. 18. quã.
1 co. æqualis 21 4/7 p̃. 3 6/7 quã.
1 co. æqualis 101 4/5 p̃. 1 4/5 quã.
80 8/35 æqualia 2 2/35 quã.
35
2008. æqualia 72. quã.
39. Valor quã.

Fig. 2.1: Cardano's construction of equations from (Cardano, 1539, f. 91^r)

The illustration is remarkable in several ways. Firstly, it shows equations where other illustrations or marginal notes by Cardano and previous authors only show polynomial expressions. As far as I know, this is the first unambiguous occurrence of an equation in print. This important fact seems to have gone completely unnoticed. Secondly, and supporting the previous claim, the illustration shows for the first time in history an operation on an equation. Cardano here multiplies the equation

$$80\frac{8}{35} = 2\frac{2}{35}y$$

by 35 to arrive at

$$2808 = 72y$$

The last line gives $39 = y$ and not 'y equals 39' which designates the implicit division of the previous equation by 72. The illustration appears both in the 1539 edition and the *Opera Omnia* (with the same misprint for 2808). As we discussed before, the term 'equation' should be used with caution in the context of early sixteenth-century practices. This case however, constitutes *the construction of an equation* in the historical as well as the conceptual sense. We have previously used an operational definition for the second unknown. Similarly, operations on an equation, as witnessed in this problem, support an operational definition of an equation. We can consider an equation, in this historical context, as a mathematical entity because it is directly operated upon by multiplication and division operators.

2.4.2 Michael Stifel introducing multiple unknowns

As a university professor in mathematics, Stifel marks a change in the typical profile of abbaco masters writing on algebra. In that respect, Cardano was a transitional figure. Cardano was taught mathematics by his father Fazio "who was well acquainted with the works of Euclid" (Cardano, 2002, 8). Although he was teaching mathematics in Milan, his professorship from 1543 was in medicine. His choice of subjects and problems fit very well within the abacus tradition. However, he did change from the vernacular of the abbaco masters to the Latin used for university textbooks. Stifel is more part of the university tradition studying Boethius and Euclid, but believed that the new art of algebra should be an integral part of arithmetic. That is why his *Complete Arithmetic* includes a large part on algebra (Stifel, 1544). Most of his problems and discussions on the cossic numbers, as he calls algebra, refer to Cardano. He concludes his systematic introduction with the chapter *De secundis radicibus*, devoted to the second unknown (ff. $251^v - 255^v$).

Several authors seem to have overlooked Cardano's use of the second unknown in the *Practica Arithmeticae*. Bosmans (1906, 66) refers to the ninth chapter of the *Ars Magna* as the source of Stifel's reference, but this must be wrong as the foreword of the *Arithmetica Integra* is dated 1543 and the *Ars Magna* was published in 1545. In fact, the influence might be in the reverse direction. Cifoletti (1993, 108) writes that "reading Stifel one wonders why the German author is so certain of having found most of his matter on the second unknown precisely in Cardano, i.e. in the *Practica Arithmeticae*. For, the *Ars Magna* would be more explicit on this topic". She gives the example of the *regula de medio* treated in chapter 51 of the *Practica Arithmeticae* (Opera, 87) and more extensively in the *Ars Magna* (Witmer, 92). She writes: "In fact, the rule Cardano gives for this case is not quite a rule for using several unknowns, but rather a special case, arising as a way to solve problems by 'iteration' of the process of assigning the unknown". However, Stifel's application of the *secundis radicibus* to linear problems unveils that he drew his inspiration from the problems in Cardano's *Questionibus* of Chapter 66, as the one discussed above. He makes no effort to conceal that:[5]

> Christoff Rudolff and Cardano treat the second unknown using the term *quantitatis*, and therefore they designate it as $1q$. This is at greater length discussed by Cardano. While Christoff Rudolff does not mention the relation of the second [unknown] with the first. On the other hand, Cardano made us acquainted with it by beautiful examples, so that I could learn them with ease.

Graciously acknowledging his sources, he adds an important innovation for the notation of the second an other unknowns. Keeping the cossic symbol ᴢℯ for the first unknown, the second is represented as $1A$, the third by $1B$, and so on, which he explains, is a shorthand notation for $1A$ᴢℯ and $1B$ᴢℯ, the square of $1A$ᴢℯ being $1A$ℨ. The use of the letters A, B and C in linear problems is common in German cossist manuscripts since the fifteenth century.[6] Although these letters are not used as unknowns, the phrasing comes very close to the full notation given by Stifel. For example, Widman writes as follows: "Do as follows, pose that C has $1x$, therefore having A 2ᴢℯ, because he has double of C, and B 3ᴢℯ, because he has triple".[7] Using Stifel's symbolism this would read as $1x$, $2Ax$ and $3Bx$. Although conceptually very different, the notation is practically the same. The familiarity with such use of letters made it an obvious choice for Stifel. Later, in his commentary on the *Coss* from Rudolff, he writes on Rudolff's use of 1ᴢℯ and $1q$., "However, I prefer to use $1A$ for $1q$.

---

[5] Stifel (1544) f. 252^r: "Christophorus et Hieronymus Cardanus tractant radices secundas sub vocabulo Quantitatis ideo eas sic signant 1 q. Latius vero eas tractavit Cardanus. Christophorus enim nihil habet de commissionibus radicum sedundarum cum primis. Eas autem Cardanus pulchris exemplis notificavit, ita ut ipsas facile didicerim", (translation AH). In the edition of Rudolff's *Coss*, he adds: "Bye dem 188 exempl lehret Christoff die Regul Quantitatis aber auss vil oben gehandelten exemplen tanstu yetzt schon wissen wie das es teyn sonderliche regel sey... Das aber Christoff und auch Cardanus in sollichen fal setzen 1 q. Das ist 1 quantitet. Daher sie diser sach den nahmen haben gegeben und nennens Regulam Quantitatis" (Stifel 1553, 307).

[6] For example, the marginal notes of the C80 manuscript written by Johannes Widman in 1481, give the following problem (C80 f. 359^r, Wappler 1899, 549): "Item sunt tres socij, scilicet A, B, C, quorum quilibet certam pecuniarum habet summam. Dicit C: A quidem duplo plus habet quam ego, B vero triplum est ad me, et cum quilibet eorum partem abiecerit, puta A 2 et B 3, et residuum vnius si ductum fuerit in residuum alterius, proveniunt 24. Queritur ergo, quod quilibet eorum habuit, scilicet A et B, et quot ego". Høyrup (2010) describes an even earlier example by Magister Wolack of 1467, note 90.

[7] Ibid.: "Fac sic et pone, quod C habet 1 x, habebit ergo A 2x, quia duplum ad C, et B 3x, quia triplum".

because sometimes we have examples with three (or more) numbers. I then use 1𝔵, $1A$, $1B$, etc.”.[8]

Distinguishing between a second and third unknown is a major step forward from Chuquet and de la Roche who used one and the same symbol for both.[9] Before Stifel, there has always been an ambiguity in the meaning of the ‘second’ unknown. From now on, the second and the third unknown can be used together as in yz, which becomes $1AB$. However, Stifel’s notation system is not free from ambiguities. For the square of A, he uses $1A$𝔷, while 𝔷B should be read as the product of x^2 and B. The product of $2x^3$ and $4y^2$, an example given by Stifel, becomes 8𝔠𝔢A𝔷. A potential problem of ambiguity arises when we multiply $3x^2$ and $4z$, also given as an example. This leads to 12𝔷B and thus it becomes very confusing that $12z^2x$ being the product of $12z^2$ and x is written as 12𝔷B while $12z^2$ would be $12B$𝔷. Given the commutativity of multiplying cossic terms, both expressions should designate the same. The problem becomes especially manifest when multiplying more than two terms together using the extended notation. Stifel seems not be aware of the problem at the time of writing the *Arithmetica integra*.

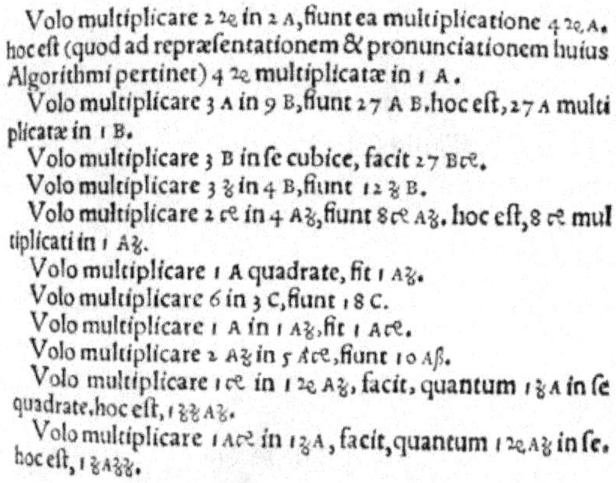
Volo multiplicare 2 𝔵 in 2 A, fiunt ea multiplicatione 4 𝔵 A. hoc eſt (quod ad repræſentationem & pronunciationem huius Algorithmi pertinet) 4 𝔵 multiplicatæ in 1 A.
Volo multiplicare 3 A in 9 B, fiunt 27 A B. hoc eſt, 27 A multiplicatæ in 1 B.
Volo multiplicare 3 B in ſe cubice, facit 27 B𝔠𝔢.
Volo multiplicare 3 𝔷 in 4 B, fiunt 12 𝔷 B.
Volo multiplicare 2 𝔠𝔢 in 4 A𝔷, fiunt 8 𝔠𝔢 A𝔷. hoc eſt, 8 𝔠𝔢 multiplicati in 1 A𝔷.
Volo multiplicare 1 A quadrate, fit 1 A𝔷.
Volo multiplicare 6 in 3 C, fiunt 18 C.
Volo multiplicare 1 A in 1 A𝔷, fit 1 A𝔠𝔢.
Volo multiplicare 2 A𝔷 in 5 A𝔠𝔢, fiunt 10 Aß.
Volo multiplicare 1 𝔠𝔢 in 1 𝔵 A𝔷, facit, quantum 1 𝔷 A in ſe quadrate, hoc eſt, 1 𝔷𝔷 A𝔷.
Volo multiplicare 1 A𝔠𝔢 in 1 𝔷 A, facit, quantum 1 𝔵 A𝔷 in ſe. hoc eſt, 1 𝔷 A𝔷𝔷.

Fig. 2.2: The rules for multiplying terms from Stifel (1545, f. 252^r)

The chapter on the *secundis radicibus* concludes with some examples of problems. Other problems, solved by several unknowns are given in *de exemplis* of the following chapters. Here we find solutions to many problems taken from Christoff Rudolff, Adam Ries and Cardano, usually including the correct ref-

[8] Stifel, 1553, f. 186^r: “Ich pfleg aber für 1q zusetzen 1A auss der ursach das zu zeyten ein exemplum wol drey (oder mehr) zalen fürgibt zu finden. Da setze ich sye also 1x, 1A, 1B etc”.
[9] For an extensive discussion of the second unknown in Chuquet, de la Roche and Rudolff and their interdependence see Heeffer (2010a).

erence. In the original sources, these problems are not necessarily treated algebraically, or by a second unknown. Let us look at one problem which he attributes to Adam Ries:[10]

> Three are in company, of which the first tells the second: if you give me half of your
> share, I have 100 fl. The second tells the third: if you give me one third of your share,
> then I have 100 fl. And the third tells the first: if you give me your sum divided by
> four, I have 100 fl. The question is how much each has.

The problem is slightly different from the example discussed above, in that the shares refer to the next one in the cycle and not to the sum of the others. The direct source of Stifel appears to be the unpublished manuscript *Die Coss* by Adam Riese, dated 1524 (Berlet 1860, 19-20). The problem is treated twice by Riese (problem 31, and repeated as problem 120). Although he uses the letters a, b and c, the problem is solved with a single unknown. Riese in turn might have learned about the problem from Fredericus Amann, who treated the problem in a manuscript of 1461, with the same values (*Cod. Lat. Monacensis* 14908, $155^r - 155^v$; transcription by Curtze, 1895, 70-1).
Stifel's version in modern notation is as follows:

$$a + \frac{b}{2} = 100$$
$$b + \frac{c}{3} = 100$$
$$c + \frac{a}{4} = 100$$

The solution is shown in Table 2.1. As a pedagogue, Stifel takes more steps than Cardano or the abacus masters before him. Line 8 is a misprint. Probably, the intention was to bring the polynomial to the same denominator as is done in step 13. This ostensibly redundant step shows the arithmetical foundation of the performed operations. Our meta-description gives the multiplication of equation (12) by 4 which makes line (13) superfluous. Stifel however, treats the polynomials as cossic numbers which he brings to the same denominator. Ten years later he will omit such operations as he acts directly on equations. The solution method is structurally not different from the one used by previous authors for similar linear problems. Note that Stifel does not use the second and third unknown in the same expression. The problem could as well be solved by two unknowns in which the second unknown is reused as by de la Roche. However, the fact that more than two unknowns are used opens up new possibilities and solution methods. How simply it may seem to the modern eye, the extension of the second unknown to multiple unknowns by Stifel was an important conceptual innovation.

---

[10] Stifel 1553, f. 296r: "Exemplum quartum capitis huius, et est Adami. Tres sunt socij, quorum primus dicit ad secundum, Si mihi dares dimidium summae tua, tunc haberem 100 fl. Et secundus dicit ad tertium: Si mihi dares summae tuae partem tertiam, tunc haberem 100 flo. Et tertius ad primum dicit: Si tu mihi dares summae tuae partem quartam, tunc haberem 100 fl. Quaestio est, quantum quisque eorum habeat".

	Symbolic	Meta description	Original text
1	$a + \frac{b}{2} = 100$	premise	Quod autem primus petit â secundo dimidium summae, quam ipse secundus habet, ut ipse primus habeat 100 fl.,
2	$x + \frac{y}{2} = 100$	choice of first and second unknown	fatis mihi indicat, aequationemen esse inter $1x + 1/2A$ et 100 florenos. Sic aût soleo ponere fracta huiusmodi $(1x + 1A)/2$ aequatae 100 fl.
3	$2x + y = 200$	multiply (2) by 2	Ergo 2x + 1A aequantur 200 fl.
4	$y = 200 - 2x$	subtract 2x from (3)	Et 1A aequantur 200 fl – 2x. Facit ergo 1A, 200fl. – 2x id quod mihi reservo loco unius A. Habuit igitur primus 1x florenorum. Et secundus 200 fl. – 2x.
5	$z = c$	choice of third unknown	Et tertius 1B flor.
6	$y + \frac{z}{3} = 100$	premise	Petit autem secundus tertiam partem summae terti socij, ut sicispe secundus habeat 100 fl.
7	$200 - 2x + \frac{z}{3} = 100$	substitute (4) in (6)	Itaque iam 200 fl. – 2x fl + 1/3 B, aequantur 100 florenis.
8	$600 - \frac{6}{3}x + z = 100$	illegal	Sic ego soleo ponere huiusmodi fractiones, ut denominator respiciat totum numeratorem. Ut 600 – 6/3 x + B aequata 100.
9	$600 - 6x + z = 300$	multiply (7) by 3	Aequantur itaque 600 – 6x + B cum 300.
10	$z = 6x - 300$	add $6x + 600$ to (9)	Atque hac aequatione vides fatis, ut 1B resolvatur in 6x – 300. Et sic primus habuit 1x florenorum. Secundus 200 fl – 2x. Tertius 6x – 300.
11	$z + \frac{x}{4} = 100$	premise	Petit autem tertius partem quartam summae, quam habet primus, ut sic ipse tertius etiam habeat centum florenos.
12	$6\frac{1}{4}x - 300 = 100$	substitute (10) in (11)	Itaque 6 x – 300 aequantur 100.
13	$\frac{25x - 1200}{4} = 100$	from (12)	Item (25x – 1200)/4 aequantur 100 fl.
14	$25x - 1200 = 400$	multiply (12) by 4	Et sic 25x – 1200 aequantur 400.
15	$25x = 1600$	add 1200 to (12)	Item 25x aequantur 1600 fl.
16	$x = 64$	divide (13) by 25	Facit 1x 64 fl.
17	$y = 200 - 128$	substitute (16) in (4)	Habuit igitur primus 1x, id est, 64 fl. Secundus habuit 200 – 2x.
18	$y = 72$	from (15)	i. 72 fl.
19	$z = 384 - 300$	substitute (18) in (10)	Et tertius habuit 6x – 300,
20	$z = 84$	from (19)	hoc est 84 fl.

Table 2.1: Stifel's exposition of the second unknown.

2.5 Cardano revisted: The first operation on two equations.

Cardano envisaged an *Opus perfectum* covering the whole of mathematics in fourteen volumes, published in stages (Cardano 1554). Soon after the publication of the *Practica arithmeticae*, he started working on the *Ars Magna*, which was to become the tenth volume in the series.[11] It was published by Johann Petreius in Nürnberg in 1545, who printed Stifel's *Arithmetica Integra* the year before as well as several other books by Cardano. We know that Cardano has seen this work and it would be interesting to determine the influence of Stifel.[12] The *Ars Magna* shows an evolution from the *Practica Arithmeticae* in several aspects. Three points are relevant for our story of the second unknown. Having learned that Tartaglia arrived at a solution to the cubic by geometrical reasoning, Cardano puts much more effort than before in delivering geometrical proofs, and this not only for the cubic equation. He also tries to be more systematical in his approach by listing all possible primitive and derivative cases of rules (which we call equations), and then by treating them separately. One of these primitive cases deals with two unknowns which he discusses in two chapters. Chapter IX is on *De secunda incognita quantitate non multiplicata* or the use of the second unknown for linear problems. Rules for solving quadratic cases are treated in Chapter X. Let us look at the first linear problem:[13]

> Three men had some money. The first man with half the other' would have had 32 *aurei*; the second with one-third the other', 28 *aurei*; and the third with one-fourth the others', 31 *aurei*. How much has each?

In modern notation the problem would be:

$$
\begin{aligned}
a + \tfrac{1}{2}(b + c) &= 32 \\
b + \tfrac{1}{3}(a + c) &= 28 \\
c + \tfrac{1}{4}(a + b) &= 31
\end{aligned}
\tag{2.2}
$$

In solving the problem Cardano introduces the two unknowns for the share of the first and the second person ("Statuemus primo rem ignotam primam,

---

[11] The dating can be deduced from the closing sentence of the Ars Magna: "Written in five years, may it last as many thousands" from Witmer (1968, 261).

[12] Cardano mentions in his biography that he is cited by Stifel in what must be the first citation index (2002, 220).

[13] Translation from Witmer (1968, 71). Witmer conscientiously uses *p* and *q* for *positio* and *quatitates* which preserves the contextual meaning. Unfortunately he leaves out most of the tables added by Cardano for clarifying the text, and replaces some of the sentences by formulas. As the illustrations and precise wording are essential for our discussion, I will use the Latin text from the *Opera Omnia* when necessary, correcting several misprints in the numerical values.

Symbolic	Meta de-scription	Original text
1 $a = x$	choice of first unknown	Statuemus primo rem ignotam primam,
2 $b = y$	choice of second unknown	secundo secundam rem ignotam
3 $c = 31 - \frac{1}{4}(x + y)$	substituting (1) and (2) in (2.2)c	tertio igitur 31 aurei, minus quarta parte rei, ac quarta parte quantitatis relicti sunt
4 $a + \frac{1}{2}(b + c) = 32$	premise	iam igitur vide, quantum habet primus, equidem si illi dimididium secundi et terti addicias, habiturus est aureos 32.
5 $a = 32 - \frac{1}{2}y - 15\frac{1}{2} + \frac{1}{8}x + \frac{1}{8}y$	substitute (2) and (3) in (4)	habet igitur per se aureos 32 m. 1/2 quan. m. 15 1/2 p. 1/8 positionis p. 1/8 quant.
6 $a = 16\frac{1}{2} - \frac{3}{8}y + \frac{1}{8}x$	from (5)	quare habebit 16 m. 3/8 quantitatis p. 1/8 pos.
7 $x = 16\frac{1}{2} - \frac{3}{8}y + \frac{1}{8}x$	substitute (1) in (6)	hoc autem sit aequale uni positioni
8 $\frac{7}{8}x + \frac{3}{8}y = 16\frac{1}{2}$	from (7)	erit 7/8 pos. et 3/8 quant. aequale 16 1/2
9 $7x + 3y = 132$	multiply (8) with 8	quare deducendo ad integra 7 pos. et 3 quant. aequabuntur 132.
10 $b + \frac{1}{3}(a + c) = 28$	premise	Rursus videamus, quantum habeat secundus, habet hic 28 si ei tertia pars primi ac tertij addatur
11 $\frac{1}{3}(a + c) = \frac{1}{3}x + 10\frac{1}{3} - \frac{1}{12}x - \frac{1}{12}y$	from (3) and (6)	ea est 1/3 pos. p. 10 2/3 m. 1/12 pos. m. 1/12 quant.
12 $\frac{1}{3}(a + c) = \frac{1}{4}x + 10\frac{1}{3} - \frac{1}{12}y$	from (11)	hoc est igitur pos. p. 10 1/3 m. 1/12 quant.
13 $b = 17\frac{2}{3} + \frac{1}{12}y - \frac{1}{4}x$	substitute (12) in (11)	abbice ex 28 relinquitur 17 2/3 p. 1/12 quant. m. pos. et tantum habet secundus.
14 $y = 17\frac{2}{3} + \frac{1}{12}y - \frac{1}{4}x$	substitute (2) in (14)	suppositum est autem habere illum quantitatem, quantitas igitur secunda, aequivalet 1/12 suimet, et 17 2/3 p. m. pos.

secundo secundam rem ignotam") (*Opera* III, 241). In the rest of the book the two unknowns are called *positio* and *quantitates*, abreviated as *pos.* and *quan.* They appear regularly throughout the later chapters, and in some cases Cardano uses *pos.* for problems solved with a single unknown.

Note how strictly Cardano switches between the role of two unknowns and the share of the first and second person by making the substitution steps of lines (7) and (14) explicit.

15	$\frac{11}{12}y + \frac{1}{4}x = 17\frac{2}{3}$	subtract $\frac{1}{12}y$ from (14) and add $\frac{1}{4}x$	abiectis communiter 1/12 quantitatis, et restituto m. alteri parti, sient 11/12 quan. p. pos aequalia 17 2/3,
16	$11y + 3x = 212$	multiply (15) by 12	quare 11 quant. p. 3 pos. aequalia erunt 212 multiplicatis partibus omnibus per 12 denominatorem.

The next part in the solution is the most significant with respect to the emerging concept of a symbolic equation. Historians have given a lot of attention to the *Ars magna* for the first published solution to the cubic equation, while this mostly is a technical achievement. We believe Cardano's work is equally important for its conceptual innovations such as the one discussed here.

The first occurrence of the second unknown for a linear problem is by an anonymous fifteenth-century abbaco master, author of Fond. prin. V.152.[14] The problem about four men buying an ox is by means of the second unknown reduced to two "linear equations", $7y = 13x + 4$ and $4y = 2x + 167$. Expressed in symbolic algebra it is obvious to us that by multiplying the two equations with the coefficients of y, we can eliminate the second unknown which leads to a direct solution. However, the author was not ready to do that, because he did not conceive the structures as equations. They are subsequently solved by the standard tool at that time, the rule of double false position. Cardano here marks a turning point in this respect. Having arrived at two equations in two unknowns Cardano gives a general method: [15]

Now raise whichever of these you like to equality with the other with respect to the number of either x or y "(in positionum aut quantitatum numero")". Thus you may decide that you wish, by some method, that in $3x + 11y = 212$, there should be $7x$. Then, by using the rule of three, there will be

$$7x + 25\frac{2}{3}y = 494\frac{2}{3}.$$

You will therefore have, as you see,

$$7x + 3y = 132 \text{ and } 7x + 25\frac{2}{3}y = 494\frac{2}{3}$$

Hence, since $7x$ is the same in both, in both the difference between the quantities of y, namely 22 2/3, will equal the difference between the numbers, which is 362 2/3.

---

[14] Franci and Pancanti, 1988, 144, ms. f. 177^r: "che tra tutti e tre gli uomeni avevano 3 oche meno 2 chose e sopra a questo agiugnerò l'ocha la quale si vole chonperare, chos aremo che tra tutti e tre gli uomeni e l'ocha saranno 4 oche meno 2 chose, dove detto fu nella quistione che tra danari ch'anno tutti e tre gli uomeni e 'l chosto del'ocha erano 176. Adunque, posiamo dire che lle 4 oche meno 2 chose si vagliano 176, chosì ài due aguagliamenti". In Heeffer (2010b) it is argued that this text is by Antonio de' Mazzinghi or based on a text by his hand.

[15] Cardano 1663, *Opera* IV, 241. I have adapted Witmer's translation to avoid the use of the terms coefficient and equation, not used by Cardano (Witmer 1968, 72).

> Divide therefore, as in the simple unknown, according to the third chapter, 362 2/3
> by 22 2/3; 16 results as the value of y and this is the second.

Using modern terms, this comes down to the following: given two linear equations in two unknowns, you can eliminate any of the unknowns by making their coefficients equal and adapting the other values in the equation. The difference between the coefficient of the remaining unknown will be equal to the difference of the numbers. Although the result is the same, the text does not phrase the procedure as a subtraction of equations. However, the table added by Cardano, which is omitted in Witmer's translation, tells a different story:

$$7x + 3y = 132$$
$$7x + 25\tfrac{2}{3}y = 494\tfrac{2}{3}$$

$$\overline{}$$

$$22\tfrac{2}{3}y = 362\tfrac{2}{3}$$

The table shows a horizontal line which designates a derivation: "from the first and the second, you may conclude the third". This table goes well beyond the description of the text and thus reads: "the first expression subtracted from the second results in the third". He previously used the same representation for the subtraction of two polynomials, also subtracting the upper line from the lower one (Cardano 1663, IV, 20). Cardano never describes the explicit subtraction of two equations in the text. Even if he did not intend to represent it that way, his peers studying the *Ars magna* will most aptly have read it as an operation on equations. As such, this is the first occurrence of an operation involving two equations, a very important step into the development of simultaneous equations and the very concept of an equation.

A second point of interest for the story of the second unknown is an addition in a later edition of the *Ars Magna* (Cardano, 1570; 1663; Witmer p. 75 note 13). Cardano added the problem of finding three so that the following conditions hold (in modern notation):[16]

$$a + b = 1\tfrac{1}{2}(a + c)$$
$$a + c = 1\tfrac{1}{2}(b + c)$$

He offers two algebraic solutions for this indeterminate problem. The second one is the most modern one, since he only manipulates equations and not polynomials. But the first solution has an interesting aspect, because we could

---

[16] Cardano, *Opera* IV, 242: "Exemplum tertium fatis accommodatum. Invenias tres quantitates quarum prima cum secunda sit sequialtera primae cum tertia et prima cum tertia sit sequialtera 2 cum tertia".

call it a derivation with two and a half unknowns. Cardano uses *positio* for the third number and *quantitates* for the second, for which we will use x and y. The sum of the first and third thus is

$$1\frac{1}{2}(x+y).$$

Subtracting the third gives the value of the first as

$$\frac{1}{2}x + 1\frac{1}{2}y.$$

Multiplying the sum of the first and third with $1\frac{1}{2}$ gives the sum of first and second as

$$2\frac{1}{4}x + 2\frac{1}{4}y.$$

Subtracting the second gives a second expression for the first as

$$2\frac{1}{4}x + 1\frac{1}{4}y.$$

As these two are equal

$$1\frac{3}{4}x = \frac{1}{4}y \text{ or } y \text{ is equal to } 7x$$

Only then, Cardano removes the indeterminism by posing that $x = 1$ leading to the solution (11, 7, 1). The interesting aspect of this fragment is that Cardano tacitly uses a third unknown which gets eliminated. As a demonstration, the reasoning can be reformulated in modern notation, with z as third unknown as follows:

$$z + y = 1\frac{1}{2}(z + x) \tag{2.3}$$

$$z + x = 1\frac{1}{2}(x + y) \tag{2.4}$$

If we subtract x from (2.4) it follows that

$$z = \frac{1}{2}x + 1\frac{1}{2}y$$

Substituting (2.4) in (2.3) gives

$$z + y = 2\frac{1}{4}x + 2\frac{1}{4}y$$

Subtracting y from this equation gives

$$z = 2\frac{1}{4}x + 1\frac{1}{4}y$$

Therefore

$$\frac{1}{2}x + 1\frac{1}{2}y = 2\frac{1}{4}x + 1\frac{1}{4}y$$

or

$$y = 7x$$

There is only a small difference between Cardano's solution and our reformulation. If only he had a symbol or alternative name for the first unknown quantity, it would have constituted an operational unknown. He seems to be aware from the implicit use of three unknowns as he concludes: "And this is a nice method because we are working with three quantities" ("Et est pulchrior modus quia operamur per tres quantitates") (*Opera*, IV, 242). It is not clear why this problem was not included in the 1545 edition. It could have been added by Cardano as a revision to the Basel edition of 1570.

A third aspect from the *Ars magna*, which reveals some evolution in Cardano's use of multiple unknowns is one of the later chapters, describing several rules, previously discussed in the *Practica aritmeticae*. Chapter 31 deals with the *Regula magna*, probably one of the most obscure chapters in the book. The rule is not described, only some examples are given. Nor does it contain any explanation why it is called *The Great Rule*. Most of these problems concern proportions which are represented by letters. Remarkably, Cardano performs operations on these letters and constructs equations using the letters such as "igitur 49 b, aequalia sunt quadrato quadrati a" (see Table 2.2). Only in the final step, as a demonstration that this solves the problem, does he switch back to regular unknown called *res*. Let us look in detail at problem 10 (Witmer 190, *Opera* IV, 276). A modern formulation of the problem is:

$$a + b = 8$$

$$\frac{a^3}{7b} = \frac{7b}{ab}$$

The text is probably the best illustration that the straightforward interpretation of the letters as unknowns is an oversimplification. If the letters would be unknowns then substituting $b = 8 - a$ in $a^4b = 49b^2$ would immediately lead to the equation. Instead, Cardano takes a detour by introducing c, d and e and then applying the magical step 5. No explanation is given, though the inference

$$\frac{a}{7} = \frac{d}{c} \text{ is correct, because } \frac{d}{c} = \frac{7b}{a^3} = \frac{a}{7}, \text{ or } \frac{7b}{a^3} = \frac{ab}{7b}$$

	Symbolic	Meta description	Original text
1	$c = a^3$	choice of unknown	Sit a minor, eius cubus c, b autem maior,
2	$ab = e$	choice of unknown	et productum b in a sit e,
3	$7b = d$	choice of unknown	et septuplum b sit d,
4	$\frac{a}{7} = \frac{e}{d}$	divide (2) by (3)	quia igitur ex b in a, sit e et ex b in 7 sit d, erit a ad 7,
5	$\frac{a}{7} = \frac{d}{c}$		ut e ad d quare a ad 7 ut d ad c
6	$ac = 7d$	multiply (5) by 7c	Igitur ex a in c, sit septuplum d
7	$a^4 = 49b$	subtitute (1) & (3) in (6)	sed est septuplum b, igitur 49 b aequalia sunt quadrato quadrati a
8	$b = \frac{1}{49}a^4$	divide (7) by 49	igitur b est aequale 1/49 quad. quadrati a
9	$a + b = 8$	premise	quia igitur a cum b est 8
10	$a + \frac{1}{49}a^4 = 8$	substitute (8) in (9)	et b est 1/49 quad. quadrati a, igitur a cum 1/49 quad. quadrati sui, aequatur 8.
11	$x + \frac{1}{49}x^4 = 8$	substitute a by x in (10)	quare res et 1/49 [quad. quadratum aequatur 8]
12	$x^4 + 49x = 392$	multiply (11) by 49	[Igitur] quad. quadratum p. 49 rebus, aequatur 392

Table 2.2: Cardano's *Regula magna* for solving linear problems

which is the reciprocal of what was given. Apparently, the fact that e is to d as d is to c, is evident to Cardano, shows how his reasoning here is inspired by proportion theory, rather than being symbolic algebra.

2.6 The improved symbolism by Stifel

From the last part of Stifel's *Coss* (1553, f. 480^r) we know that he has read the *Ars magna*. He cites Cardano on the discovery of Scipio del Ferro (f. 482^r) and adds a chapter on the cubic equation. The influence between Cardano and Stifel is therefore bidirectional. At several instances he discusses the second unknown from a methodological standpoint, as Cardano did in the *Ars magna*. Although Rudolff does use the second unknown in the original 1525 edition for several problems, in other examples Stifel recommends the *regula quantitatis* as a superior method to the ones given by Rudolff ("Christoff setzet vier operation oder practicirung auff diss exemplum. Ich will eine setzen ist besser und richtiger zu lernen und zu behalten denn seyne vier practicirung", 223v). He notes that there is nothing magical about the second unknown. For him, it is basically not different from the traditional *coss*: "Den im grund ist regula Quantitatis nichts anders denn Regula von 1𝔢." (Stifel 1553, ff. 223^v − −224^r). While we can only wonder why it has not been done before, for Stifel it seems natural to use multiple unknowns for the typical shares or values expressed in linear problems: "Man kan auch die Regulam (welche sye nennen) Quantitatis nicht besser verstehn den durch sollische exempla [i.e. linear problems] Weyl sye doch nichts anders ist denn da man 1 𝔢 setzt under einem andern zeychen" (Stifel 1553, f. 277^v). He considers arithmetical operations on shares not fundamentally different from algebraic operations on unknowns: "Der Cossischen zeychen halb darffest du dich auch nicht hart bekumern. Denn wie 3 fl. un 4 fl. machen 7 fl., also auch 3𝔢 und 4𝔢 machen 7𝔢" (1553, f. 489^r).

After treating over 400 problems from Rudolff, Stifel adds a chapter with some examples of his own. Half of the 24 problems added are solved by two unknowns. Interestingly, he silently switches to another notation system for quadratic problems involving multiple unknowns, thus avoiding the ambiguities of his original system. The improved symbolism is well illustrated with the following example:[17]

> Find two numbers, so that the sum of both multiplied by the sum of their squares equals 539200. However, when the difference of the same two numbers is multiplied by the difference of their squares this results in 78400. What are these numbers?

This is a paraphrase of Stifel's solution: Using 1𝔢 and 1A for the two numbers, their sum is 1𝔢 + 1A. Their difference is 1𝔢 − 1A. Their squares 1ℨ and 1AA. The sum of the squares 1ℨ + 1AA. The difference between the squares 1ℨ − 1AA. So multiplying 1𝔢 + 1A with 1ℨ + 1AA gives ℭ + 1ℨA + 1𝔢AA + 1AAA which equals 539200. Then I multiply also 1𝔢 − 1A. with 1ℨ − 1AA. This gives ℭ − 1ℨA − 𝔢AA + 1AAA and that product equals 539200.

---

[17] Stifel 1553, ff. 469^r − 470^v, translation mine.

So Stifel now uses AA for the square and AAA for the third power of A. He thus eliminates the ambiguities discussed before. Now that $A\tilde{z}$ becomes AA, the product of the square of A with $1\tilde{e}$ can be expressed as $AA\tilde{e}$ and the product of the square of $1\tilde{e}$ with A as $A\tilde{z}$ or $\tilde{z}A$ – thus also removing the ambiguity of multiplying cossic terms together. As such, algebraic symbolism is functionally complete with respect to to the representation of multiple unknowns and powers of unknowns. What is still missing, as keenly observed by Serfati (2010), is that this does not allow to represent the square of a polynomial. In order to represent the square of $1\tilde{z} + 1\tilde{e} + 2$, for example, Stifel has to perform the calculation. Also, the lack of symbols for the coefficients does not yet allow that every expression of seventeenth-century Cartesian algebra can be written unambiguously in Stifel's symbolism. This was later introduced by Viète. However, the important improvement by Stifel in his *Coss*, was an important step necessary for the development of algebraic symbolism, and has been overlooked by many historians.[18] Having shown that Stifel resolved the ambiguities in the interpretation of multiplied cossic terms, we will further replace the cossic signs for *coss*, *census* and cube by x, x^2 and x^3.

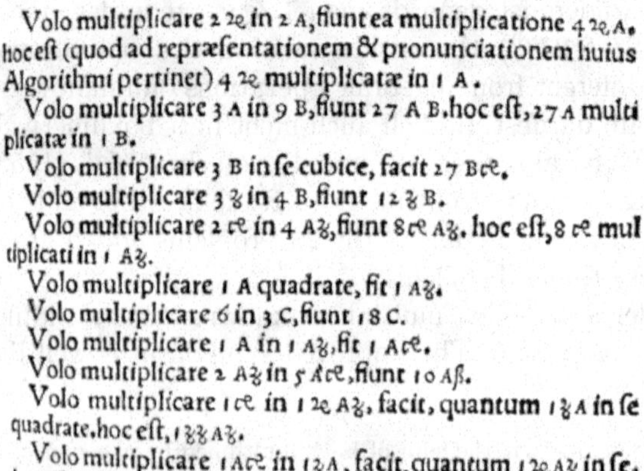

Volo multiplicare 2 ℞ in 2 A, fiunt ea multiplicatione 4 ℞ A. hoc eft (quod ad repræfentationem & pronunciationem huius Algorithmi pertinet) 4 ℞ multiplicatæ in 1 A .
   Volo multiplicare 3 A in 9 B, fiunt 27 A B. hoc eft, 27 A multiplicatæ in 1 B.
   Volo multiplicare 3 B in fe cubice, facit 27 Bℂ.
   Volo multiplicare 3 ℥ in 4 B, fiunt 12 ℥ B.
   Volo multiplicare 2 ℂ in 4 A℥, fiunt 8 ℂ A℥. hoc eft, 8 ℂ multiplicati in 1 A℥.
   Volo multiplicare 1 A quadrate, fit 1 A℥.
   Volo multiplicare 6 in 3 C, fiunt 18 C.
   Volo multiplicare 1 A in 1 A℥, fit 1 Aℂ.
   Volo multiplicare 2 A℥ in 5 Aℂ, fiunt 10 Aß.
   Volo multiplicare 1 ℂ in 1 ℞ A℥, facit, quantum 1 ℥ A in fe quadrate, hoc eft, 1 ℥℥ A℥.
   Volo multiplicare 1 Aℂ in 1 ℥A, facit, quantum 1 ℞ A℥ in fe, hoc eft, 1 ℥ A℥℥.

Fig. 2.3: The improved symbolism by Stifel (1553, f. 469^r)

Next, Stifel eliminates terms from the equation by systematically adding, subtracting, multiplying and dividing the equations, not seen before in his *Arithmetica Integra* of 1544 (Stifel 1553, 469^v):

---

[18] The symbolism introduced by Stifel in the *Arithmetica integra* is discussed by Bosmans (1905-6), Russo (1959), Tropfke (1980, 285, 377), Gericke (1992, 249-50), Cifoletti (1993) chapter 3, appendix 1 and 2. With the exception of Cajori (1928-9, I, 144-146) who mentions Stifel's innovation as "another notation", none of these authors discuss the significance of the

Multiply the two equations in a cross as you can see below:

$$x^3 + 1x^2A + 1xAA + 1AAA = 539200$$
$$x^3 - 1x^2A - 1xAA + 1AAA = 78400$$

But dividing these numbers by their GCD ("yhre kleynste zalen") gives 337 and 49 and so we arrive at the two sums:

$$49x^3 + 49x^2A + 49xAA + 49AAA$$
$$337x^3 - 337x^2A - 337xAA + 337AAA$$

and these two sums are equal to each other. If we now add $337x^2A + 337xAA$ to each side so, this result in

$$337x^3 + 337AAA = 49x^3 + 386x^2A + 386xAA + 49AAA$$

Now subtract $49x^3 + 49AAA$ from each side, this will give

$$386x^2A + 386xAA = 288x^3 + 288AAA$$

Divide each side by $2x + 2A$, this results in

$$193xA = 144x^2 - 144xA + 144AA \qquad (2.5)$$

Next (as you can extract the square root from each side) subtract from each side $144xA$

$$49xA = 144x^2 - 288xA + 144AA$$

Extract from each side the square root, which becomes $\sqrt{49xA} = 12x - 12A$. This we keep for a moment.

Here, operations on equations are remarkably extended to root extraction. Although not fully correct, this can be considered a 'natural' step from previous extensions. Because the alternative solutions are imaginary they are not recognized as such. Only in the seventeenth century we will see the full appreciation of double solutions to quadratic equations. Now Stifel returns to the equation (2.5) ("Ich widerhole yetzt die obgesetzte vergleychung").

Add to each side [of this equation] as much as is needed to extract the root of each side. This is 3 times $144xA$, namely $432xA$. So becomes

$$144x^2 + 288xA + 144A^2 = 625xA$$

Extract again from each side the square root, so will be

$$\sqrt{625xA} = 12x + 12A$$

And before I have found that $\sqrt{49xA} = 12x - 12A$. From these two equations I will make one through addition. Hence

improvements of 1553. Eneström (1906-7, 55) spends one page on the improved symbolism discussing Cantor's *Vorlesungen* (1892, 441, 445).

$$24x = \sqrt{1024xA}$$

Next I will square each side, which results in $576x^2 = 1024xA$ and then I divide each side with $576x$. Thus

$$1x = 1\frac{7}{9}A \text{ or } 1A = \frac{9}{16}x$$

Having formulated both unknowns in terms of the other, one of them can be eliminated, or in Stifel's wording *resolved*. He reformulates the original problem in x and $9/16\ x$, which leads to a cubic expression with solution 64.

We have previously shown that Cardano's operations on equations are implicit in the illustrations but are not rhetorically phrased as such. In this text by Stifel we have a very explicit reference to the construction from one equation by the addition of two others: "From these two equations I make one equation by addition" ("Aufs desen zweyen vergleychungen mach ich ein einige vergleychung mit addiren"). This is certainly an important step forward from the *Arithmetica Integra*, and from then on, operations on equations will be more common during the sixteenth century.

We have here an unique opportunity to compare two works, separated by a decade of development in Stifel's conceptions of algebra. It gives us a privileged insight into subtle changes of the basic concepts of algebra, in particular that of a symbolic equation. As an illustration, let us look at one problem with three numbers in geometric progression. The same problem is presented in Latin in the *Arithmetica Integra* and in German in the Stifel edition of Rudolff's *Coss*, though with different values. The problem is solved using two unknowns in essentially the same way, but there are some delicate differences which are very important from a conceptual point of view. As Stifel presents the problem in a section with "additional problems by his own", we can assume that he constructed the problem himself. In any case, it does not appear in previous writings. In modern formulation the problem has the following structure:

$$a : b = b : c$$
$$(a + c)(a + c - b) = d$$
$$(a + c - b)(a + b + c) = e$$

with respectively (4335, 6069) and (90720, 117936) for d and e. The start of the solution is identical in the Latin and German text, except that the choice of the first and second unknowns are reversed (see Table 2.3).

In both cases Stifel arrives at two equations in two unknowns. These compares very well with those from Fond. prin. V.152 and the example of Cardano's *Ars Magna*, except that we now have a quadratic expression. If we swap back the two unknowns in the German text, the equations compare as follows:

Stifel 1544, f. 313r	Problem 24, Stifel 1553, f. 474r
Quaeritur tres numeri continue proportionales, ita ut multiplicatio duorum extremorum, per differentiam, quam habent extremi simul, ultra numerum medium, faciant 4335. Et multiplicatio eiusdem differentiae, in summam, omnium trium faciat 6069.	Es sind drey zalen continue proportionales so ich das aggregat der ersten, und dritten, multiplicir mit der differentz dess selbigen aggregatis uber die mittel zal, so kommen 90720. Und so ich die selbige differentz multiplicir in die summa aller dreyer zalen, so kommen 117936. Welche zalen sinds?
1A + 1x est summa extremorum 1A − 1x est summa medij 2A est summa omnium trium 2x est differentia quam habent extremi ultra medium.	Die drey zalen seyen in einer summa 2x. Die zurlege ich also in zwo summ 1x + 1A, 1x − 1A Nu last ich 1x − 1A die mittel zal seyn so muss 1x + 1A die summa seyn der ersten und dritten zalen. Und also sind 2A die differentz dess selbigen aggregats uber die mittel zal.
Itaque 2x multiplicatae in summam extremorum, id est, in 1A + 1x faciunt 2xA + 2x2 aequata 4335.	Drumb multiplicir ich 2A in 1x + 1A facit 2xA + 2AA gleych 90720.
Deinde 2x multiplicatae in 2A seu in summam omnium, faciunt 4xA aequata 6096.	So ich aber 2A multiplicir in die summ aller dreyer zalen, nemlich in 2x, so kommen 4xA die sind gleych 117936.

Table 2.3: Two ways how Stifel solves structurally the same problem.

$$2xy + 2x^2 = 4335 \qquad 2xy + 2x^2 = 90720$$
$$4xy = 6096 \qquad 4xy = 117936$$

The next step is to eliminate one unknown from the two equations. We have seen that Cardano was the first to do this by multiplying one equation to equal the coefficients of one term in both equations and then to subtract the equations, albeit implicitly. In this respect, the later text deviates from the former (see Table 2.4).

The method in the Latin text articulates the value of xy from the two expressions and *compares* the resulting values. The text only states that their values are equal. Although Stifel writes "Confer iam duas aequationes illas", this should be understood as "now match those two equal terms", *aequationes* being the acts of comparing. So from the first expression we can infer that the value of xy is $(4335 - 2x2)/2$. From the second we can know that the value is $6069/4$. Thus, $(4335 - 2x2)/2$ must be equal to $6069/4$, from which we can deduce the value of x. The reasoning here is typical for the abacus and early cossist tradition were the solution is based on the manipulation and equation of polynomials expressions. In the German text, a decade later, Stifel distinctly moves to the manipulation of equations. He literally says: "Now double the equation above" and "from this [equation] I will now subtract the numbers

Stifel 1544, f. 313r	Problem 24, Stifel 1553, f. 474v
Confer iam duas aequationes illas. Nam ex priore sequitur quod $1xA$ faciat $(4335 - 2x^2)/2$.	So duplir ich nu die obern vergleychung, fa. $4xA + 4AA$ gleych 181440.
Ex posteriore autem sequitur quod $1xA$ faciat $6069/4$. Sequitur ergo quod $(4335 - 2x^2)/2$ et $6069/4$ inter se aequentur. Quia quae uni et eidem sunt aequalia, etiam sibi invicem sunt aequalia. Ergo (per reductionem) $17340 - 8x^2$ aequantur 12138 facit $1x^2 \cdot 650\frac{1}{4}$.	Da von subtrahir ich yetzt die zalen diser yetzt gefundnen vergelychung. Nemlich $4xA$ gleych 117936 so bleyben $4AA$ gleych 63504.
Et $1x$ facit $25\frac{1}{2}$.	Also extrahir ich auff yeder seyten die quadrat wurzel, so werden $2A$ gleych 252 und ist die differentz dess aggregats uber die mittel zal. So in nu $1A$ gleych 126.

Table 2.4: Two ways how Stifel solves structurally the same problem.

of the newly found equation", thus eliminating the second unknown. The last step also shows a clear evolution. In the Latin text he reduces the expression to the square of the unknown $1x^2$ and then extracts the root. In the later text he "extracts the square root of each side [of the equation]". The rest of the problem is to reformulate the original problem using the value of the second unknown. This is done in similar ways.

The example shows how the road to the concept of a symbolic equation is completed in a crucial decade of algebraic practice of the mid-sixteenth century. We have witnessed this evolution within a single author. The French algebraists from the second half of the sixteenth century will extend this evolution to a system of simultaneous linear equations.

2.7 Towards an aggregate of equations by Peletier

Stifel's edition of the *Coss* was published in Köningsberg in 1553, his foreword is dated 1552. Peletier's postscript ends the *Algèbre* with the date July 28, 1554. The printer's permit allows him to print and sell the book for three years from June 15, 1554. So, while Peletier might have seen Stifel's edition of the *Coss*, it does not show in his book. He certainly has studied the *Arithmetica Integra* well.

Jacques Peletier spends one quarter of the first book on the second unknown which he calls *les racines secondes* (pp. 95-117), a direct translation of Stifel's *de secundis radicibus* (Stifel 1544, f. 251v). He introduces Stifel's notation by way of the problem of finding two numbers, such that, in modern formulation (Peletier 1554, 96):

$$x^2 + y^2 = 340$$
$$xy = \frac{6}{7}x^2$$

If we would use the same name for the unknown for both numbers, this would lead to confusion, he argues. He therefore adopts Stifel's notation of $1A$, $1B$ for the second and third unknown in addition to his own sign for the first unknown. He then discusses the operations with multiple unknowns: addition, subtraction, multiplication and division, as was done with polynomials in his introductory chapters. He retains Stifel's ambiguity from the *Arithmetica Integra* that xy cannot be differentiated from yx.

Peletier has selected this example, instead of the one used by Stifel, because that problem can easily be solved in one unknown ("Car il est facile par une seule posicion sans l'eide des secondes racines", Peletier 1554, 102).

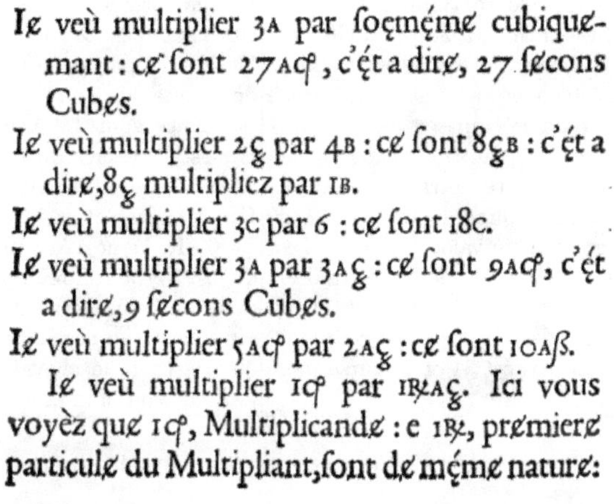

Iɇ veù multiplier 3A par foɇmɇmɇ cubiquɇ-
mant : cɇ font 27Aɋ , c'ɇt a dirɇ, 27 fɇcons
Cubɇs.
Iɇ veù multiplier 2ɕ par 4B : cɇ font 8ɕB : c'ɇt a
dirɇ,8ɕ multiplicz par 1B.
Iɇ veù multiplier 3c par 6 : cɇ font 18c.
Iɇ veù multiplier 3A par 3Aɕ : cɇ font 9Aɋ, c'ɇt
a dirɇ,9 fɇcons Cubɇs.
Iɇ veù multiplier 5Aɋ par 2Aɕ : cɇ font 10Aβ.
Iɇ veù multiplier 1ɋ par 1℞Aɕ. Ici vous
voyèz quɇ 1ɋ, Multiplicandɇ : e 1℞, prɇmierɇ
particulɇ du Multipliant,font dɇ mɇmɇ naturɇ:

Fig. 2.4: The rules for multiplying terms with multiple unknowns from Peletier (1554, 98). Compare these with Stifel (1545, f. 252^r)

Using x for the larger number and y for the smaller one he squares the second equation to

$$x^2y^2 = \frac{36}{49}x^4 \text{ which leads to } 49y^2 = 36x^2$$

Because $y^2 = 340 - x^2$ this can be rewritten as $y^2 = 340 - \frac{49}{36}y^2$. Then the second unknown can be expressed as

$$2\frac{13}{36}y^2 = 340 \text{ or } y^2 = 144,$$

leading to the solution 12 and 14.

Peletier gives four other problems solved with multiple unknowns. The first two are taken from Cardano's *De Quaestionibus Arithmeticis* in the *Practica arithmeticae*, problem 97 and 98 (Cardano, Opera III, 168-9), the third is the problem from Cardano's *Ars magna* discussed above (2.2). The fourth is one from Stifel (1544, f. 310^v), reproducing the geometric proof. This shows that Peletier was well acquainted with the most important algebraic treatises of his time. In fact, Peletier's example III (1554, 105-7) and its solution, is a literal translation from Cardano's, only using the symbolism from Stifel. The problem is structurally similar to problem 41 from Pacioli discussed earlier and follows the method by Pacioli. Compare the following text fragments:

Cardano, 1539, ff. HH.vir - HH.viv	Peletier, 1554, p. 106
Igit*ur* per p*r*aecedente*m* iunge summa*m* eoru*m* sit 3 quan. m. 31/30 co. divide p*er* 1 m. numero hominu*m* q*uod* est 2 exit 1 quan. m. 31/60 co. et haec est summa quae debet aequari valori equi sed aequus valet 1. quan. igit*ur* 1 quan. m. 31/60 co. ae-quantur 1 quan. quare detrahe 1 quan. ex 1 quan. remanebit quan. equivale*n*s 31/60 co. igitur 1 quan. aequivalet duplo q*uod* est 31/30 co. igitur dabis ex hoc fracto valore*m* denominatoris q*ui* es*t* 30 [sic] ad co. et nu-meratore*m* ad quan. igitur valor co. es*t* 30 et valor qua*n*titatis est 31 et in bursa fuere 30.	Par la precedente, assemblez les troes sommes: ce sont 3A m. 31/30 R. Divisez par un nombre moindre de 1 que les hommes, savoer est par 2: ce sont 1 A m. 31/60 R. E c'est la valuer du cheval. Donq, 1A est egale a 1 A m. 31/60 R. E par souttraction, A est egale a 31/60 R. Donc 1A, vaut la double, qui est 31/30 R. Meintenant, prenez pour 1A, le numerateur, que est 31, e pour 1R prenez le denominateur 30. Partant, le cheval valoet 31 e l'argant commun etoest 30.

Table 2.5: The dependence of Peletier on Cardano's *Practica Arithmeticae*.

Peletier thus literally translated Cardano's text only changing 1 *quan.* in 1*A* and reformulating the common sum as the value of a horse. We included this fragment to show how strongly Peletier bases his algebra on Cardano while Cifoletti attributes to him an important role in the development towards a symbolic algebra. Nonetheless, Peletier introduces some interesting new aspects in the next linear problem taken from *Ars Magna*. He first gives a literal translation of Cardano's solution calling the problem text *proposition* and the solution *disposition*. Interestingly he leaves out the substitution steps from Cardano, lines (7) and (14). Cardano considered these important for a demonstration, but apparently Peletier does not. Then he introduces a solution of his own ("trop plus facile que l'autre"). Starting from the same formulation (2.2), Peletier adapts Cardano's solution method by means of Stifel's symbolism for multiple unknowns.

	Symbolic	Meta description	Original text
1	$a = x$	choice of first unknown	Le premier à 1R
2	$b = y$	choice of second unknown	Le second 1A
3	$c = z$	choice of third unknown	Le tiers 1B.
4	$a + \frac{1}{2}(b+c) = 32$	premise	E par ce que le premier avec $\frac{1}{2}$ des deus autres, an à 32:
5	$x + \frac{1}{2}(y+z) = 32$	substitute (1), (2) and (3) in (4)	1R p. (1A p. 1B)/2 seront egales a 32.
6	$2x + y + z = 64$	multiply (5) by 2	E par reduccion, e due transposicion: 2R p. 1A p. 1B sont egales a 64, qui sera la premiere equacion.
7	$b + \frac{1}{3}(a+c) = 28$	premise	Secondemant, par ce que le second, avec 1/3 partie des deus autres an à 28:
8	$y + \frac{1}{3}(x+z) = 28$	substitute (1), (2) and (3) in (6)	ce sont 1A p. (1R p. 1B)/3 egales a 28:
9	$x + z + 3y = 84$	multiply (8) by 3	E par reduccion, 1A p. 1B p. 3A seront egales a 84, qui sera la seconde equacion.
10	$c + \frac{1}{4}(a+b) = 31$	premise	Pour le tiers (lequel avec $\frac{1}{4}$ partie des deus autres an à 31),
11	$z + \frac{1}{4}(x+y) = 31$	substitute (1), (2) and (3) in (10)	nous aurons 1B p. (1R p. 1A)/4, egales a 31.
12	$x + y + 4z = 124$	multiply (11) by 4	e par samblable reduccion, 1R p. 1A p 4B seront egales a 124. Voela, noz troes equacions principales.

Table 2.6: Peletier solving a problem by multiple unknowns.

Having arrived at three equations in three unknowns there seems to be little innovation up to this point. All operations and the use of three unknowns have been done before by Stifel. However, we can discern two subtle differences. Firstly, the last line (12) suggests that Peletier considers the three equations as an aggregate. In the rest of the problem solving process he explicitly acts on this aggregate of equations ("disposons donq nos troes equacions an cete sorte"). Secondly, he identifies the equations by a number. In fact, he is the

first one in history to do so, a practice which is still in use today.[19] The
identification of equations, as structures which you can manipulate, facilitates
the rhetorical structure of the *disposition*. This becomes evident in the final
part (see Table 2.7).

	Symbolic	Meta description	Original text
13	$2x + 4y + 5z = 208$	add (9) and (12)	Ajoutons la seconde e la tierce, ce seront, pour quatrieme equacion 2R p. 4A p. 5B egales a 208
14	$3y + 4z = 144$	subtract (6) from (13)	Donq an la conferant a la premier equacion, par ce que 2R sont tant d'une part que d'autre, la differance de 64 a 208 (qui est 144) sera egale avec la differance de 1A p. 1B a 4A p. 5B: Donq, an otant 1A p. 1B de 4A p. 5B, nous aurons pour la cinquieme equacion 3A p. 4B egales a 144
15	$3x + 4y + 2z = 148$	add (6) and (9)	ajoutons la premiere e la seconde: nous aurons pour la sizieme equacion 3R p. 4A p. 2B egales a 148.
16	$3x + 2y + 5z = 188$	add (6) and (12)	ajoutons la premiere e la tierce: nous aurons pour la sesttieme equacion 3R p. 2A p. 5B egale a 188.
17	$6x + 6y + 7z = 336$	add (15) and (16)	ajoutons ces deus dernieres: nous aurons, pour la huitieme equacion 6R p. 6A p. 7B egales a 336.
18	$6x + 6y + 24z = 744$	multiply (12) by 6	Finablemant, multiplions la tierce par 6 (pour sere les racines egales, de ces deus dernieres equacions) e nous aurons, pour la neuvieme equacion 6R p. 6A p. 24B egales a 744.

Table 2.7: Peletier eliminating unknowns by adding and subtracting equations.

Peletier succeeded in manipulating the equations in such a way that he arrives
at two equations in which two of the unknowns have the same coefficients, or
in his terms, "equal roots". Subtracting the two gives $17z = 408$ arriving at
the value 24 for z. The other values can then easily be determined as 12 and
16. Comparing his method with Cardano's, it is not shorter or more concise.
Cardano takes 16 steps to arrive at two equations in which one unknown can
be eliminated, Peletier takes 18 steps to the elimination of two unknowns. But

---

[19] The classic work by Cajori (1928-9) on the history of mathematical notations, does not
include the topic of equation numbering or referencing. I have seen no use of equation

Peletier does not use the argument of length, instead he considers his method easier and clearer, thus emphasizing the argumentative structure. Indeed, as can be seen from the table, the actual text fits our meta-description very well. Peletier systematically uses operations on equations and applies addition and subtraction of equations to eliminate unknowns. Moreover, he explicitly formulates the operations as such: "add the second [equation] to the third, this leads us to a fourth equation". Although we have seen such operations performed implicitly in Cardano's illustration, the use of the terminology in the argumentation is an important contribution. The use of multiple unknowns, the symbolism and the argumentation, referring to operations on structures, called equations, makes this an important entrance into symbolic algebra.

2.8 Valentin Mennher (1556)

Valentin Mennher, a reckoning master from Antwerp, introduces the rule in between problems 254 and 255 as *regle de la quantité, ou seconde radice* in his *Arithmétique seconde* (Mennher, 1556, f. Qi^v; 1565, f. FFi^r) as a "rule which exceeds all other rules and without which many examples would otherwise be unsolvable". He refers to Stifel for the origin of the rule and adopts Stifel's notation.[20] From problem 267, it becomes clear that he has used Stifel's edition of Rudolff (1553) as he also uses the improved notation AA for the square of the second unknown (1556, ff. $Qvi^r - Qvi^v$; 1565, ff. $Ffviii^r - Ffviii^v$). We will give one example from Mennher, though the method does not differ from Stifel's solution to problem 193 of Rudolff's *Coss*. The problem is about four persons having a debt, with the four sums of three given. The problem is known from early Indian sources. Stifel uses four unknowns while Rudolff originally reuses the second unknown. Mennher adopts Stifels method with different values and slightly changing the unknowns. Mennher uses the values:

$$a + b + c = 18$$
$$b + c + d = 25$$
$$a + c + d = 23$$
$$a + b + d = 21$$

With the unknowns x, A, B and C for d, a, b and c respectively, he expresses the sum of all four as $18 + x$, $25 + A$, $23 + B$ and $21 + C$.

---

numbers prior to Peletier's.

[20] Mennher, clearly learned the use of letters from Stifel, as he writes: "tout ainsi comme M. Stiffelius l'enseigne, en posant apres le x pour la seconde position A, et pour la troisiesme B, et pout la quatriesme C." (Mennher, 1556, Qi^v; 1565 $Ffi^r - Ffi^v$).

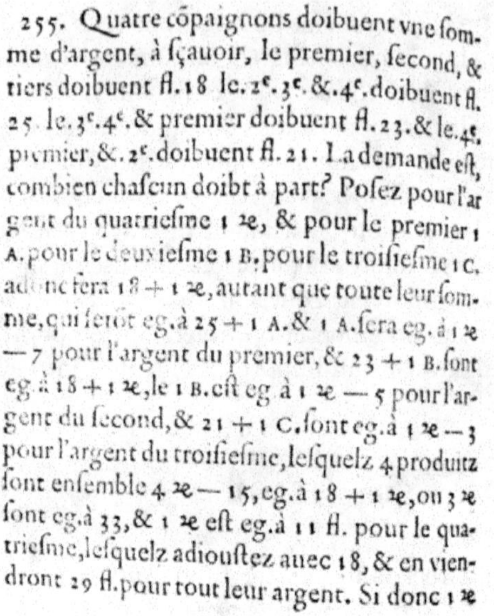

255. Quatre cōpaignons doibuent vne fom-
me d'argent, à fçauoir, le premier, fecond, &
tiers doibuent fl. 18 le. 2ᶜ. 3ᶜ. &. 4ᶜ. doibuent fl.
25 le. 3ᶜ. 4ᶜ. & premier doibuent fl. 23. & le. 4ᶜ.
premier, &. 2ᶜ. doibuent fl. 21. La demande eſt,
combien chaſcun doibt à part? Poſez pour l'ar
gent du quatrieſme 1 æ, & pour le premier,
A. pour le deuſieſme 1 B. pour le troiſieſme 1 C.
adōne fera 18 + 1 æ, autant que toute leur fom-
me, qui ſerōt eg. à 25 + 1 A. & 1 A. fera eg. à 1 æ
— 7 pour l'argent du premier, & 23 + 1 B. font
eg. à 18 + 1 æ, le 1 B. eſt eg à 1 æ — 5 pour l'argent du fecond, & 21 + 1 C. font eg. à 1 æ — 3
pour l'argent du troiſieſme, leſquelz 4 produitz
font enfemble 4 æ — 15, eg. à 18 + 1 æ, ou 3 æ
font eg. à 33, & 1 æ eſt eg. à 11 fl. pour le quatrieſme, leſquelz adiouſtez auec 18, & en vien-
dront 29 fl. pour tout leur argent. Si donc 1 æ

Fig. 2.5: The use of the second unknown by Mennher (1556, f. Ffi^r).

As these four expressions have the same value, the debts of the first three can be restated in terms of x, namely $x - 7$, $x - 5$, and $x - 3$ respectively.

Adding the three together with x leads to the sum of all four $4x - 15$, which is equal to $18 + x$. From this it follows that x is 11, and the other debts are 4, 6 and 8. Most of the last twenty problems in the book are solved using several unknowns.

2.9 Kaspar Peucer (1556)

The humanist Caspar Peucer wrote, among other works on medicine and philosophy, a Latin algebra with the name *Logistice Regulae Arithmeticae*. The book contributed little to the works published by Stifel and had little influence. Except for a recent paper by Meißner and Deschauer (2005), Peucer seems to be forgotten. He discusses the *regula quantitatis* by the term *radicibus secundis* and provides four examples (Peucer 1556, ff. Tvir-Viir). He refers to Rudolff, Stifel and Cardano for the origin of the method. His first example is the ass and mule problems from the Greek epigrams, creating the indeterminate equation $1x+1 = 1A-1$. The other problems are linear ones involving multiple unknowns. The symbolism is taken from Stifel (1544).

2.10 Towards a system of simultaneous equations

2.10.1 Buteo (1559)

Jean Borel, better known under his Latinized name Buteo, is an underestimated as an author of mathematical works during the sixteenth century. He started publishing only after he became sixty. His *Logistica* of 1559 is a natural extension of the ideas of Peletier. Though Peletier was the first to consider an aggregate of equations, Buteo improved on Peletier and raised the method to what we now call solving a system of simultaneous linear equations. The naming of his book by the Greek term of *logistics* is an implicit denial of the Arab contributions to Renaissance algebra. This position is shared by several humanist writers of the sixteenth century.

Buteo introduces the second unknown in the third book on algebra in a section *De regula quantitatis* (Buteo 1559, f. 189^r). For the origin of the rule he cites Pacioli and de la Roche (by the name Stephano). While the name of the rule is indeed derived from de la Roche, Buteo remains quiet about his main source, his rival Peletier.[21]

After an explanation of the method by means of four examples he solves many linear problems by multiple unknowns in the fifth book. He introduces some new symbols but he had too little influence on his peers to be followed in this. Where Peletier and Mennher still used the radix or cossic sign for the first unknown, Buteo assigns the letter A to the first unknown and continues with B, C, .. for the other unknowns. Ommitting the cossic signs all together, Buteo takes a major step into the "representation of compound concepts", a necessary step towards algebraic symbolism according to Serfati (2010). The next step would be the use of exponents as introduced by Descartes in the *Regulæ*. Buteo further uses a comma for addition, the letter M for subtraction and a left square bracket for an equation. Thus the linear equation

$$6x + 3y + 2c = 84$$

is written as

$$6A, 3B, 2C\lceil 84$$

---

[21] Apart from a theoretical dispute on the angles of contact, in which Buteo's *Apologia* of 1562 pursues a refutation of Peletier, there existed a real hostility between them.

Once an equation is resolved in one unknown, he uses two brackets as in

$$5C[60] \text{ for } 5z = 60$$

A fragment of the fourth example is shown in Figure 2.6.

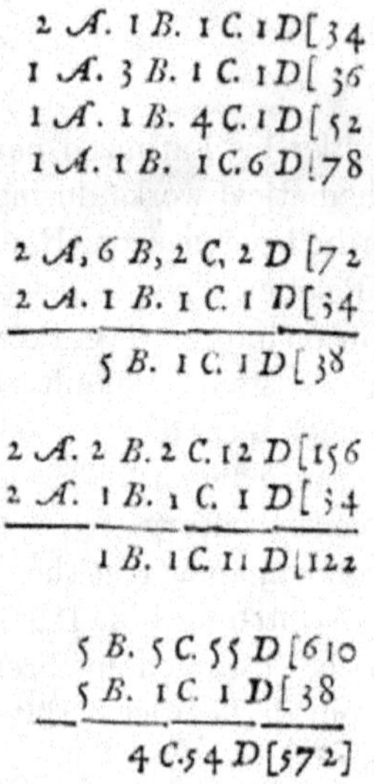

$$
\begin{aligned}
&2\,A.\ 1\,B.\ 1\,C.\ 1\,D[34 \\
&1\,A.\ 3\,B.\ 1\,C.\ 1\,D[36 \\
&1\,A.\ 1\,B.\ 4\,C.\ 1\,D[52 \\
&1\,A.\ 1\,B.\ 1\,C.6\,D[78
\end{aligned}
$$

$$
\begin{aligned}
&2\,A,6\,B,2\,C,2\,D\ [72 \\
&2\,A.\ 1\,B.\ 1\,C.\ 1\,D[54 \\
\hline
&\quad 5\,B.\ 1\,C.\ 1\,D[38
\end{aligned}
$$

$$
\begin{aligned}
&2\,A.\ 2\,B.\ 2\,C.\ 12\,D[156 \\
&2\,A.\ 1\,B.\ 1\,C.\ 1\,D[34 \\
\hline
&\quad 1\,B.\ 1\,C.\ 11\,D[122
\end{aligned}
$$

$$
\begin{aligned}
&5\,B.\ 5\,C.\ 55\,D[610 \\
&5\,B.\ 1\,C.\ 1\,D[38 \\
\hline
&\quad 4\,C.54\,D[572]
\end{aligned}
$$

Fig. 2.6: Systematic elimination of unknowns by Buteo (1559, 194)

Buteo refers to equations, not by numbers as Peletier but at least by their order. As an example let us look at question 30 (Buteo 1559, 357-8). His commentary is very terse (see Table 2.8).

With this and other examples, Buteo systematically manipulates equations to eliminate unknowns. His explanation refers explicitly to the multiplication of equations and the operations of adding or subtracting two equations. The idea of substitution is implicitly present, but is not performed as such, as can be seen from the missing commentaries for steps (13) and (16).

2.10.2 Pedro Nunes criticizing the second unknown

Although from Portugese origin, Nunes wrote his treatise on algebra in Spanish and published it in Antwerp.[22] Because his *Algebra* was published in 1567, it could appear that Nunes did not take advantage of the significant advances

---

[22] His name is therefore often written in the Spanish form Pedro Nuñez.

	Symbolic	Meta description	Original text
1	$x + \frac{y}{2} + \frac{z}{3} = 14$	premise	Huius solution secundum quantitatis regulam investigabitur, hoc modo. Pone Biremes esse $1A$, Triremes $1B$, Liburnicas $1C$. Erit igitur $1A$, B, 1/3 C [14. Item $1B$, 1/3 A, C [13. Et $1C$, 1/6 A, 1/8 B [14.
2	$\frac{x}{3} + y + \frac{z}{4} = 13$	premise	
3	$\frac{x}{6} + \frac{y}{8} + z = 14$	premise	
4	$6x + 3y + 2z = 84$	multiply (1) by 6	
5	$4x + 12y + 3z = 156$	multiply (2) by 12	
6	$4x + 3y + 24z = 336$	multiply (3) by 24	
7	$24x + 12y + 8z = 336$	multiply (4) by 4	multiplica aequationem (4) in 4
8	$20x + 5z = 180$	subtract (5) from (7)	auser (5) restat
9	$10x + 15y + 5z = 240$	add (4) and (5)	adde (4) (5)
10	$10x + 60 = 15y$	subtract (9) from (8)	Inter duas equationem postremas que sunt (8) et (9) differentia est (10)
11	$5z = 60$	subtract (10) from (8)	qua sublata ex (10) restat (11)
12	$z = 12$	divide (11) by 5	Partire in 5 provenit (12)
13	$20x + 60 = 180$	substitute in (12) in (8)	
14	$20x = 180 - 60$	resolves (13)	habeas Biremes ex aequatione ubi est 180 auser 60
15	$x = 6$	divide (14) by 20	partire (14) in 20
16	$2 + y + 3 = 13$	substitute (15), (12) in (2)	
17	$y = 8$	resolves (16)	et Trimeres erunt 8
			Quod erat quaesitum.

Table 2.8: Buteo's handling of a system of linear equations.

in symbolic algebra established during the decades before him. However, in the introduction, Nunes explains that he wrote most of the book over thirty years ago.[23] He chose to base much of the problems treated in his book on the *Summa* by Pacioli (1494). He questions some innovations that he learned from Pacioli, such as the use of the second unknown. Nunes discusses the problem of three men comparing their money as treated by Pacioli in distinction 9, treatise 9, paragraph 26 (1494, f. $191^v - 192^r$). However, the values of the problem are not those of Pacioli but are identical to the problem of Cardano, which we discussed above (2.2). Nunes does not reduce the problem

---

[23] John Martyn discovered a manuscript in 1990, the Cod. cxiii/1-10 at Municipal Library of Évora, Portugal. This Portugese text, written in 1533, contains an algebra which he attributed to Pedro Nunes. The date corresponds well with this thirty years of time difference. Martyn (1996) published an English translation and put much effort in the demonstration of the similarities with the Spanish text of 1567. The attribution of this text to Nunez has recently been refuted by Leitão (2002).

to two linear equations in two unknowns to be resolved by manipulating the equations as did Cardano (1545). Instead he follows the solution method in two unknowns from Pacioli.[24] He then provides a solution of his own, using a single unknown and concludes with the following observation:[25]

> But having treated the same example, that is case 51, we solved this with much ease, and more concise by the single unknown, without the use of the absolute quantity. And all the cases that Father Lucas solved with the [rule of] quantity, we solved by the rules of the unknown, without the aid of this last quantity.

Nunes is not very impressed by the *regula quantitatis* in which others saw "a more beautiful" way for solving problems or even "a perfection of algebra". He believes that most (linear) problems can be solved easier and shorter by a single unknown.

Similar criticism was formulated by other authors. Bosmans discovered a copy of the *Arithmetica Integra* by Stifel (1544) with marginal annotations from Gemma Frisius. The book, kept at the Louvain university library, has unfortunately been destroyed during World War I. Bosmans (1905-6, 168) reports three occasions in which Frisius critizes Stifel for using the second unknown: "Haec quaestio non requirit secundas radices" (f. 252v), "hic quoque secundis radicibus non est opus" (f. 253r), "et haec quastio secundis radicibus non est opus" (f. 253v) and "et haec quaestio secundis radicibus absolve potest" (f. 255r). This demonstrates that the use of the second unknown was still controversial during the mid-sixteenth century.

One could blame Frisius and Nunes for a reactionary view point. Bosmans (1908a, 159) quotes Nunes with some examples in which he rejects negative solutions and zero as a solution to an equation. However, Nunes had a very modern approach to algebra. As pointed out by Bosmans (1908a, 163), he can be credited as being the first who investigates the relationship of the following product with the structure of the equations (Nunes 1567, f. 125^v):

$$(x+1)(x+1), (x+1)(x+2), (x+1)(x+3)\dots$$
$$(2x+1)(x+1), (2x+1)(x+2), (2x+1)(x+3)\dots$$

As we now known from further developments, such investigations were important to raise sixteenth century algebra from arithmetical problem solving to the study of more abstract algebraic structures and relations. This leads us to the last author before Viète writing on the *Regula quantitatis*.

---

[24] We omit the solution here because a complete transcription of the problem with a symbolic translation is provided by Bosmans (1908b, 21-2).

[25] Nunes 1567, f. 225v: "Pero nos avemos tratado esto mismo exemplo, que es el caso 51, y lo practicamos muy facilmente, y brevemente por la cosa, sin usar de la quantidad absoluta. Y todos los casos que Fray Lucas practica por la quantidad, practicamos nos por las reglas de la cosa, sin ayuda deste termino quantidad".

2.10.3 Gosselin (1577)

Guillaume Gosselin's *De Arte Magna* is our last link connecting the achievements of Cardano, Stifel, and Buteo using the second unknown with Viète's study of the structure of equations in his *Isagoge*. Cifoletti (1993) has rightly pointed out the importance of this French tradition to the further development of symbolic algebra.

Gosselin is rather idiosyncratic in his notation system and seems to ignore most of what was used before him. For the arithmetical operators, addition and subtraction he uses the letters P and M, rather than + and – as was commonly used in Germany and the Low Countries at that time and also adopted by Ramus in France. However, five years later in *de Ratione* (Gosselin, 1583) he did use the + and – sign. The letter 'L' (from *latus*) is used for the unknown; the square becomes 'Q' and the cube 'C'. In some cases he refers to the second unknown by 'q', as did Cardano. For a linear problems with several unknowns he switches to the letters A, B, C, as Buteo, but evidently leading to ambiguities with the sign for x^3. Even more confusing is the use of 'L' for the root of a number, such as

$$\text{L9 for } \sqrt{9} \text{ and LC8 for } \sqrt[3]{8}$$

Accepting isolated negative terms, the letter 'M' is also used as M8L for $-8x$. Gosselin follows Buteo with equations to zero as in '3QM24L aequalia nihilo', for $3x^2 - 24x = 0$ (Gosselin 1577, f. 73^v). The symbolism adopted by Gosselin can be illustrated with an example of the multiplication of two polynomials (ibid. f. 45^v):

	4 L M 6 Q P 7
	3 Q P 4 L M 5
	12 C M 18 QQ P 21 Q
Producta	16 Q M 24 C P 28 L
	M 20 L P 30 Q M 35
Summa	67 Q P 8 L M 12 C M 18 QQ M 35

The major part of book IV deals with the second unknown, though his terminology is rather puzzling. Chapter II is titled *De quantitate absoluta* (f. 80^r) and chapter III (misnumbered as II) as *De quantitate surda* (f. 84^r). In both these chapters Gosselin solves linear problems with several unknown quantities. So what is the difference? Gosselin gives no clue as he leaves out any definitions of the terms. However, we have previously seen that 'abso-

lute quantity' is used by Nunes and *quantita sorda* by Pacioli and Cardano.[26]
From a comparison of the five problems solved by 'absolute quantities' with
the four solved by the *quantita surda* it becomes apparent that Gosselin places
the distinction between multiple unknowns and the second unknown. Thus the
'absolute quantities' correspond with the symbolic unknowns A, B, C, .. as
used by Buteo. Gosselin leaves out the primary unknown of Stifel or Peletier,
as was previously done by Buteo. The *quantita surda* corresponds with the
quan. of Cardano (1545), for which Gosselin uses the symbol q. The *positio* of
Cardano becomes the *latus* for Gosselin.

$$\tfrac{1}{2}y + 2 + x = \tfrac{9}{2}y - 18$$
$$x + 20 = 4y$$
$$y = \tfrac{1}{4}x + 5$$

$$\tfrac{1}{3}x + 3 + \tfrac{1}{4}x + 5 = 2x - 9$$
$$\tfrac{17}{12}x = 17$$

> Sit prior numerus 1 L, fecundus 1 q,
> atque fic $\frac{1}{2}$ q P 2 P 1 L æqualia funt re-
> fidui noncuplo nempe $\frac{2}{2}$ q M 18, &
> addito quod deficit fubductoque fu-
> perfluo 1 L P 20 æqualia 4 q, fit 1 q,
> $\frac{1}{4}$ L P 5, iam prior vt fupra fit 1 L, fe-
> cundus erit $\frac{1}{4}$ L P 5, atque adeo $\frac{1}{4}$ L P
> 3 P $\frac{1}{4}$ L P 5 æqualia 2 L M 9, & addi-
> to quod deficit fubductoque fuper-
> fluo $\frac{7}{12}$ L æquales 17, fit vnum latus
> 12, & tantus eft prior numerus, fecun-
> dus $\frac{1}{4}$ L P 5 hoc eft 8.

$$x = 12, y = 8$$

Table 2.9: Gosselin's use of the *quantita surda* (Gosselin, 1577, f. 84*v*)

Cifoletti (1993, 138-9) concludes on Gosselin that

> it is true that this innovation originates with Borrel [Buteo], but Gosselin uses it with
> a new skill that permits him to more easily solve the same problems proposed by
> Borel. It seems reasonable to think that Viète took his symbol as point of departure to
> arrive at his A, E. Gosselin could also be a source for the notation used by Descartes,
> who in the *Regulae* proposes to designate the known term with lower-case letters and
> the unknown with capitals.

---

[26] Cifoletti (1993, 136) is wrong in claiming that "Cardano does not use the word *surda*
in this sense". Furthermore, she translates the *quantita surda* as the surd quantity and
speculates on irrational quantities. However, the Italian term *sorda*, as used by Pacioli,
means 'mute' in Italian. Thus *quantitate sorda* may simply refer to the voiceless consonant

	Symbolic	Meta description	Original text
1	$x + \frac{y}{2} + \frac{z}{2} + \frac{u}{2} = 17$	premise	1ABCD aequalia 17
2	$\frac{x}{3} + y + \frac{z}{3} + \frac{u}{3} = 12$	premise	1B1/3A1/3C1/3D aequalia 12
3	$\frac{x}{4} + \frac{y}{4} + z + \frac{u}{4} = 13$	premise	1CABD aequalia 13
4	$\frac{x}{6} + \frac{y}{6} + \frac{z}{6} + u = 13$	premise	1D1/6A1/6B1/6C aequalia 13
5	$2x + y + z + u = 34$	multiply (1) by 2	revocentur ad integros numeros, existent 2A1B1C1D aequalia 34
6	$x + 3y + z + u = 36$	multiply (2) by 3	1A3B1C1D aequalia 36
7	$x + y + 4z + u = 52$	multiply (3) by 4	1A1B4C1D aequalia 52
8	$x + y + z + 6u = 78$	multiply (4) by 6	1A1B1C6D aequalia 78
9	$2x + 2y + 5z + 7u = 130$	add (7) and (8)	addamus duas ultimas aequationes, tertiam scilicet et quartam, existent 2A2B5C7D aequalia 130
10	$y + 4z + 6u = 96$	subtract (5) from (9)	tollamus hinc primam, restabunt 1B4C6D aequalia 96
11	$2x + 4y + 2z + 7u = 114$	add (6) and (8)	addamus quartam et secundam, fient 2A4B2C7D aequalia 114
12	$3y + z + 6u = 80$	subtract (5) from (11)	tollamus hinc primam, supererunt 3B1C6D aequalia 80
13	$2x + 4y + 5z + 2u = 88$	add (6) and (7)	addamus secundam et tertiam aequationem, fient 2A4B5C2D aequalia 88
14	$3y + 4z + u = 54$	subtract (5) from (13)	tollamus primam, restabunt 3B4C1D aequalia 54
15	$3y + 12z + 18u = 288$	multiply (10) by 3	iam vero triplicemus 1B4C6D quae fuerunt aequalia 96 fient 3B12C18D aequalia 288
16	$11z + 12u = 208$	subtract (12) from (15)	tollamus hinc 3B1C6D aequalia 80, restabunt 11C12D aequalia 20
17	$8z + 17u = 234$	subtract (14) from (15)	subducamus iterum ex eadem triplicata aequatione 3B4C1D eaqualia 54, restabunt 8C17D aequalia 234
18	$88z + 187u = 2574$	multiply (17) by 11	multiplicemus hanc aequationem in 11, fient 88C187D aequalia 2574
19	$88z + 96u = 1664$	multiply (16) by 8	ducamus etiam 11C12D aequalia 208, in 8, existent 88C96D aequalia 1664
20	$91u = 910$	subtract (19) from (18)	tollamus 88C96D aequalia 1664 ex 88C187D aequalibus 2574, restabunt 91D aequalia 910 sicque stat aequatio
21	$u = 10$	divide (20) by 91	partiemur 910 in 91, quotus erit 10 valor D, est ergo 10 ultimus numerus ex quaesitis
22	$11z + 120 = 208$	substitute (21) in (16)	et quoniam 11C12D erant aequalia 208,

Table 2.10: Gosselin's solution to a problem from Buteo.

q representing 'quantity'. In English a voiceless consonant is also called a surd.

We believe that the influence on Viète and Descartes attributed to Gosselin by Cifoletti is too much of an honour for Gosselin. The many ambiguities in Gosselin's system of symbols are clearly a departure from the achievements by Stifel (1553). As to the superior way of solving problems with multiple unknowns let us look at the fifth problem which Gosselin solves by 'absolute quantities'. The problem text and its meta-description is as follows (Gosselin 1577, f. 82v):

> Quatuor numeros invenire, quorum primus cum semisse reliquorum faciat 17. Secundus cum aliorum triente 12. Tertius cum aliorum quadrante 13. Quartus cum aliorum sextante faciat 13.

$$x + \frac{y}{2} + \frac{z}{3} + \frac{u}{2} = 17$$
$$\frac{x}{3} + y + \frac{z}{3} + \frac{u}{3} = 12$$
$$\frac{x}{4} + \frac{y}{4} + z + \frac{u}{4} = 13$$
$$\frac{x}{6} + \frac{y}{6} + \frac{z}{6} + u = 13$$

This is the very same problem of Buteo (1559, 193-6) shown in Figure 2.6. Bosmans (1906, 64) writes that here "Gosselin triumphs over Buteo who gets confused in solving the problem". Let us first look at Gosselin's solution in Table 2.10.

23	$11z = 88$	subtract 120 from (22)	tollamus 12D hoc est 120, restabunt 88 aequalia 11C
24	$z = 8$	divide (23) by 11	dividemus 88 in 11, quotus erit 8, valor C et tertius numerus
25	$3y + 10 + 32 = 54$	substitute (21) and (24) in (14)	sed etiam 3B4C1D aequalia sunt 54,
26	$3y = 12$	subtract 42 from (25)	tollamus hinc 4C1D, hoc est 10 et 32, nempe 42, restabunt 12 aequalia 3B
27	$y = 4$	divide (26) by 3	estque B et secundus numerus 4
28	$2x + 4 + 8 + 10 = 34$	substitute (21), (24) and (27) in (5)	iam vero 2A1B1C1D aequantur 34,
29	$2x = 12$	subtract 22 from (28)	tollamus 1B, nempe 4, 1C 8, 1D 10, hoc est 22, restabunt 12 aequalia 2A
30	$x = 6$	divide (28) by 2	quare 1A et primus numerus est 6

Table 2.11: Final part of Gosselin's solution to a problem from Buteo.

Buteo provides three different but correct solutions to the problem. In the first he reduces the number of equations by multiplication and subtraction to eliminate an unknown in every subtraction step. Gosselin's method may be somewhat more resourceful but there is little conceptual difference between both with regards to equations and the possible operations on equations. Remark that the solution text is close to identical with our meta-description.

This signifies the completion of an important phase towards the emergence of symbolic algebra.

2.11 Simon Stevin (1585)

In his *L'arithmetique*, Simon Stevin (1585) employs the second unknown for several problems. From questions 25 to 27 it becomes obvious that he used Cardano's *Ars Magna* for his use of the second unknown. Although not original in its method, Stevin's use of symbolism is quite novel (see Figure 2.7). Let us look at question 27 asking for three numbers in GP with the sum given and the condition that the square of the middle term is equal to twice the product of the two smaller numbers plus six times the smaller number (Stevin 1585, 402-404). In modern symbolism the structure of the problems is:

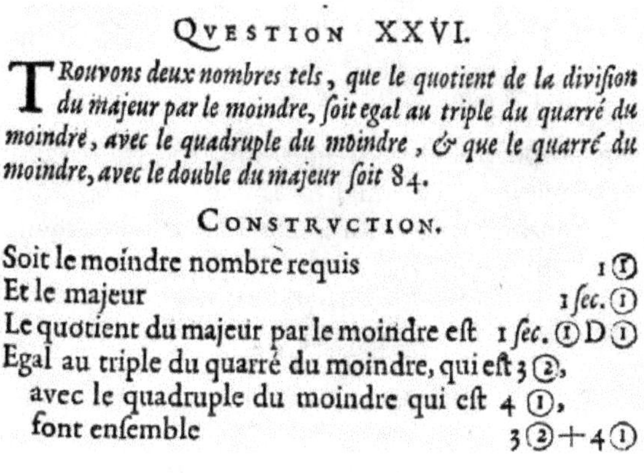

QVESTION XXVI.

Trouvons deux nombres tels, que le quotient de la division du majeur par le moindre, soit egal au triple du quarré du moindre, avec le quadruple du moindre, & que le quarré du moindre, avec le double du majeur soit 84.

CONSTRVCTION.

Soit le moindre nombre requis   1 ①
Et le majeur   1 sec. ①
Le quotient du majeur par le moindre est   1 sec. ① D ①
Egal au triple du quarré du moindre, qui est 3 ②,
     avec le quadruple du moindre qui est 4 ①,
     font ensemble   3 ②+4 ①

Fig. 2.7: Simon Stevin's symbolism for the second unknown (from Stevin 1585, 401)

$$a : b = b : c$$
$$a + b + c = d$$
$$eab + fa = b^2$$

Cardano discusses the problems with (20; 2, 4) for the values of (d; e, f). Stevin writes that he has the problem from Cardano and changes the values to (26; 2, 6). Stevin calls his solution a construction (of an equation) and starts by using the first unknown for the middle term and the second for the lower extreme, for which we will use x and y. An unknown is represented by

Stevin as a number within a circle. The number inside denotes the power of the unknown. Thus

$$① \text{ stands for } x \text{ and } ② \text{ for } x^2$$

To differentiate the second unknown from the first the power of the unknown is preceded by sec., for example

$$5y^2 \text{ becomes } 5sec. ②$$

For multiplication, Stevin used the letter M, thus

$$5xy^2 \text{ would be } 5 ①Msec. ②$$

Remark that if this system would be extended to *pri.* and *ter.*, the circled numbers correspond to our exponents and the Stevin's symbolism becomes very similar with the one adopted by Descartes in 1637.

Stevin proceeds by formulating the condition in terms of the two unknowns as $x^2 = 2xy + 6y$, or using his notation, as

$$1② \text{ égale à } 2 ①Msec. ① + 6sec.②$$

As x is the mean proportional between y and the third number c, $x^2 = yc$ and the larger extreme must be equal to $2x + 6$. Thus, y, x and $2x + 6$ are in continuous proportion and their sum is 26. This allows Stevin to express the value of the second unknown as:

$$-3① + 20$$

Substituting $(-3x + 20)$ as the value of y in $x^2 + 2xy + 6y$ leads to

$$x^2 = -6x^2 + 22x + 120$$

for which Stevin gives the root of 6 leading to the solution $(2, 6, 18)$.

2.12 Conclusion

We have treated the development of symbolism with regards to the second unknown from 1539 to 1585, the period preceding Viète's *Isagoge* (1591). We have argued that the search – or we might even say, the struggle – towards a satisfactory system for representing multiple unknowns has lead to the creation of a new mathematical object: the symbolic equation. The solution to linear problems by means of the second unknown initiated, for the first time, operations on equations (in Cardano's *Practica Arithmeticae*) and operations between equations (in Cardano's *Ars Magna*). Once operations on equations became possible, the symbolic equation became a mathematical object of its own and hence required a new concept. Algebraic practice before Cardano consisted mostly of problem solving by means of the manipulation of polynomials – on the condition that they were kept equal – in order to arrive at a format for which a standard rule could be applied. We therefore use the term 'co-equal polynomials' for these structures rather than "equations" in the modern sense. Half a century of algebra textbooks marked the transition from algebra as a practice of problem solving (the abbaco and cossic tradition) to algebra as the study of equations. These authors, and especially Cardano and Stifel paved the Royal road for Viète, Harriot, and Descartes, to use algebra as an analytic tool within the wider context of mathematics. In order to study the structure of equations, the concept of a symbolic equation had to be established.

References

1. Arrighi, Gino (ed.), 1967. Antonio de' Mazzinghi, *Trattato di Fioretti* nella trascelta a cura di Maestro Benedetto secondo la lezione del Codice L.IV.21 (sec. XV) della Biblioteca degl'Intronati di Siena. Siena: Domus Galilaeana.
2. Berlet, Bruno, 1860. *Die Coss von Adam Riese*, Siebzehnter Bericht über die Progymnasial- und Realschulanstalt zu Annaberg, Annaberg-Buchholz.
3. Bosmans, Henri, 1905-6. "Le commentaire de Gemma Frisius sur l'*Arithmetica Integra* de Stifel", *Annales Société Scientifique*, 30, pp. 165-8.
4. Bosmans, Henri, 1906. "Le *De Arte Magna* de Guillaume Gosselin", *Bibliotheca Mathematica*, 3 (7), pp. 46-66.
5. Bosmans, Henri, 1907. "L'Algèbre de Jacques Peletier du Mans (XVIe siècle)", *Revue des questions scientifiques*, 61, pp. 117-73.
6. Bosmans, Henri, 1908a. "Sur le libro de algebra de Pedro Nuñez", *Bibliotheca Mathematica*, 3 (8), pp. 154-69.
7. Bosmans, Henri, 1908b. "L'algèbre de Pedro Nuñez", *Annaes Scientificos da Academia Polytechnica do Porto*, 3 (4), pp. 222-71.
8. Bosmans, Henri, 1926. "La théorie des équations dans l'Invention nouvelle en l'algèbre d'Albert Girard", *Mathesis*, 41, pp. 59-67, 100-9, 145-55.
9. Buteo, Ioannes, 1559. *Logistica, quae et arithmetica vulgo dicitur, in libros quinque digesta... Ejusdem ad locum Vitruvii corruptum restitutio...* Lyon: apud G. Rovillium.
10. Cardano, Girolamo, 1539. *Hieronimi C. Cardani medici Mediolanensis, Practica arithmetice, & mensurandi singularis. In qua que preter alias cõtinentur, versa pagina demnstrabit..* Milan: Io. Antonins Castellioneus medidani imprimebat, impensis Bernardini calusci.
11. Cardano, Girolamo, 1545. *Artis magnae; sive, De regulis algebraicis, Lib. unus. Qui & totius operis de Arithmetica, quod opus perfectvm inscripsit, est in ordine decimus.* Nurnberg: Johann Petreius.
12. Cardano, Girolamo, 1663. *Hieronymi Cardani Mediolanensis Opera omnia tam hactenvs excvsa: hic tamen aucta & emendata: quàm nunquam alias visa, ac primum ex auctoris ipsius autographis eruta cura Caroli Sponii..* (10 vols). Vol IV: *Operum*

Tomvs Qvartus quo continentvr Arithmetica, Geometrica, Mvsica. Part II: *Practica Arithmeticae generalis omnium copiosissima & vtilissima.* 14-215. Part IV: *Artis magnae sive de Regulis Algebraicis.* 222-302. Lyon: Sumptibus Ioannis Antonii Hvgetan, & Marci Antonii Ravaud.

13. Cardano, Girolamo, 1643. *De vita propria,* Paris:. Translated by Jean Stoner, *The Book of My Life,* Toronto (reprint with introduction by A. Grafton, New York, 2002).

14. Cajori, Florian, 1928. *A History of Mathematical Notations.* I. *Notations in Elementary mathematics.* II. *Notations Mainly in Higher Mathematics.* La Salle, Illinois: Open Court, 1928–29 (Dover edition, 1993).

15. Cantor, Moritz, 1880-1908. *Vorlesungen über Geschichte der Mathematik* (4 vols.), Vol. I (1880): "Von den ltesten Zeiten bis zum Jahre 1200 n. Chr.", Vol. II (1892): "Von 1200-1668", 2nd ed. Leipzig: Teubner, 1894, 1900.

16. Cifoletti, Giovanna, 1993. *Mathematics and Rhetoric: Peletier and Gosselin and the Making of the French Algebraic Tradition,* PhD Diss., Princeton University (0181, *Dissertation Abstracts International.* 53 (1993): 4061-A).

17. Curtze, Maximilian, 1895. "Ein Beitrag zur Geschichte der Algebra in Deutschland im fünfzehnten Jahrhundert", *Abhandlungen zur Geschichte der Mathematik,* (supplement of *Zeitschrift für Mathematik und Physik*) 5, pp. 31-74.

18. Eneström, Gustav, 1906-7a. "Kleine Bemerkungen zur zweiten Auflage von Cantors *Vorlesungen über Geschichte der Mathematik*", *Bibliotheca Mathematica,* 3 (7), p. 204.

19. Franci, Raffaella, and Marisa Pancanti (eds), 1988. Anonimo (sec. XIV), *Il trattato d'algibra* dal manoscritto Fond. Prin. II. V. 152 della Biblioteca Nazionale di Firenze. (Quaderni del Centro Studi della Matematica Medioevale, 18). Siena: Servizio Editoriale dell'Università di Siena.

20. Gericke, Helmut, 1992. *Mathematik im Abendland. Von den römischen Feldmessern bis zu Descartes,* Wiesbaden: Fourier.

21. Gosselin, Guillaume, 1577. *Gvlielmi Gosselini Cadomensis Bellocassii. De arte magna, seu de occulta parte numerorum, quae & algebra, & almucabala vulgo dicitur, libri qvatvor. Libri Qvatvor. In quibus explicantur aequationes Diophanti, regulae quantitatis simplicis, [et] quantitatis surdae.* Paris: Aegidium Beys.

22. Gosselin, Guillaume, 1583. *Gulielmi Gosselini Cadomensis Issaei de Ratione discendæ docendque mathematices repetita prælectio.* Ad Ioannem Chandonium et Carolum Bocherium supplicum libellorum in regia Magistros.

23. Heeffer, Albrecht, 2008. "A Conceptual Analysis of Early Arabic Algebra". In S. Rahman, T. Street and H. Tahiri (eds.) *The Unity of Science in the Arabic Tradition: Science, Logic, Epistemology and their Interactions.* Heidelberg: Springer, pp. 89–126.

24. Heeffer, Albrecht, 2010a. "Estienne de la Roche's appropriation of Chuquet (1484)". In *Pluralité ou unité de l'algèbre à?* Études rassemblées par Maria-Rosa Massa Estève,, Sabine Rommevaux, Maryvonne Spiesser, *Archives internationales d'histoire des sciences,* 2010 (to appear).

25. Høyrup, Jens, 2010. "Hesitating progress — the slow development toward algebraic symbolization in abbacus - and related manuscripts, c. 1300 to c. 1550". in A. Heeffer and M. Van Dyck (eds.) *Philosophical Aspects of Symbolic Reasoning in Early Modern Mathematics,* Studies in Logic 26, London: College Publications, 2010 (this volume, chapter 1).

26. Heeffer, Albrecht, 2010b. "Algebraic partitioning problems from Luca Pacioli's Perugia manuscript (Vat. Lat. 3129)". *Sources and Commentaries in Exact Sciences,* (2010), 11, pp. 3-52.

27. Hughes, Barnabas, 2001. "A Treatise on Problem Solving from Early Medieval Latin Europe", *Mediaeval Studies,* 63, pp. 107-41.

28. Leitão, Henrique, 2002. "Sobre as 'Notas de Álgebra' atribuídas a Pedro Nunes (ms. évora, BP, Cod. CXIII/1-10)", *Euphrosyne,* 30, pp. 407-16.

29. Martyn, John, R. C. (ed., trans.), 1996. *Pedro Nunes (1502-1578) His Lost Algebra and Other Discoveries*. New York: Peter Lang.
30. Mennher, Valentin, 1556. *Arithmétique seconde*, Anwterp: Jan van der Loë.
31. Oaks, Jeffrey A. and Haitham M. Alkhateeb, 2007. "Simplifying equations in Arabic algebra", *Historia Mathematica*, 34 (1), February 2007, pp. 45-61.
32. Peletier, Jacques, 1554. *L'algebre de Iaques Peletier dv Mans, departie an deus liures*. Lyon: Jean de Tournes.
33. Picutti, Ettore, 1983. "Il 'Flos' di Leonardo Pisano dal codice E.75 P. Sup. della Biblioteca Ambrosiana di Milano", *Physis* 25, pp. 293-387.
34. Romero Vallhonesta, Fàtima, 2010. "The second quantity in the Spanish algebras in the 16th century". In *Pluralité ou unité de l'algèbre à la Renaissance?* Études rassemblées par Maria-Rosa Massa Estève, Sabine Rommevaux, Maryvonne Spiesser, *Archives internationales d'histoire des sciences*, 2010 (to appear).
35. Russo F., 1959. "La consititution de l'algebre au XVIe siècle. Etude de la structure d'une évolution", *Revue d'histoire des sciences*, 12 (3), 193-208.
36. Serfati, Michel, 2010. "Symbolic revolution, scientific revolution: mathematical and philosophical aspects". In A. Heeffer and M. Van Dyck (eds.) *Philosophical Aspects of Symbolic Reasoning in Early Modern Mathematics*, Studies in Logic 26, London: College Publications, 2010 (this volume, chapter 3).
37. Simi, Annalisa (ed.), 1994. Anonimo (sec. XIV), *Trattato dell'alcibra amuchabile* dal Codice Ricc. 2263 della Biblioteca Riccardiana di Firenze. (Quaderni del Centro Studi della Matematica Medioevale, 22). Siena: Servizio Editoriale dell'Università di Siena.
38. Singmaster, David, 2004. *Sources in Recreational Mathematics, An Annotated Bibliography*, Eighth Preliminary Edition (unpublished, electronic copy from the author).
39. Stevin, Simon, 1585. *L'arithmetique de Simon Stevin de Brvges: contenant les computations des nombres Arithmetiques ou vulgaires: aussi l'Algebre, auec les equations de cinc quantitez. Ensemble les quatre premiers liures d'Algebre de Diophante d'Alexandrie, maintenant premierement traduicts en françois. Encore un liure particulier de la Pratique d'arithmetique, contenant entre autres, Les Tables d'Interest, La Disme; Et vn traicté des Incommensurables grandeurs: Avec l'explication du Dixiesme liure d'Euclide*. A Leyde: De l'imprimerie de Christophe Plantin, M.D. LXXXV.
40. Stifel, Michael, 1544. *Arithmetica integra*. Nürnberg: Petreius.
41. Stifel, Michael, 1553. *Die Coss Christoffe Ludolffs mit schönen Exempeln der Coss* / Zu Königsperg in Preussen: Gedrückt durch Alexandrum Lutomyslensem.
42. Tropfke, Johannes, 1980. *Geschichte der Elementar-Mathematik in systematischer Darstellung mit besonderer Berücksichtigung der Fachwrter*, Vol. I: *Arithmetik und Algebra*, revised by Kurt Vogel, Karin Reisch and Helmuth Gericke (eds.), Berlin: Walter de Gruyter.

Chapter 3
Symbolic revolution, scientific revolution: mathematical and philosophical aspects

Michel Serfati

Abstract This paper is devoted to showing how the introduction of symbolic writing in the seventeenth century was a true revolution in thought patterns that instituted a powerful tool for the creation of mathematical objects, without equivalent in natural language. I first comment on the constitution of symbolic representation, the main protagonists being Vieta and Descartes. The final institution of symbolic representation involved six patterns, which turn out to be constitutive. I give a short account of these six patterns and discuss two of them in more details, first the dialectic of indeterminacy, then the representation of compound concepts. All these innovations are found gathered together in the *Géométrie* of 1637. I conclude with two important points. First, the introduction of substitution in a symbolic text: with Leibniz substitutability actually became an essential, everyday element. One can discover here one of the creative aspects of symbolism. Second, the description of an important aspect of symbolism: the emergence of a specific scheme in three phases for creating mathematical objects from it.

Key words: Descartes, exponential, Leibniz, representation, symbolic.

3.1 Symbolic revolution, scientific revolution

A fundamental chapter in the history of humanity, the scientific revolution in seventeenth century Europe was a time of rupture with old Greek and scholastic visions of the world, and that of the establishment of a new scien-

IREM. Université Paris VII-Denis Diderot.

tific world — the dawn of modern science. Already abundantly commented upon, this period of upheavals is the subject of numerous specific historical studies, as well by discipline (physics, cosmology, philosophy, mathematics), as by protagonist (Galileo, Descartes, Kepler or Newton for example). Among the components of these upheavals, there is one, the new mathematical symbolism, which has remained almost entirely without comment, however. An essential chapter in the philosophy of language, the constitution of mathematical writing indeed has suffered an almost completely neglect in theoretical studies. *In fine* I shall return to the platonic origins of this disinterest.

This paper is devoted to showing how the introduction of symbolic writing in the seventeenth century was a true revolution in thought patterns, and that it instituted a powerful tool for the creation of mathematical objects, without equivalent in natural language.

I will first comment on the constitution of symbolic representation, the main protagonists being Vieta and Descartes. One might be surprised that I have the culmination of this formative period coincide with Descartes's *Géométrie* (1637) and not, for example, with Frege's *Begriffsschrift* (1879). My position is pragmatic: most of the constitutive and decisive elements were in place by 1637, although a theory about them had not yet been elaborated. So for two and a half centuries, mathematicians went on inventing, without being able to refer to a theory of symbolism. I am mostly interested in this aspect of creation, which is too often neglected by the philosophy of mathematics.

3.2 From Cardano to Descartes

One can introduce the question in a natural way by comparing two celebrated texts. On the one hand, a page of Cardano's *Ars Magna* (1545) (Figure 3.1) that appears unreadable today, an archaic text; but a major representative work of sixteenth century mathematics. On the other hand, a quasi-modern excerpt (Adam and Tannery, VI, 473) of Descartes's *Géométrie* (1637) using the usual symbols of algebra (letters and the sign of the square root) as we do today (Figure 3.2).

Both texts deal with basically the same subject: equations of the third degree. The first contains no symbols other than numbers and abbreviations, and we are able to read it only after a prior study and with great effort. Below, I shall comment more accurately on the reason why it is unreadable. The second is perfectly intelligible, because it uses the appropriate symbols, which have at that time acquired their final form.

More precisely, we fix the time of the split between 1591 (Vieta's *Isagoge*) and 1637 (*La Géométrie*). After 1637, mathematical writing would certainly

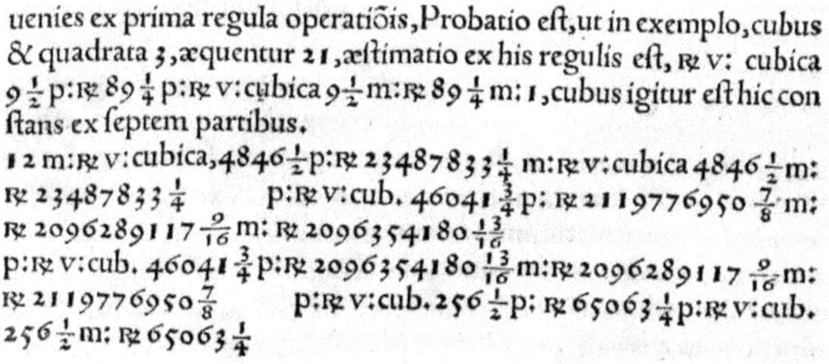

uenies ex prima regula operatiõis,Probatio est,ut in exemplo,cubus & quadrata 3,æquentur 21,æstimatio ex his regulis est, ℞ v: cubica 9 1/2 p:℞ 89 1/4 p:℞ v:cubica 9 1/2 m:℞ 89 1/4 m: 1,cubus igitur est hic constans ex septem partibus.

12 m:℞ v:cubica,4846 1/2 p:℞ 23487833 1/4 m:℞ v:cubica 4846 1/2 m: ℞ 23487833 1/4 p:℞ v:cub. 46041 3/4 p: ℞ 2119776950 7/8 m: ℞ 2096289117 9/16 m: ℞ 2096354180 13/16 p:℞ v:cub. 46041 3/4 p:℞ 2096354180 13/16 m:℞ 2096289117 9/16 m: ℞ 2119776950 7/8 p:℞ v:cub. 256 1/2 p: ℞ 65063 1/4 p:℞ v:cub. 256 1/2 m: ℞ 65063 1/4

Fig. 3.1: Cardano's *Ars Magna* of 1545, Chapter XV, fol. 34^v

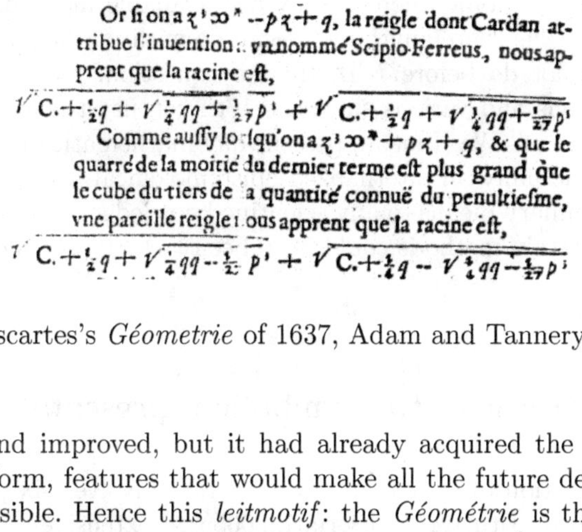

Or si on a z³ ∞ * -- pz + q, la reigle dont Cardan attribue l'inuention à vn nommé Scipio Ferreus, nous apprent que la racine est,

√C.+ ½q + √¼qq + 1/27 p³ + √C.+ ½q + √¼qq + 1/27 p³

Comme aussy lorsqu'on a z³ ∞ * + pz + q, & que le quarré de la moitié du dernier terme est plus grand que le cube du tiers de la quantité connuë du penultiesme, vne pareille reigle nous apprent que la racine est,

√C.+ ½q + √¼qq -- 1/27 p³ + √C.+ ½q -- √¼qq -- 1/27 p³

Fig. 3.2: Descartes's *Géometrie* of 1637, Adam and Tannery, VI, p. 473

be amended and improved, but it had already acquired the main features of its present form, features that would make all the future developments of symbolism possible. Hence this *leitmotif*: the *Géométrie* is the first text in history directly readable by present-day mathematicians.[1]

How and why was such a reversal possible? We find ourselves here at a moment in historical time (in the seventeenth century in Europe), at the birth of a new written language, the mathematical language, with many texts and authors. How did it happen? That is, what was it about the specific material aspects of the text at this time, which we can now regard as the major differences between Cardano and Descartes? I will propose epistemological answers to that question, because the study of the constitution of symbolic representation must go beyond a merely historical approach that would limit itself to describing and listing various occurrences of signs. This was a task remarkably

[1] On the structure of the *Géométrie*, see (Serfati 2005b).

carried out by Florian Cajori, without this work offering any opening towards epistemological interpretations, however.

Rather, this constitution directly concerns philosophy. First of all because analyzing the progressive constitution of a new written language is doubtless a genuinely philosophical task, not to be carried out in reference to a scheme of eternal "mathematical idealities", supposed to exist abstractedly and in themselves, but always with an eye to the concrete interests and motivations of the mathematicians of the time and, equally important, the constraints that reality opposed to them.[2] This part of the analysis brings to light a form, empirical and real, of the "necessity" of the system. Above all, the advent of symbolism appears not as a mere "change of notation" against a (supposedly) unchanged mathematical background, but on the contrary, as a decisive conceptual revolution, in particular with regard to the creation of objects. (This aspect of the history of ideas is even more philosophical.) In order to *prove* this point, there is only one available method: to ask always and everywhere: "Is there anything we can do with mathematical symbolism that we could not do before?" To this key question, *positive* answers have been so numerous and various since the 17th century that it would be hard even to list them all. We (I mean, present-day mathematicians) are so used to symbolism — actually an internalized epistemological frame, and the necessary and preliminary means to any scientific knowledge — that we can hardly imagine how certain methods could have been lacking or could have taken so long to emerge.

3.3 The patterns of the symbolic representation

The final institution of symbolic representation involves six patterns, which turn out to be constitutive. In (Serfati, 2005a), I extensively developed these various points. Here, I shall give a short account of these patterns and discuss two of them in more detail.

1. **The representation of the "unknown" or "required"**. Simple as it may appear, this involves surprising features, in Diophantus's work for instance.
2. **The dialectic of indeterminancy** (the representation of the "given"). It is for example the use of a in $y = ax$. This aspect is discussed below in a certain detail.

---

[2] These constraints, particularly of a symbolic nature, are imposed by the need for the mathematician to produce objects that are both consistent and that can be operated on. For example the representation of the lineage of powers mentioned below in Section 3.5.

3. **Structuration by assemblers** (the representation of elementary operative instructions: addition, division, extraction of roots, etc.), with the signs $+$, $\div$, $\cdot$, $-$. Thus the (very simple) writing of: $2 \cdot x + 1$.

4. **The ambiguity of order** (the representation of the succession and *entanglement* of instructions). This pattern is coextensive to both dividing and gathering signs in the symbolic text, *via* brackets or *vinculum* for instance, that is via *punctuation* signs, also denoted *aggregation* signs. Thus the concatenation of signs $2 \cdot x + 1$ above is naturally ambiguous in reading, and its significance must be specified according to $(2 \cdot x) + 1$ or $2 \cdot (x + 1)$. This part of the study concludes first with stressing the conceptual importance of a tree-structure underlying any symbolic text (it structures the mathematical thought). Thus, with the two possible readings of the preceding example, one can associate two tree-structures as shown in Figure 3.3.

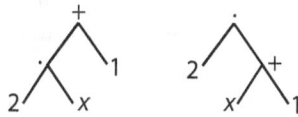

Fig. 3.3: Parsing trees for two interpretations of the ambiguous $2 \cdot x + 1$

One can then distinguish two ways of traversing the tree (from the root or from the leaves), which I eventually identified as two (theoretical) epistemological positions of the mathematical subject, namely the author's and the reader's.

This example is very simple, however, undoubtedly too simple. It is interesting to carry out the same analysis on the solution (developed by Descartes in the *Géométrie* in solving Pappus' problem) of a certain quadratic equation of the second degree, which leads him to the formula:

$$m - \frac{n}{\xi} + \sqrt{m \cdot m + o \cdot x - \frac{p}{m} x \cdot x}$$

This formula leads to the tree as in Figure 3.4:

We can conclude that the employment in the symbolic text of signs of aggregation, like the brackets, progressively opened up the possibility of the successive execution of a significant number of various instructions, a complex operation which could not be considered rhetorically before. Of course, the 17th century geometers did not draw the tree of a symbolic expression any more than do contemporary mathematicians wishing to write or deci-

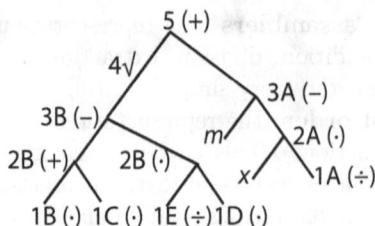

Fig. 3.4: Parsing tree for a quadratic equation of the second degree

pher a symbolic text! It remains that this tree-like organization structures mathematical thought.

Cardano's text above is definitely not punctuated, and that's why it is unreadable. A line of calculation from the end of the 17th century could, on the contrary, usually display twenty-five partial results, with nevertheless an easy deciphering, by the use of a full set of implicit hierarchical conventions.

This fourth of the six "patterns of the representation" constitutive of the new writing, therefore represented a major methodological step. Apparently resulting from a mere quantitative modification aiming at the number of instructions to describe, it in fact found an extension such as to become a difference of kind.

Note incidentally that this part of the study (4th pattern) also stresses a clearly possible distinction in the text between two spontaneous interpretations of symbolism: as a *procedure* or as an *object*.

5. **Representation of relationship** (for example by means of the signs = (Recorde), > (Harriot), and ≅ (Leibniz)). In the case of equality, this was established by Recorde's work (Recorde 1557), then Descartes's (*Géométrie*, 1637). It was a late-coming representation, which nevertheless made it impossible to maintain the syntax of natural language in the text. In effect, a new symbolism expressing ideal interchangeability succeeded the predicate structure of rhetorical expression.

6. **Representation of compound concepts**, based on the example of the representation of the sequence of powers (squares, cubes, sursolids or quadratics, etc.). This point is also discussed below.

3.4 The dialectic of indeterminacy (second pattern)

The symbolic representation of the "given" was a major innovation in the late sixteenth century, due to François Vieta, who introduced a new system of signs, entirely made up of letters, and whose true function consisted, in the last analysis — I anticipate my conclusions here — to combine within the symbolic system two concepts hitherto considered as opposites, the arbitrary and the fixed, or the one and the multiple or even, maybe more significantly, the unspecified and the singular.

In Vieta's time, and since Antiquity, geometrical figures had been considered "arbitrary", i.e., generic or emblematic of some given geometrical situation — the figure was some particular example of its type, but its particularity was not significant.[3] There was no comparable representation of arbitrary numbers. To represent an unknown, a non-numeric symbol (sometimes a letter) was needed, precisely because it was unknown. In his *Isagoge*, Vieta (1591) introduced the use of letters to represent the *given* as well. The letters used were of different types, according to the concept: vowels for the unknowns (which were less numerous), and consonants for the given.

In his preliminary considerations, Vieta is very careful to find an "obvious and durable symbol" to distinguish "the given and required magnitudes". But this definition, insofar as his contemporaries understood it according to the rules of the time, might contain a contradiction due to this simple fact: in any calculation considered valid at that time, the "given" was just what could be explicitly represented by numbers. Saying, as Vieta did, that the consonant B represented a *given* magnitude, meant that the author of the text knew its value. But, as thus represented, the reader had no knowledge of it! How could Vieta call B the sign of a given? Objections to such a conception of knowledge and representation cannot be overlooked, and bear a resemblance to those later exchanged (circa 1905) by proponents and opponents of the Axiom of Choice in their celebrated correspondence. How is the author certain to always "think" of the same element? As Lebesgue wrote to Zermelo, it was not a question of contradiction, but of self-consistence: how could Zermelo be sure that Zermelo would always "think" of the same element, since he did not characterize it in any other way *per se* ? (Cf. our discussion on this point in (Serfati, 1995).) These objections to Vieta's symbolism were actually considerable epistemological obstacles, so that many centuries were needed to understand and (in some way) go beyond them. One can thus clearly understand why, although it is widespread knowledge that the representation of givens by letters was a conclusive element in the development of mathematics, this major discovery was not made until thirteen centuries after Diophantus.

In those conditions, however, the withdrawal of numbers as explicit symbolizations of given data led directly to a new power and a new obligation:

---

[3] All the commentary on the ancient Greek mathematics recognizes this Platonic conception. Heath e.g. notes that the figure actually represented a class of figures "The conclusion can, of course, be stated in as general terms as the enunciation, since it does not depend on the particular figure drawn; that figure is only an illustration, a type of the class of figure and it is legitimate therefore, in stating the conclusion, to pass from the particular to the general" (Heath, 1921, I, 370).

to consider the given as arbitrary. If in fact the only information provided by the letter was to indicate a convention relating to the category of some represented entity (namely the "given") and not to make its value explicit, then this one, although fixed, was free to be arbitrarily selected. In other words, Vieta's letter B was doubtless the sign of a given quantity, but of an *arbitrary* one. But how was it possible for a quantity to be altogether arbitrary and yet fixed, fixed and moving, specific and general? Such statements appear as contradictions in natural language, so that what Vieta actually asked his reader to do was to accept a convention: on the one hand, "arbitrary and fixed" entities exist; on the other hand, the warrant of this kind of existence is symbolic language (not natural) — thus there is a "dialectic" here.

Immediately accepted by the mathematical community, in particular by Descartes, Vieta's system of letters spread in Europe. No matter what was its legitimacy, it authorized the use of literal formulas for the resolution of the problems. Since then, the formulas came to replace the rhetorical counting rhymes or pieces of poetry which, from the Middle Ages to the Renaissance, had described the resolution of problems in natural language. A little later, the concept (pseudo-concept?) of "indetermined" was equipped with a convenient but ambiguous term, which appeared at the end of the seventeenth century with Leibniz, the "variable". Opposed to "constant", the term once knew (and still has) a considerable success, due to the fact that it is accompanied by a strong naive cinematic connotation, that of a quantity that could, by the assumption of some imaginary displacement, "take all its values" inside a certain field.

This issue of the status of the letter, so central, remained (at least: almost) unchanged until the early twentieth century when it rebounded in very important discussions between mathematicians, logicians and philosophers such as Frege, Russell, Hilbert and Gödel. Thus Russell and Frege were vigorously opposed on the character of the fundamental contradiction, whether it was reducible in the natural language or not (Serfati, 2005a, 189-193). Meanwhile Hilbert and Gödel believed being able to evacuate the question, i.e. to eliminate it as a question, by simply stating the vanity of any interpretation — each one in his way, however, formalist for the first, logicist for the second. It is well known that this refusal of a certain constructivism led to a well located metamathematical split between interpretations and formalism. In everyday practice, the question is still not settled among mathematicians at work, for example Bourbaki.

To repeat: we must interpret Vieta's approach as an implicit request to accept a radical change in the *level* of convention, legitimizing the "arbitrary but fixed", whose existence is secured only by the symbolic. We must then consider that since Vieta (and perhaps, in a sense, up to Hilbert and Gödel) only those are recognized as mathematicians who have agreed to enroll in this agreement — an assent which has always been counted as a sign of a strong implicit adherence to the mathematical community, testifying once again that the symbolic is not the transcript, nor the reflection in signs, of writings in natural language.

3.5 The representation of compound concepts (sixth pattern)

Faced with an unknown quantity, such as the "Thing" of the cossic system or Diophantius' *Arithmos*, calculators saw very quickly that it was useful to conceive the "Square", and also the "Cube". Beyond these, there were higher species called the "Square-Square" (or "Biquadratic", according to the authors), then the "Sursolid", and finally a whole lineage of powers. As yet the simplest of calculations interspersed "Thing", "Square" and "Cube", it became necessary to represent symbolically the whole lineage. The solutions proposed were remarkably different, despite the apparent simplicity of the question, not only in terms of choosing the type of signs (number, letter, or "figure"), but mostly in the procedure of representation itself. In (Serfati, 2005a, chap. VIII) we have given some epistemologically significant examples (e.g. Vieta's and Bombelli's) of the proposed systems, from Diophantus to Descartes. There were other good attempts from others writers, recorded and reviewed by Cajori.[4] Descartes's exponent put an end to centuries of scattered notations, not yet operationally completed before him. One will better understand the central conceptual importance of the representation of powers by ultimately pointing out that being a basic condition in terms of mathematical technique, it was simultaneously the first, historically speaking, which led to the representation in symbolic writing of a compound concept.[5]

For brevity, we limit ourselves here to two symbolic systems prior to Descartes, first Diophantus's (the first symbolic algebra) and secondly, the cossic system, which immediately preceded Descartes. Clavius, who imposed the reform of mathematics in Jesuit schools, was one of its last practitioners and the young Descartes, who was one of his followers at La Flèche, still used it in its early texts (e.g. the *Cogitationes Privatæ* of 1619-1621 (Adam and Tannery, X, 213-256)). Countless other systems were employed, both in Europe (like in Lucas Pacioli's the *Summa di Aritmetica*,[6] or Cardano's Italian system in the sixteenth century) as well as in Arabic algebra.[7] On these important

---

[4] "Signs of Powers" in (Cajori, 1928, I, 335–360).

[5] I developed some of these conclusions in (Serfati, 1998).

[6] See for instance Høyrup (2010) who gives a reproduction of Pacioli's scheme, showing the signs for powers with root names.

[7] See for instance (Cajori, 1928, I), *Symbols in Arithmetic and Algebra* (Elementary Part), 71–400. Also (Woepcke, 1954, 352–353).

historical points, we refer the reader to two papers quoted in the references, to Høyrup (2009) for Europe, for the Middle Ages and the Renaissance, and to Heeffer (2008) for Arabic algebra.

First, we briefly investigate Diophantus's system, noting that a specificity of his symbolism lies in the coexistence of a sign for ς for *arithmos* (i.e the unknown) together with another, radically different, for the Square, Δ^{v}, and also for the Cube K^{v}. As one can observe, it cannot be deduced from the symbolism that Δ^{v} and K^{v} denoted the square and the cube of an unknown denoted by ς. A battery of different signs then completed the representation of the "Square-Square" $\Delta^{v}\Delta$, the "Square-Cube" ΔK^{v}, and the "Cube-Cube" $K^{v}K$. Diophantus also exhibited a new sign ς', for what we today call the inverse of the unknown, and finally another $\Delta^{v'}$ for the square of the latter.

A quite similar inventory was in use in the cossic system, with signs different from the previous ones, however: the "Thing" (the unknown) had $\mathfrak{r}$ for symbol, the "Square" (or Census) was usually represented (in Stifel or Rudolff for instance) by $\mathfrak{z}$, and the "Cube" by $\mathfrak{c}$. Similarly, the "Biquadratic" had $\mathfrak{zz}$ for denotation. Rudolff's book exposes one of the first inventories of such signs, with ten levels (Cajori, 1928, I, 134):

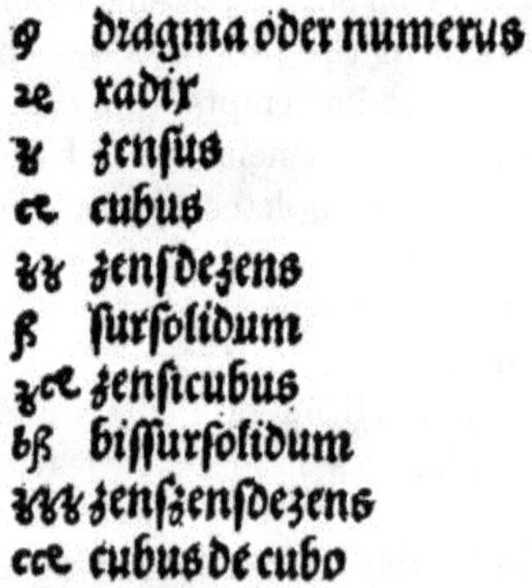
φ dragma oder numerus
𝔯 radix
𝔷 zensus
𝔠𝔢 cubus
𝔷𝔷 zensdezens
ß sursolidum
𝔷𝔠𝔢 zensicubus
bß bissursolidum
𝔷𝔷𝔷 zenszensdezens
𝔠𝔠𝔢 cubus de cubo

Fig. 3.5: Different powers of the unknown (from Rudolff fol. D^{iiij})

At first glance, this might seem a satisfactory system, then as now! We will however describe its significant structural deficiencies in an example (an equation in Stifel). In the *Arithmetica Integra*, Stifel proposes to solve

$$1\mathfrak{zz} + 2\mathfrak{c} + 6\mathfrak{z} + 5\mathfrak{r} + 6 \text{ æqu. } 5550$$

Weaving together "Thing", "Square", "Cube" and "Biquadratic", the problem is what is anachronistically called (since Descartes!) a numerical equation of the fourth degree.[8] To resolve, Stifel introduced the new expression:

$$1 \maltese + 1 \maltese + 2$$

and found that if he added it to its own square, he got the first member of the original equation.[9] This requires comment: one first has to calculate the square of $1 \maltese + 1 \maltese + 2$. Now this square *cannot be written symbolically* in the cossic system, since representation by the sign $\maltese$ allows writing the square of the initial term only. To evoke the square of the new expression, Stifel was forced to abandon the symbolic representation of direct powers and multiply the expression by itself. Under these conditions, however, and to continue the calculation, he would then necessarily have to known how to deal with expressions such as $1 \maltese \cdot 1 \maltese$ for example. He has therefore been obliged to involve four basic rules governing the multiplication of the cossic signs. Here are the first two. First:

$$1 \maltese \cdot 1 \maltese \text{ equal to } 1 \maltese$$

or

("Thing" *multiplied by* "Thing" *makes* "Square")

which, written in Latin, was the maxim, famous in its time, *Res in Rem fit Census*. Also:

$$1 \maltese \cdot 1 \maltese \text{ equal to } 1 \maltese$$

or

("Thing" *multiplied by* "Square" *makes* "Cube")

This situation was very similar to what we described above in Diophantus's algebra. And since Diophantus, and up to Vieta, a list of these rules had to be memorized and was, in various forms (especially as counting-out-rhymes), part of the baggage of any mathematician (Serfati, 2005a, 208). Note that the protagonists of the time did not perceive them as we would naturally regard them today, namely as instances of a simple multiplication table.

Note also incidentally that such equation would seem difficult to solve at a time when Ferrari's method for equations of the fourth degree was not yet available. A careful examination of Stifel's statement of the question (a gambling problem coming from Cardano's *Arithmetica* (1539), chapter 51)

---

[8] From Stifel's *Arithmetica Integra* f. 307^v. The example is reproduced in (Cajori, 1928, I, 139-140).

[9] $(x^2 + x + 2)^2 + (x^2 + x + 2) = x^4 + 2x^3 + 6x^2 + 5x + 6.$

makes us understand how its mode of elaboration is genetic, meaning that this equation is entirely *ad hoc:* to the unknown (say x), the author adds first its square, then the number 2. To the sum thus obtained, he adds its own square and wants the latter sum to be equal to 5550. He is then seeking the value of x. In modern terms (completely anachronistic) the proposed equation of the fourth degree thus decomposes into the successive resolution of two equations of degree 2:

$$A^2 + A = 5550 \tag{3.1}$$

followed by

$$x^2 + x + 2 = A \tag{3.2}$$

This is obviously not a general method for the resolution of an equation of the fourth degree! Actually Stifel made the analysis of what he himself has synthesized.[10] One must therefore not anachronistically consider the system (3.1) and (3.2) as an indetermined change of variables (see below).

Let us repeat: Stifel could write ⅔ as the square of the "Thing". However, he could not write the square of 1⅔+1ᶻℯ+2 in a structurally analogous manner, as we do today in the Cartesian system, very simply by replacing A with A^2. In other words, the cossic expressions ᶻℯ and 1⅔+1ᶻℯ+2 could not be freely substituted for one another in the symbolic expression of a square. Naturally, this was also impossible for an expression such as 3ᶻℯ+5. In fact, the Cossic could never allow the symbolic text to represent the square of an *arbitrary* expression. This point is crucial. Therefore Stifel could evoke the square of 1⅔+1ᶻℯ+2 in the text — which was essential — only by developing it, i.e. by computation.

Developing, however, is not representing; developing requires the calculator to use various memorizing methods, and thus to appeal to elements of meaning foreign to the symbolic system. Actually, such "square", since it is unrepresentable in the cossic system is not capable of being individuated, that is, it cannot be *objectified*. In other words, it is *inconceivable* as an object in the system. Admittedly, some calculation may give a specific value to a specific symbolic expression, but such expression is definitely the product of two other symbolic expressions, and not a "square" *per se.*

---

[10] A careful analysis of the complete resolution shows to what even greater extent the solution is ad hoc. Actually, the two quadratic equations of the system have the same specific form, namely a product of two consecutive numbers, $Z(Z+1) = C$. First $A(A+1) = 5550$ is satisfied by $A = +74$, the other root, (obviously -75) must be rejected as "false". Then Stifel had to solve $x(x+1) + 2 = 74$, that is $x(x+1) = 72$, satisfied similarly by $x = 8$ (the other root being -9). 8 is therefore the single solution of the problem. I think it highly probable that Stifel actually began synthetically, that is to say in the opposite direction, by considering first the number 8, then forming $8 \cdot 9$, then the equation $x(x+1) + 2 = 74$, etc.

This example also illustrates the inability of the mathematician of the time to make an *arbitrary* change of unknowns *within* the system. One could argue that in Cardano's *Ars Magna* (Chapter XXXIX for example) there are changes of unknowns (e.g.: $y = x + \frac{b}{3}$) which remove the "square" term from the cubic equation $x^3 + bx^2 + cx + d = 0$ so as to get a standard reduced form (Vieta used a similar device). This interpretation, however, appears anachronistic. In fact, a careful analysis of Cardano's text shows that it is but a straightforward verification: just like in the equation of the second degree $x^2 + ax + b = 0$ the first two terms $x^2 + ax$ had long been regarded as the beginning of a square (in order to reach the standard canonical form), so the block $x^3 + bx^2$ was considered by Cardano as the beginning of a cube, so that the whole calculation was equivalent to a simple check. Epistemologically speaking, Cardano's technique (just like Vieta's) was adventitious: it was not an arbitrary (indetermined) change of unknown and could therefore hardly be transposed to an example of an even slightly different nature. The first actual examples of arbitrary changes of unknowns appear later with Tschirnaus's transformations under the rules of the new (post Cartesian) symbolic writing.[11]

Thus our initial conclusions can be summarized as a list of deficiencies and drawbacks inherent to the cossic system, similar to those of Diophantus: inability of an arbitrary change of unknowns, necessity of the use of rhymes to make any calculation.

The analysis of these disadvantages stresses (retrospectively) the major importance of two predicates: substance and relation (e.g., in the $2a^3$ of Descartes's *Regulae*[12] where a and 3 are respectively signs for substance and relation). Should a symbolic system have 'decided" to represent them both, the univocity rule would necessitate two signs, not just one, and the represented concept would therefore have been considered compound. And this is what Descartes was the first to do in Rule XVI of *Regulae*, where a is the sign of the substance and 3 that of the relation. For various reasons the Cossic "decided" to use only one sign, thereby implying that the concept was simple. Thus the Cossic for all intents and purposes represented *neither* of the two predicates. This was not, however, a logical fault. Following a (supposedly)

---

[11] An indeterminate change of unknowns to solve an algebraic equation $F(x) = 0$ is a mapping g of the form $x = g(t,a)$ where t is the new unknown, and a an indeterminate parameter, such as the new equation, known as the transformed equation, namely $Ga(t)(= F(g(t,a))) = 0$ becomes simpler by a suitable choice of the parameter a. This is the simplest case. It may happen that the process involves several parameters with the same function. A paradigmatic example is Tschirnaus' method for equations of the third degree which considers the equation $F(x) = x^3 + bx^2 + cx + d = 0$ and sets $y = ax^2 + bx + c$ (thus there are 3 parameters) so as to reduce it to the "binomial" form $y^3 = K$. Lagrange studied the method and its inherent limitations. Many attempts to extend Tschirnaus' method to higher degrees were developed in the nineteenth and twentieth centuries. See for instance (Adamchik and Jeffrey, 2003).

[12] The first exponential in history appears in Rule XVI of Descartes's *Regulae'*.

natural approach, the Cossic worked in effect as a listing operator: each new concept was given a distinct representation. But as far as mathematics is concerned, such a procedure has no future. Descartes's exponent put an end to the Cossic.

The introduction of the Cartesian exponent marked thus the disappearance of the "diophanto-cossic" symbolism which for centuries had ruled mathematical thinking on the issue of powers. With it disappeared its main limitations. We can make ours Cajori's conclusion: "There is perhaps no symbolism in ordinary algebra which has been as well chosen and is as elastic as the Cartesian exponents" (Cajori, 1928, I, 360).

It was indispensable to the advancement of mathematics to represent the original lineage of powers by two signs, and not just one: so we can summarize the lesson of this first part of the story of powers. Thus the representation of powers was historically the first which led to today's mode of representation of a compound concept in universal symbolic writing (assembler and open places). At the end of the century, one of the epistemological lessons drawn by posterity from the outcome of the question of "powers", was undoubtedly the analogous creation of Leibniz's "New Calculus" (Serfati, 2005a, 274).

3.6 Descartes's *Géométrie* or the "Rosetta Stone"

Out of the effective representation of the six above-mentioned concepts emerged the essence of the new symbolic system. Thus, from Vieta to Descartes, mathematical symbolic writing was constituted, taking on the principal aspects of its current structure. The role of Leibniz, on which I will expand a bit in what follows, was quite as capital, but different. (On the various contributions of the three protagonists, cf. Serfati (2005a, 386)).

All these innovations are found gathered together in the *Géométrie* of 1637. This text took center-stage because of the richness of its mathematical content as well as Descartes's authority as a philosopher; and despite the fact that Descartes gave no explicit instructions on symbolism, it served during the 17th century as the model for deciphering new symbolic texts (according to the so-called "principle of the Rosetta Stone").

3.7 The introduction of substitution

From the above discussion of the sixth pattern (compound concepts) it is clear that the substitution of

$$A = x^2 + 1 \text{ in } Z = 2A^3 - 5A$$

so simple to write and perform today, remained an *inconceivable* operation for the medieval rhetorical writing of mathematics. But with Leibniz, substitutability became an essential, everyday element. One can discover one of the creative aspects of symbolism here. The successive creation of first fractional exponents, then irrational and real exponents, is a good elementary example.

Fractional exponents was the object of a famous letter from Newton to Leibniz in June 1676, the *Epistola Prior* (Leibniz, 1899, 179). Nothing in Leibniz' preliminary experience at that time, nor in the Cartesian definition of exponential (the only one he knew) allowed him to understand what meaning Newton could bring to symbolic forms as $3^{\frac{1}{2}}$, $(x+3)^{\frac{1}{2}}$, $5^{\frac{2}{3}}$, or $(\sqrt{2})^{-\frac{6}{7}}$. Any attempt of rhetoric translation "à la Descartes's led to nonsense: if the procedure of the form "3^5" can be described by

"Multiply the number of sign 3 five times by itself",

which significance could be reasonably allotted, with respect to $3^{\frac{1}{2}}$

"Multiply the number of sign 3 half-time by itself"?

There was thus for Leibniz, faced with symbolic forms without significance, an epistemologically crucial time of incomprehension (a momentary, "logical" time), which was dissipated by following the letter. In the *Epistola Posterior* (Leibniz, 1899, 225) that followed Newton persevered, now introducing irrational exponents, as in:

$$\frac{\sqrt{3^{\frac{2}{3}}}}{x^{\sqrt{2}} + x^{\sqrt{7}}}$$

That the first version of the Newtonian exponential, the so-called *broken* one (i.e. with fractional exponents) in the *Epistola Prior*, was precisely defined, while the second, *surd* (quadratic irrational) in the *Epistola Posterior* was not at all, hardly worried Leibniz. On the contrary, capturing the essence of the Newtonian process, Leibniz worked at that time to build – by imitation – a new exponential, with sign a^x or y^x, whose importance, as Leibniz does naively repeat, would exceed both Descartes's and Newton's. The question was of course: what could at that time and for him be the meaning of a symbolic form where at the place of the exponent was a sign of an arbitrary (indeterminate) number? An exponential, which was for him one of the three aspects of what he then named the transcendent in the mathematical sense.[13]

As it is known, this method is more explored today, for example by substituting a complex number, or an endomorphism, or else a square matrix,

---

[13] Leibniz was the first to import in mathematics the concept of transcendence, with diverse meanings of the word. See for instance (Breger, 1986) and (Serfati, 1992).

defining the exponential A^B of two arbitrary objects A and B of an (almost) arbitrary category C.[14]

Far from appearing strange or arbitrary, these creations of objects by substitution were each time recognized as legitimate and fertile by the mathematical community. The authentic reasons for such a universal approval constitute a capital epistemological fact and will be analyzed in another place. Here, I will consider an instance in the last section, however.

Other examples in Leibniz reinforced the importance of substitutability without any equivalent in natural language. Thus, in his demonstration of *Arithmetic Squaring of the Circle*, he used, with modifications, the demonstration that Mercator had given in his *Logarithmotechnia* (Mercator, 1668) for squaring the hyperbola. Mercator developed $\frac{1}{1+x}$ as a power series, and then integrated term by term. In order to "square the circle", Leibniz substituted x^2 for x, then integrated the development of $\frac{1}{1+x^2}$ term by term (cf. for instance (Leibniz, 2004, 208)), a substitution that goes without saying for us, but that was amazing at the time. Today one can hardly imagine the difficulties faced by scholars of the time, whose minds were filled with geometrical truths, in grasping such a substitution that involves only symbolism. It is hardly necessary to specify to which extent the procedure requires the use of the symbolic writing to be conceived!

3.8 Symbolic notation and the creation of mathematical objects

I will conclude with an important aspect of symbolism, describing the emergence of a process of creating objects from it. I will analyze it *in statu nascendi* in the correspondence of 1676 between Leibniz and Newton mentioned above, by returning to this question: how to provide meaning to the symbolic form $a^{\frac{1}{2}}$ which for Descartes certainly had no significance? The reconstruction of the method is the following: first, the geometer chooses a formula for the exponential (Cartesian), in the stock of all those that it was known to validate. This will be for example the so-called "multiplicative formula"

$$(a^r)^s = a^{r \cdot s}$$

valid if r and s are the signs of any natural numbers, and a the sign of any positive (real) number.

This formula will be known as *elective* in what follows (it is the mathematician who chooses it). If however r is interpreted as an unspecified rational,

---

[14] The category C must be "cartesian closed". This definition (due to F. W. Lawvere) legitimately points out the reference to Descartes.

say $r = \frac{p}{q}$, then $a^{\frac{p}{q}}$ is without significance: a and $\frac{p}{q}$ being separately both equipped with significance, it is the assembler, that is the exponential copula which is deprived from it, and the nonsense comes from an inadequacy of categories. The method then consists in the definition, if possible, of the values of a^r and a^s as numbers such that the same formula remains true for any value of the couple of the rational numbers, of signs r and s. With this intention, one will start by affirming the validity in the particular case of it where r is the inverse of a natural number, that is to say $r = \frac{1}{q}$. One then shows simply by stages that the only possible value for $a^{\frac{p}{q}}$ is $(\sqrt[q]{a})^p$, i.e. the one proposed by Newton in his letter — for instance $a^{\frac{1}{2}} = \sqrt{a}$, a value which satisfies the elective formula:

$$\left(a^{\frac{1}{2}}\right)^2 = a^{\frac{1}{2}\cdot 2} = a^1 = a$$

In this modest example, the calculator can undoubtedly believe to have won on all counts: he provided significance to a symbolic form which did not have any. Thus he uncovers a rational, scientific management of nonsense — ruptures of meaning here being consubstantial with (mathematical) creation. At the same time, he extended *extra muros* (i.e. to the rationals) the field of validity of the multiplicative formula, which can thus appear as depositary of a higher form of truth, intrinsic and enlarged, an hypostasis of some (alleged) general concept of "Exponential". The mechanism thus consolidates the (Platonic!) feeling of the protagonists to be in the presence of a "natural" concept, with the simple acceptation of "which one finds in nature". I will close here this short parenthesis of philosophy of psychology to note that this form of illusion is the result of the occultation of the fact that the seemingly natural significance is the exact counterpart of the method of the elective formula.

I am well aware of the modesty of this example. It is however decisive for the abstraction of a method. One can indeed show that this same pattern in three phases, meaningless forms, elective formulas and analogical extensions, was at work in creation of many objects, both in the eighteenth century with Euler (e.g. complex exponential) and more recent (Moore-Penrose's pseudo-inverses of matrices, derivation in the sense of distributions, the elective formula being here integration by parts, etc.). Let us quote another example, of major importance in mathematical Analysis, highlighted by J.P. Kahane,[15] the couple of Fourier's relations

$$c_n = \int f(\ldots) \text{ and } f(\ldots) = \sum_{n \in \mathbb{Z}} c_n \ldots$$

that institute, in his terminology, a "program", composed of two elective formulas, registered in a structural canonical duality between series and integrals.

---

[15] Intervention at the 6 May 2008 Meeting of the Académie des Sciences: "La vertu créatrice

This term of "program" means that the concerned formulas (Fourier's relations) are conceived, not as completed data, but as objectives to be reached, and this in various situations, by *defining* each time *ad hoc* objects to satisfy them. In other words, the above formulas "require" new objects for their own satisfaction.

Epistemologically speaking, the scheme is a pattern for the creation of objects by analogical extension and permanence of symbolic forms in a mode that is entirely specific to mathematics and consubstantial with the notation. It thus shows one of the aspects under which the advent of the symbolic system contributed, from the seventeenth century onwards, to invention in mathematics, thus helping to clarify the intimate nature of this "power to create" by mathematicians — actually the fundamental subject here — evoked by Dedekind and stressed by Cavaillès (Cavaillès, 1981, 57).

In fine, one can philosophically comment on the psychological aspects of the scheme, which, notwithstanding its essentially *ad hoc* nature, has an ambiguous effect: it indeed helps to "spontaneously" reinforce the Platonic (or realistic) vision of mathematics. Admittedly, the scheme is purely constructive, and the product of human activities. Admittedly, it is based on the desire of permanence of some symbolic forms — a central point in the philosophy of psychology, of which one will certainly have to discuss both the origin and the relevance, but not the historical reality which is undeniable. Nevertheless, once in the presence of the scheme, the mathematician can believe that he does nothing but put his steps in the way of a discovery traced by others that uncovers idealities, i.e. objects and formulas eternal as well as transcendent to the human subject of knowledge. Note also that a primitive form of the scheme has been uncovered by George Peacock in the 1830's, under the name of "permanence of equivalent symbolic forms" (Peacock, 1830), but with simple examples only and without that he perceived the universality and systematicity of the schema, as well as its deep rationality.[16]

Thus, after the introduction of the symbolic writing system, nothing in mathematics was anymore like before. The outcome was, strictly speaking, a (symbolic) revolution, one of the major components of the scientific revolution. An epistemologist therefore cannot fail to wonder about the reasons for the already noted, persistent absence of any study of the subject. Our analysis delivers some astonishing conclusions here. One indeed uncovers underground epistemological obstacles attached to unquestioned beliefs. Overall, the pregnant Platonic conception that since mathematical objects are suppos-

---

du symbolisme mathématique".

[16] Heeffer (2010, 521) rightly emphasizes the creative role of Peacock's principle in the early history of numbers. He thus writes: "We should like to demonstrate that Peacock's principle of the permanence of equivalent forms is a fruitful framework for studying changes in the history of numbers". He gives examples of expansions of the number concept in Maestro Dardi and Cardano.

edly ideal, contingent representations would not matter; with as corollary the naive belief, as widespread as false, that the symbolic system would only be shorthand, the reflection in signs of natural writing.[17] A conception which, as I showed here, cannot face the reality of mathematical developments since the seventeenth century, in the very first place because of the question of substitutability — a major point of division between natural language and scientific (mathematical) language.

References

1. Adamchik, Victor S. and David J. Jeffrey, 2003. "Polynomial Transformations of Tschirnhaus, Bring and Jerrard", ACM SIGSAM Bulletin, 37, **3**, September 2003, 90-94.
2. Breger, Herbert, 1986. "Leibniz Einfürhung der Transzendenten", *300 Jahre "Nova Methodus" von G.-W. Leibniz (1684-1984)* in Studia Leibnitiana, Sonderheft XIV.
3. Cajori, Florian, 1928. *A History of Mathematical Notations. I. Notations in Elementary mathematics. II. Notations Mainly in Higher Mathematics.* La Salle, Illinois: Open Court, 1928–29 (Dover edition, 1993).
4. Cavaillès, Jean, 1981. *Méthode axiomatique et formalisme. Essai sur le problème du fondement des mathématiques.* Paris: Hermann, 1981.
5. Dascal, Marcelo, 1978. *La sémiologie de Leibniz.* Paris: Aubier-Montaigne, 1978.
6. Adam, Charles and Paul Tannery (eds.) 1897-1910. *René Descartes, Œuvres complètes* (13 vol.), édition Adam-Tannery. Paris: Léopold Cerf. (Reprint of the first 11 volumes from 1964. Paris: Vrin, pocket edition from 1996).
7. Descartes, René, 1637. *La Géométrie* In C. Adam, Charles and P. Tannery (eds.) 1897-1910. VI, 367-485. English translation: *The Geometry of René Descartes*, With a Facsimile of the First Edition. Trans. D.E Smith and M. L. Latham. Dover. New-York. 1954.
8. Frege, Gottlob, 1879. *Begriffsschrift: eine der arithmetischen nachgebildete Formelsprache des reinen Denkens.* Halle: Nebert, 1879.
9. Heath, Thomas, 1921. *A History of Greek Mathematics* (2 vol.). Oxford: Clarendon Press, 1921 (Reprint: New York: Dover, 1981).
10. Heeffer, Albrecht, 2008. "A Conceptual Analysis of Early Arabic Algebra". In S. Rahman, T. Street and H. Tahiri (eds.) *The Unity of Science in the Arabic Tradition: Science, Logic, Epistemology and their Interactions.* Heidelberg: Springer, pp. 89–126.
11. Heeffer, Albrecht, 2010. "The Symbolic Model for Algebra: Functions and Mechanisms". L. Magnani, W. Carnielli and C. Pizzi (eds.) *Model-Based Reasoning in Science and Technology, Abduction, Logic, and Computational Discovery.* Heidelberg: Springer, 2010. pp. 519–532.
12. Høyrup, Jens, 2010. "Hesitating progress — the slow development toward algebraic symbolization in abbacus - and related manuscripts, c. 1300 to c. 1550". in A. Heeffer and M. Van Dyck (eds.) *Philosophical Aspects of Symbolic Reasoning in Early Modern Mathematics*, Studies in Logic 26, London: College Publications, 2010 (this volume, chapter 1).

---

[17] The point is also noted by Heeffer (2010, 524) in the context of early symbolic texts:

13. Leibniz, Gottfried Wilhelm, 1899. *Der Briefwechsel von G. W. Leibniz mit Mathematik-ern.* Berlin: C.I. Gerhardt. (Reprint Hildesheim: Olms, 1962).
14. Leibniz, Gottfried Wilhelm, 2004. *Quadrature arithmétique du cercle, de l'ellipse et de l'hyperbole et la trigonométrie sans tables trigonométriques qui en est le corollaire.* Introduction, translation and notes from M. Parmentier. Latin text from E. Knobloch. Paris: Vrin.
15. Mercator, Nicolaus, 1668. *Logarithmotechnia: Sive Methodus Construendi Logarithmos,*
16. Peacock, George, 1830. *A Treatise on Algebra.* 2 vols. (I. Arithmetical Algebra, II. Symbolical Algebra). Cambridge: J. Smith. (Reprint New York: Dover, 2005).
17. Recorde, Robert, 1557. *The whetstone of witte whiche is the seconde parte of Arith-metike: containyng theextraction of rootes: the cossike practise, with the rule of equa-tion: and the woorkes of surde nombers. Though many stones doe beare greate price, the whetstone is for exersice ... and to your self be not vnkinde,* London: By Ihon Kyngston.
18. Serfati, Michel, 1992. *Quadrature du cercle, fractions continues, et autres contes.* Paris: Éditions de l'Association des Professeurs de Mathématiques.
19. Serfati, Michel, 1995. "Infini 'nouveau'. Principes de choix effectifs". In F. Monnoyeur (ed.)*Infini des philosophes, infini des astronomes* Paris: Belin. 1995, pp. 207-238.
20. Serfati, Michel, 1998. "Descartes et la constitution de l'écriture symbolique mathématique". *Revue d'Histoire des Sciences* **51**, 237–289.
21. Serfati, Michel, 2005a. *La révolution symbolique. La constitution de l'écriture symbol-ique mathématique.* Paris: Éditions Petra.
22. Serfati, Michel, 2005b, René Descartes, Géométrie, Latin edition (1649), French edi-tion (1637). In I. Grattan-Guinness (ed.) *Landmark Writings in Western Mathematics 1640-1940.* Amsterdam: Elsevier, 2005.
23. Serfati, Michel, 2008a. "Symbolic inventiveness and irrationalist' practices in Leib-niz' mathematics". In M. Dascal (ed.) *Leibniz: What kind of rationalist?* Heidelberg: Springer, pp. 125-139.
24. Serfati, Michel, 2008b. "L'avènement de l'écriture symbolique mathématique. Symbol-isme et création d'objets". In Lettre de l'Académie des Sciences **24** (automne 2008), pp. 25-27.
25. Stifel, Michael, 1544. *Arithmetica integra.* Nürnberg: Petreius.
26. Viète, François, 1591. *In artem analyticem isagoge. Seorsim excussa ab Opere restitu-aeæ mathematicaeæ analyseos, seu algebra nova.* Tournon: apud Iametium Mettayer typographum regium.
27. Viète, François, 1646. *Opera Mathematica.* Edition by Frans Van Schooten. Leyden. (Reprint Hildesheim: Olms, 1970).
28. Woepcke, Franz, 1854. "Recherches sur l'histoire des sciences mathématiques chez les Orientaux, d'après des traités inédits arabes et persans. Premier article. Notice sur des notations algébriques employées par les Arabes". *Journal Asiatique*, 5e série **4**, 348-384.

"The objection marks our most important critique (...): symbols are not just abbreviations or practical short-hand notations".

Part II
The interplay between diagrams and symbolism

Chapter 4
Translating Euclid's diagrams into English, 1551–1571

Michael J. Barany

Abstract The years 1551–1571 saw the first published translations of Euclid's *Elements* in the English language. Euclid's first English translators had to translate not just his words, but his entire system of geometry for a vernacular public unversed in a method and study hitherto 'locked up in straunge tongues.' Throughout its written and printed history, diagrams have been crucial features of Euclid's text. This paper considers the variety of diagrammatic approaches used in these first English translations, arguing that the strategic inclusion and exclusion of points, lines, letters, and labels, along with depictions of surveying instruments and landscapes, played crucial roles in establishing the authors' voices, vocabularies, methodologies, and mathematical philosophies. Using simple but polysemic objects such as points and lines, and appealing to familiar practices such as drawing, using a compass, and surveying a field, Euclid's translators projected and enforced an image of a geometry which could be seen to be already present and meaningful. Their diagrams, rather than being *mere* illustrations, played indispensable roles in establishing the new English geometry.

Key words: Euclid's *Elements*; Robert Recorde, John Dee, Henry Billingsley, Leonard Digges, Thomas Digges, Geometry, Translation, Diagrams, Representation

4.1 Translating Diagrams

In 1551, Robert Recorde published England's first surviving vernacular textbook on the principles of Euclidean geometry, *Pathway to Knowledg*.[1] Recorde's

Princeton University Program in History of Science, mbarany@princeton.edu

[1] Taylor (1954, 14–15, 312) discusses earlier surveying texts and a possible prior translation

text presents the definitions, axioms, postulates, and propositions of the first four books of Euclid's *Elements* in such a way that "the simple reader might not justly complain of hardnes or obscuritee."[2] Making Euclid's ancient geometry newly accessible to his vernacular readers, Recorde's translation involved more than mere linguistic substitution and coinage. For Recorde was not merely translating between languages, but across concepts, idioms, places, times, social positions, and professions. Euclid's new tongue grew out of a translation in the fullest sense of the word.

Henry Billingsley completed his own edition of the *Elements* in 1570. The volume has been identified by Heath (1956) as "the first and most important translation" of Euclid's *Elements* into English (109). Billingsley's text incorporated "Scholies, Annotations, and Inventions, of the best Mathematiciens, both of time past, and in this our age" and a "very fruitfull Præface" by his collaborator John Dee.[3] The text was produced for, as Dee writes, "unlatined people, and not Universitie Scholers," and included all fifteen books then attributed to Euclid of Megara.[4] The following year, Thomas Digges posthumously published his father Leonard's tripartite geometrical practice, *Pantometria*, appending a preface and his own discourse on geometrical solids. *Pantometria* emphasizes surveying and the military arts, while Thomas's contribution concerns "matters only new, rare and difficile."[5]

This paper focuses on one aspect of these five authors' translations: their diagrams. Geometric diagrams and figures had been present even in the first print editions of the *Elements*, dating to the printer Erhard Ratdolt's 1482 volume, and the inclusion of illustrative diagrams was a standard feature for geometric texts.[6] Indeed, Euclid's *Elements* can scarcely be understood without the aid of geometric illustrations, and the visual vocabulary of the *Elements* had long been established as a central feature of geometric learning. Even so, the sheer variety of diagrammatic approaches used in the first English vernacular translations of the *Elements* indicates that the choice of how to illustrate one's text was no trivial matter.

I use the word 'diagrams' here in an unusual and anachronistic sense, but one which seems to me the most justifiable for the discussion that will fol-

---

of Euclid.

[2] Sig.*a*1[r]. Page citations are according to Gaskell (1972). I use Johnson and Larkey's (1935) convention of preserving spellings while sometimes modernizing typography by, for example, expanding contractions and converting 'u's to 'v's where appropriate. Emphasis in quotations is the quoted author's.

[3] Sig.[fist]1[r].

[4] Sig.*A*3[v]. The *Elements* are now typically attributed to Euclid of Alexandria, instead of Megara, and only the first thirteen of the books included in Billingsley's translation are considered to be of Euclid's authorship (Heath 1956, 3–5).

[5] Sig.*S*4[v].

[6] On manuscript geometric diagrams, see Keller (2005) and De Young (2005, 2009).

low. The diagrams considered here comprise all manner of visual para-text – including illustrations of definitions, constructions, proof figures, and drawings of instruments – designed to facilitate geometric understanding in these translations of the *Elements*. Such an expansive view is necessary for two principal reasons. The first is that by construing diagrams broadly one is better able to account for the diversity of illustrative approaches used by the different authors. Drawing from the same representational traditions found in past editions of the *Elements*, each translator took a different approach to rendering the visual and geometric meaning in those texts for his vernacular readers.

More importantly, a detailed reading of the visual vocabularies in these starkly varied texts cannot help but undermine any narrow circumscription of what counts as a diagram. One finds in our texts a variety of meanings for terms such as 'figure' and 'example' as well as a variety of uses for illustrations of different sorts. As all of these works were produced in print, the safest delineation seems to be that between conventional alphabetical text and other printed illustrations, including their captions. There, the extra work of producing figures and arranging the rest of the text around them suggests a special place for such images in our consideration of these translations. In these figures, we shall see the junctures where our authors found, for a diversity of reasons, that words did not suffice.[7]

Diagrams, for our authors, were integral means of establishing a new English geometry which was simultaneously comprehensible, even familiar, to its vernacular readers and a part of an ancient mathematical tradition. The next section provides some further necessary context for the authors and texts under consideration, exploring what it means to translate the *Elements*. I then consider Recorde and Billingsley's uses of diagrams, first in turn and then in comparison, and contrast these uses to those of Dee and Leonard and Thomas Digges. Finally, I synthesize these observations by comparing how each author establishes the definitions for parallel lines and the simple geometric point. In these texts, I argue, the strategic inclusion and exclusion of points, lines, letters, and labels, along with the depiction of instruments and landscapes, figured crucially in the establishment of the authors' voices, vocabularies, methodologies, and mathematical philosophies.

4.2 Translating Euclid

Between the first geometrical writings of Recorde, Billingsley, Dee, and Leonard and Thomas Digges, one finds the foundations of English vernac-

---

[7] This is not to discount the crucial role of printers and engravers in preparing diagrams. The texts engage sufficiently with their illustrations that it is safe to presume substantial involvement on the part of the authors, but there are also signs (particularly in errors or omissions) that suggest the limits of such involvement. For the remainder of this paper, I

ular geometry. Billingsley was a wealthy merchant and translator of several genres. His collaboration with Dee connected him to a closely interlinked Tudor tetrumvirate of English mathematics. Leonard and Thomas Digges both use elements of Recorde's terminology in their work, and Dee had worked directly with Recorde's arithmetical text *The Ground of Artes*.[8] Dee and the elder Digges knew each other personally, and the younger Digges was a pupil of Dee, who became Thomas's "second mathematical father" after Leonard's death.[9]

While Billingsley's is the only work of these five authors typically counted among translations of Euclid's *Elements*, and is certainly the most complete and literal of the group, the texts of each played pivotal roles in shaping Euclid's reception in England. *Pathway* offered many new geometric terms for geometry's new language, and, following Proclus, was the first modern text to classify Euclid's propositions as either constructions or theorems.[10] Dee's preface presented a taxonomy of the mathematical sciences and was among the most influential mathematical texts of the late sixteenth century.[11] *Pantometria* offered a definitive bridge between practical and theoretical geometry from an author already widely read by practical users of the art.[12] Beginning with "Elementes of Geometrie, or Diffinitions," its Euclidean allusions and aspirations permeate the text.[13] Thomas Digges's treatise on geometric solids was the first of many works securing his place as one of England's most eminent mathematicians.[14] All five authors incorporate Euclid's style and content, implicitly or explicitly, into their own.

In this respect, all five should be counted among the first English translators of Euclid's *Elements*.[15] One must remember that before the work of these translators there was no geometry, as such, outside the universities in England.[16] Our authors realized the novelty and significance of what they were creating.[17] "For nother is there anie matter more straunge in the englishe tungue," Recorde explains, "then this where of never booke was written be-

---

will attribute the constellation of authorships underlying the diagrams to the works' official authors.

[8] Johnson (1944, 132), Roberts (2004), Johnston (2004a, 2004b), Easton (1967, 515), Heninger (1969, 109).

[9] Digges (1573, Sig.$A2^r$), Johnston (2004b), Johnson (1936, 398–399), Taylor (1954, 166).

[10] Johnson and Larkey (1935, 68), Johnson (1944). Johnson identifies 'straight line' as the only one of Recorde's coinages to have survived to the present day.

[11] Roberts (2004, 200), Taylor (1954, 170–171, 320).   [12] McRae (1993, 345).   [13] Sig.$B1^r$.

[14] Johnston (2004b).

[15] Euclid's work was first printed in Latin in 1482 and its first full vernacular rendering was in Italian some sixty years later. Heath (1956, 97, 106). On English translations of the *Elements*, see Barrow-Green (2006, esp. 3–7).

[16] Feingold (1984, 178), for instance, credits these very authors with introducing higher mathematics to "London's practitioners as well as its scholars."

[17] Bennett (1986, 10–11); Hill (1998, 253).

fore now, in that tungue."[18] In his 1556 *Tectonicon*, an elementary technical treatise, Leonard Digges explains his intention to write his forthcoming *Pantometria* for "all maner men of this realme" and for making accessible "those rules hidde, and as it were locked up in straunge tongues."[19] Billingsley desired that "our Englishe tounge shall no lesse be enriched with good Authors, then are other straunge tounges."[20]

But these authors did not aim merely to reproduce the same obscure knowledge in a different language. Rather, as Thomas Digges writes of his father's intentions, works were "compiled in the Englishe tong, desiring rather with plaine and profitable conclusions to store his native language and benefite his Countrey men, than by publishing in the Latin rare and curiouse demonstrations, to purchase fame among straungers."[21] In the second edition of *Pantometria*, Digges further declares his resolve to publish "onely in my Native Language: Aswell to make the benefite thereof the more private to my Countreymen, as also to make thereby other Nations to affect as much our Language as my selfe have desired to learne the Highe Dutche."[22] Recorde and Thomas Digges both saw mathematics and a mathematically literate public as important elements of statecraft.[23] Dee adds the aims that this *"Englishe Geometrie"* would occupy those with sharp wits but lacking philosophical inclinations and simultaneously serve to increase the prestige of university mathematics among the general public.[24]

The geometry our authors made was a local geometry, valid in particular ways for its particular users. It was also, however, a global geometry which, though newly minted, could be traced in the authors' prefaces as far back as Archimedes's defense of Syracuse.[25] Our authors rendered geometric truth for English vernacular audiences by appealing in multiple ways to their audiences' situated and local experiences of the art. Their task was to render as geometry the multifarious knowledges and practices brought to bear by their vernacular readers.

It need hardly be mentioned that the translations did not emerge in a vacuum. Indeed, the annotated text from which Billingsley prepared the bulk of his translation survives to this day – a 1548 edition of Zamberti's Latin translation of Theon's version of the *Elements*.[26] One can infer from the others' writings that they were well read in the mathematics of their contemporaries,

---

[18] Sig.*a*2^r. [19] Sig.π2^r. [20] Sig.[fist]3^r. [21] Sig.*A*4^v.

[22] Digges (1591, 176). Recorde, too, published only in English. Johnson and Larkey (1935, 85).

[23] Easton (1966, 354–355; 1967, 523), Hill (1998, 256), Feingold (1984, 186, 206–207).

[24] Sig.*A*4^r. [25] Recorde Sig.†2^v–†3^v, Digges Sig.*A*3^v, Dee Sig.*C*4^v.

[26] Archibald (1950) reviews the history and historiography of Billingsley's sources. See also Feingold (1984, 158).

both in England and on the Continent. Recorde, for instance, refers in his text to claims made by German near-contemporary Albrecht Dürer.[27] Moreover, the personal and scholarly ties joining the translators makes it reasonable to assume that they had access to a similar corpus of mathematical works. Beyond that, however, it is difficult to untangle the variety of sources upon which they drew for their geometric works – certainly, such a task is beyond the scope of the present paper.

Even in the case of their diagrams, which could presumably be transfered more recognizably from their various sources, one can assert little beyond the observation that, broadly speaking, there is little that is particularly innovative about the representational strategies employed in the English translations. Features identified in the discussion below can, with few exceptions, be found in prior works in other languages, both from the overtly Euclidean volumes[28] and from other geometric texts.[29] Each of the English translations combines images and motifs identifiable in multiple prior works. In light of the common visual vocabulary upon which our five translators could draw, along with the convergence in pedagogical intent among all but Thomas Digges, it is all the more remarkable how different their works appear.

My analysis takes these authors' aims and dispositions, drawn largely from the writings of the authors themselves, as its starting point. My goal is not to evaluate the success or failure of these authors, nor to assess their influence and influences, nor even to address the surely complicated matters of authorship and responsibility for the various words and figures of their respective texts. Taking the works' attributed authors at their word, the ensuing analysis explores what can be learned by contrasting the different representations as they stand before us.

Each author set out to fashion a new English geometry on the back of Euclid's *Elements*. I shall interrogate their texts in order to shed light on what they deemed necessary in order to accomplish such a monumental task. The result will not account for the images, nor will it be simply an accounting of them. Rather, it will comprise a first inquiry into how the images account for geometry. My question shall be how Euclid's diagrams, in our broad sense of the word, were translated for an English vernacular readership. How, in other words, was the visual vocabulary of the *Elements* made meaningful for this (at least purportedly) new audience? The different strategies employed by the texts under consideration offer a view of the choices each author made and the range of strategies each author set aside in his attempt to render a vernacular geometry.

---

[27] Sig.$A4^v$.   [28] For example, Ratdolt (1482), Pacioli (1523), Grynaeus (1533), Hirschvogel (1543), Benedetti (1553), or Xylander (1562).
[29] For example, Nemorarius (1533), Fine (1544), Cardano (1554), or Frisius (1557).

4.3 Recorde's English Geometry

Robert Recorde's *Pathway to Knowledg* presented a geometry steeped in the familiar trades and practices of men in all walks of life, whether or not they were potential readers of his book. To help establish that geometry truly was everywhere, his preface lists no fewer than sixteen 'unlearned' professions which, he claimed, already relied on the subject.[30] Users of geometry include the commoner, the deity, the contemporary, the ancient, and (implicitly) everyone in between. "Ceres and Pallas," for instance, join a congregation of Ancient figures who "were called goddes" for teaching little more than geometry's applications, and Galen "coulde never cure well a rounde ulcere, till reason geometricall didde teache it him."[31]

Yet Recorde's geometry consisted of a mass of terms, methods, and ways of organizing knowledge which had never before appeared in the English language. To bridge this gap between theoretical knowledge and purported practice, Recorde enlisted both words and images. Diagrams and illustrations in the early pages of Recorde's exposition are laden with extra contextualizing details. Thus, 'A twiste line' is shown wrapped about a column and a right angle in a construction is shown against a drafting square. Even abstract shapes are drawn with hatching in order to give a sense of depth and form (figures 4.1(a)–(c)).

Definitions are illustrated with figures that can also stand alone without the expository text surrounding them. Typographically differentiated terms from Recorde's exposition match copious labels attached to the figured objects being defined.[32] Such typographical cues create an explicit link between text and figure, and in so doing they establish parallel functions for the textual definition and its associated diagram. Thus, the components of the figure are not just semantically but also structurally mapped onto the components of the definition. The structural authority thus acquired by the definitional diagram makes it a credible stand-in for its textual counterpart. Particularly in a setting where geometry's rhetorical formulae had not entered the vernacular, such an elevated role for diagrams offered a crucial conceptual bridge for Recorde's readers.

Nor did Recorde's diagrams stop at merely illustrating individual terms. In many cases, definitional figures show the multiplicity and variety encompassed by the term or terms in question – something the written text would be hard-pressed to do without distractingly wordy descriptions. Individual concepts are instantiated, in Recorde's illustrations, by a sometimes-vast variety of

---

[30] Sig.[ez]4^v–†1^v.   [31] Sig.†1^v–2^r.
[32] See figures 4.1(a)(b)(d) and 4.27 (below). In 4.1(d), the label for 'A corde' was included without the cord even being drawn into the figure.

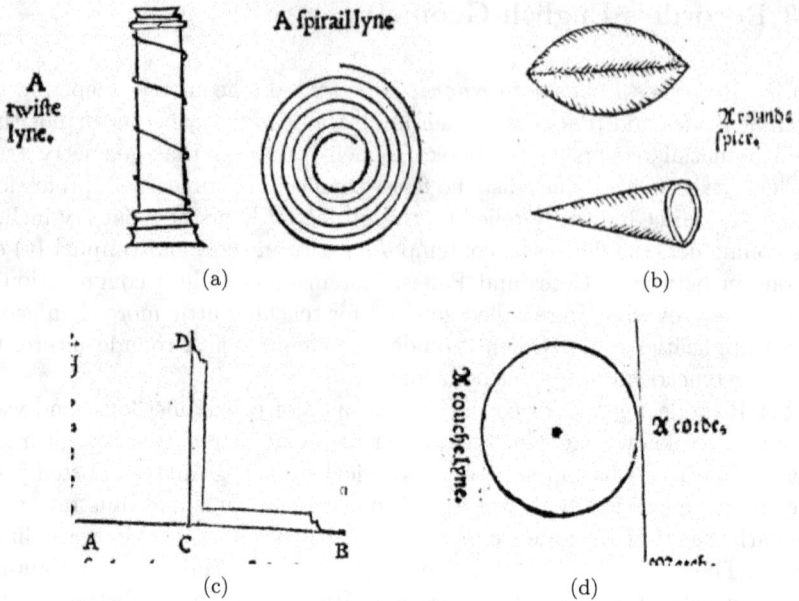

A fpiraillyne

A twifte lyne.

Xrounds fpicr.

(a)   (b)

X touchelyne

X co:dc,

D

A   C   B

(c)   (d)

Fig. 4.1: Definition and construction figures with contextualizing details and internal captions from Recorde (1551): (a) twist and spiral lines, Sig.$A4^v$; (b) two three-dimensional shapes, Sig.$C1^r$; (c) a construction using a drafting square, Sig.$D1^v$; (d) a tangent ('touche') line, Sig.$B1^r$

cases. In some places, images are reused to illustrate multiple phenomena, as when Recorde's exemplary 'spirail line' joins a dizzying array of 'croked' ones (figure 4.2). Here, expediency for the printer reinforces the mathematical principle that the same object can belong to many geometric classes.

According to Recorde, the diagrams establish abstract geometric concepts on the basis of "such undowbtfull and sensible principles."[33] It is important that this approach is emphasized at an early stage in the text. On the first page of his definitions, for example, Recorde explains that a line is composed of points[34] by saying that "if you with your pen will set in more other prickes betweene everye two of these [in the dotted line above], then wil it be a line."[35] His demonstration purports to explain the composition of both the particular dotted and solid lines on the page and the general concept of a geometric line. He thus transforms the familiar action of drawing points and

---

[33] Sig.$c1^r$.

[34] See Alexander (1995, 580–581) for contemporary disputes over the composition of the continuum.

[35] Sig.$A1^r$.

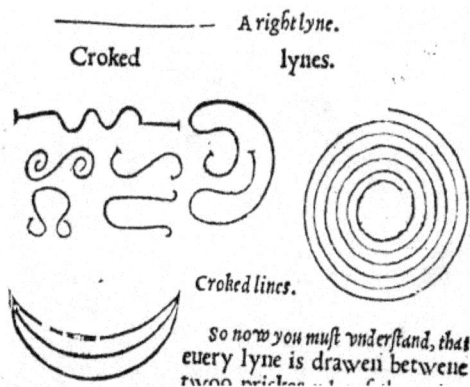

Fig. 4.2: Definition figure for crooked lines from Recorde (1551, Sig.$A1^v$)

lines from the mere production of marks on the page into the production of geometry. The reader can see that all lines are composed of points through a practice so intuitive that it need not actually be performed. That such practices need only be performed hypothetically becomes important later on, where assertions which cannot be verified with a few dots of a pen nonetheless inherit the same degree of clarity and obviousness as those which can.

This same reliance on what was readily sensible led Recorde to exercise caution where the diagrams might introduce ambiguities. He explains the dangers of deceptive figures in his discussion of right angles (figure 4.3). Though "angles (as you see) are made partly of streght lines, partly of croked lines, and partly of both together," his illustrations of right angles show only straight lines, "because it would muche trouble a lerner to judge them: for their true judgment doth appertaine... rather to reason then to sense."[36] Indeed, geometry was, for Recorde, a foundation for reason, not something reason could teach. As the more fundamental subject, geometry was to be grounded in the already-meaningful and already-obvious – that is, it was to be grounded in the purely sensible.

Recorde's use of an excess of exemplars in order to define geometric entities was and remains a widely employed practice. But Recorde does not stop his parade of multiplicity as he embarks on constructions, which in other contexts are illustrated and discussed in just one putatively generic case. Emphasizing the development of an intuitive comprehension of geometric truths over their rigorous proof-based establishment, Recorde frequently offers multiple constructions for a single proposition or problem. His diagrams show how the construction should be applied in different situations and imply that the ge-

---

[36] Sig.$A1^v–A2^v$.

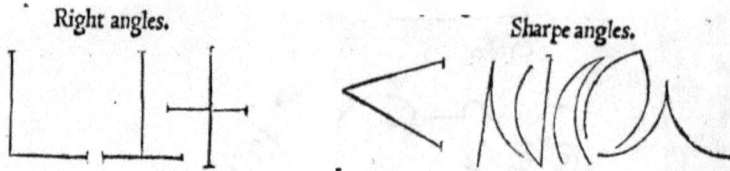

Fig. 4.3: Definition figures for right and sharp angles from Recorde (1551, Sig.$A2^r$)

ometric conclusion is no mere byproduct of an overly facile case study while suggesting how other geometric case studies might be applied beyond the cases in which Recorde considers them explicitly.

Pathway also extends the familiarity-granting depiction of the everyday objects of practical geometry to the constructions later in the text. Thus, Recorde depicts a window arch with a hanging plumbline to accompany three constructions for bisecting a semicircle using different surveying tools, including compasses and drafting squares. He illustrates Euclid's petition to construct a circle from a point and a radius with a picture of a compass, rather than the series of embedded circles, sometimes accompanied by radii, used in nearly all of the other Euclidean texts of his period (figures 4.1(c), 4.4; cf. figure 4.10).

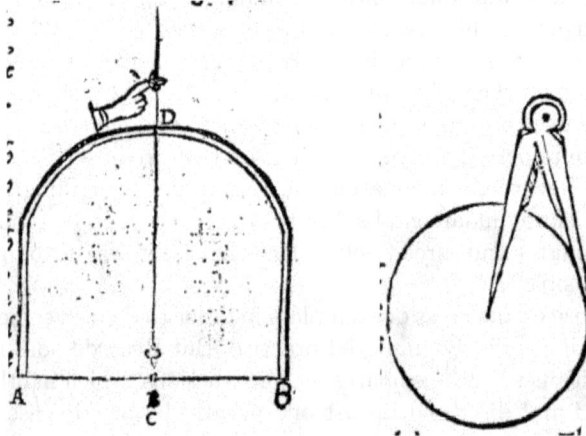

Fig. 4.4: Constructions using, respectively, a plumb line and a compass from Recorde (1551, Sig.$D1^v$, $b1^r$)

Diagrams can, moreover, indicate information that is omitted in the text, as where Recorde's diagrams show that he intends his theorems about triangles to apply only to ones formed of straight lines, even though his definitions stress that triangles can also be formed using curved, or crooked, lines. Recorde's diagrams rarely include numbers, and while some numbers appear to correspond with their associated measures in the diagram,[37] others correspond hardly at all to the proportions of the drawn figure itself (e.g. figure 4.5). With his proofs, Recorde is explicit about having

> drawen in the Linearic examples many times more lines, than be spoken of in the explication of them, whiche is doone to this intent, that if any manne list to learne the demonstrations by harte,...those same men should finde the Linearic exaumples to serve for this purpose, and to want no thing needefull to the juste proofe....[38]

This practice is evident in his figure for the Pythagorean theorem, discussed below (page 141), but we shall also see more examples where pluralistic considerations lead to the inclusion of many more diagram elements than would be used in more conventional Euclidean proofs such as Billingsley's (see figures 4.7 and 4.11).

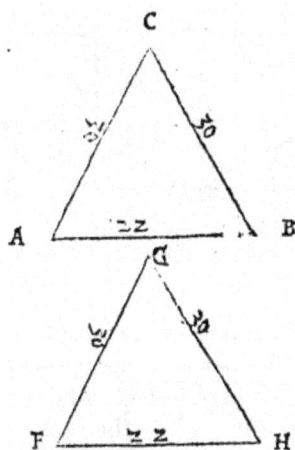

Fig. 4.5: Triangles with specified measures from Recorde (1551, Sig.$c1^v$)

Pedagogical through and through, Recorde's text works by guiding the reader through concepts using explicitly exemplary situations. For Euclid's common notions relating to equalities of magnitude, Recorde uses the areas of rectangles and triangles as his case studies (figure 4.6). For the sixth common notion, that two doubles of the same thing are equal to each other, Recorde's diagram includes two copies of the doubled rectangle. These are

---

[37] e.g. Sig. $b2^v$ and $b3^v$, both of which show tick-marks, and $e4^r$, for the Pythagorean theorem, discussed below.

[38] Sig $a3^r$–$a3^v$.

arranged alongside two larger rectangles formed by joining the smaller ones along different edges, showing both that geometric objects can be doubled in multiple ways and underscoring that either doubling produces the same new area. As with the definitions discussed above, the diagrams for Recorde's common notions aim to establish not just the legitimacy of the claim but also its scope and purpose. Recorde's readers had to be convinced that it was meaningful to compare different ways of doubling an object before examining those comparisons, just as it was necessary to exhibit a multitude of angles and shapes before embarking on their systematic classification.

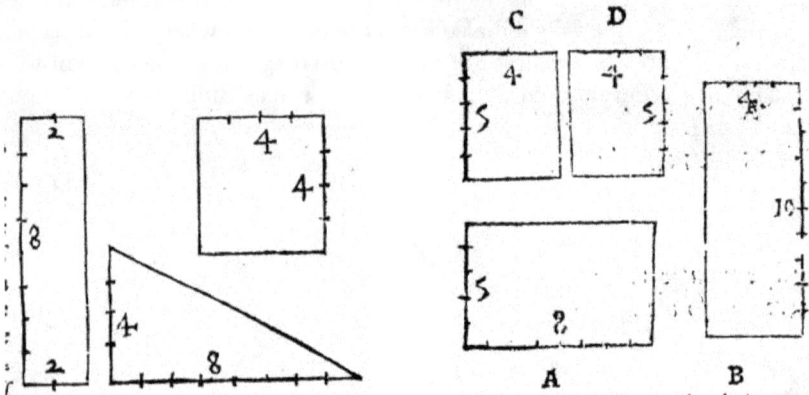

Fig. 4.6: Figures for two common notions from Recorde (1551, Sig.$b2^v$, $b3^v$)

Elsewhere, Recorde puts his images to multiple use by illustrating a method of partitioning polygons using parallel lines in a construction involving triangulation (figure 4.7). One diagram indicates how to triangulate simple polygons of increasingly many edges while the other shows a large selection of more complicated polygons which suggests the general applicability of the construction and a practical means of applying it. Finally, Recorde takes care to show why some possible exceptions to his geometric principles are not so. This involves showing variations on a theorem which fail to hold (see figure 4.11 below) and demonstrating how two straight lines cannot enclose a region by showing a regions and non-regions made of different combinations of two curved or straight lines (figure 4.8).

The diagrams in Recorde's *Pathway to Knowledg* are thus made to perform a variety of functions as a crucial supplement to the text. His images establish the legitimacy, meaningfulness, and familiarity of everything from simple geometric objects to relatively complex assertions, constructions, and theorems. Recorde exhibits a geometry addressed to a dazzling array of shapes and objects from both the geometric world and the everyday one. His fig-

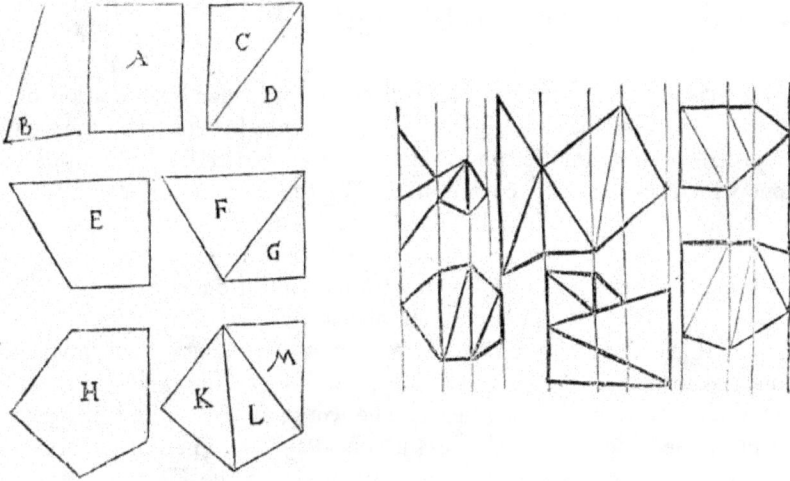

Fig. 4.7: Constructions involving triangulation from Recorde (1551, Sig.$E1^v$–$E2^r$)

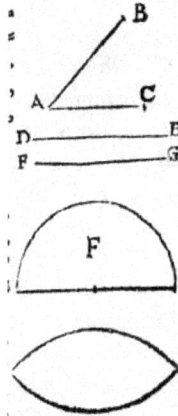

Fig. 4.8: Pairs of lines from Recorde (1551, Sig.$b2^r$)

ures justify geometry through its contexts while simultaneously showing how such contexts are to be translated and manipulated according to geometric conventions. Many of these aspects appear in different ways in the figures of subsequent texts, and Recorde's work offers a rich template against which to set later English geometries.

4.4 Euclid According to Billingsley

At first glance, the illustrations in Billingsley's compendious edition of the *Elements* are unremarkable. He holds close to what by that time were highly standardized diagrammatic conventions in the manuscript and even print traditions which preceded his contribution. For the purposes of this essay, and without aiming to describe Billingsley's many influences, it should suffice to note that the 'look and feel' of Billingsley's (albeit exceedingly thorough and well-appointed) text does not depart dramatically from other authoritative versions of the *Elements* in circulation at his time.

The text does not arouse our interest for its representational innovations so much as for the means by which it deploys its very unoriginal illustrations to serve an utterly original audience. The work's diagrams and figures must have been the object of much careful consideration. The author attests to the "charge & great travaile" incurred in translating the *Elements*, stating in the text's frontmatter that "I have added easie and plaine declarations and examples by figures, of the definitions."[39] The book is copiously illustrated, and no cost was spared in annexing images to proofs, definitions, scholia, examples, and other textual features. Where proofs span a page-turn, their corresponding diagrams are typically copied over so that they are always visible when following the proof.[40] To help the reader grasp three-dimensional shapes, Billingsley adds to his two-dimensional diagrams a parallel set which use fold-out flaps so that the shapes literally pop out of the page. This latter was perhaps the most distinctive of Billingsley's arsenal of illustrative tools.

Billingsley's representational strategies are best seen in contrast to Recorde's. Although they claim in their prefaces to be writing for similar audiences and to similar ends, it is not hard to see where their purposes diverge. The difference goes all the way down to what sort of geometry they would have their readers learn. As a case in point, contrast Recorde's approach to multiple representations of a triangulation procedure (figure 4.7 above) to Billingsley's treatment of the proof for Euclid's second proposition, which concerns the reproduction of a line segment at a new location.[41] Recorde shows a suitably representative variety of case studies for his procedure, addressing it to increasingly complex polygons in order to inspire confidence in the method's general applicability and point to how it might be so applied. Billingsley, by contrast, exhausts all of the logical possibilities for the proof's diagram, showing how the respective diagram for each of four cases is related to the proof's text (figure 4.9). Recorde prizes instructiveness, Billingsley completeness.

At the same time, Billingsley's completeness is necessarily a qualified one. Previous editions of the *Elements* in other languages break proofs down by

---

[39] Sig.[fist]2^v.   [40] This feature can also be found, albeit less frequently, as far back as the 1482 Ratdolt text.   [41] Fol.10^r–11^r.

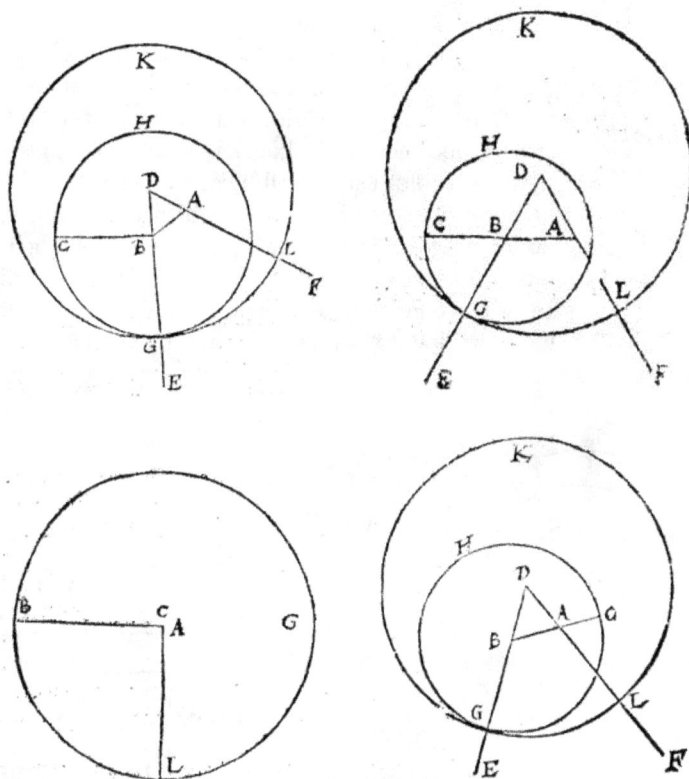

Fig. 4.9: Billingsley's (1570) four cases for Proposition 2. Fol.10^v–11^r

their possible diagrams in similar ways, but Billingsley appears particularly zealous in treating proposition I.2 in this manner. Just as Recorde uses demonstrations by simple manual practices such as drawing early in his text where such demonstrations are still simple and plausible, Billingsley can only afford to be exhaustive with such proofs at a relatively early stage. Thus, Billingsley's detailed demonstration of proposition I.2 manages to stand in for the great range of demonstrations where such a consideration would be prohibitively impractical. He shows how one diagram and argument can stand for many in this simple case so as to avoid having to do so for later ones.

Recorde and Billingsley's different approaches to instructiveness and completeness play out dramatically in the different illustrations they attach to Euclid's common notions. Where Recorde instantiates the principle with a specific example which illustrates and justifies the claim (as in figure 4.6 above), Billingsley aspires to the most abstract possible representation by joining the claim to images of appropriately related line segments (figure 4.10). Here, Billingsley appropriates a convention found in many prior Euclidean texts from later books of the *Elements* where quantities are depicted as linear magnitudes, often arranged in series next to each other for ease of comparison. The segments become, for Billingsley, standard representatives for any type of quantity, and could just as well be of any length or dimension. In Billingsley's metonymy of magnitude, "a line, which is the first kynde of quantitie,"[42] stands in for all geometric quantities. The letters labeling their lengths reinforce this point, as well as the convention in geometric diagrams of using such labels to produce general geometric arguments. Recorde's rectangles, on the other hand, work only as rectangles and have specific numerical sizes associated to them.

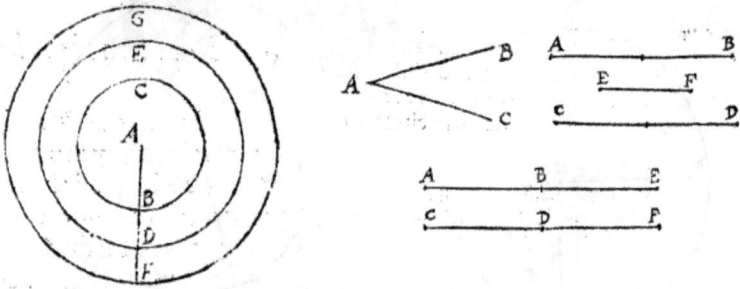

Fig. 4.10: Illustrations of postulates and common notions from Billingsley (1570) Fol.6^r–7^r

Similarly, Recorde depicts a compass to show how circles of arbitrary center and radius may be made (figure 4.4 above), whereas Billingsley asserts their multiplicity by drawing a nested collection of three circles with a common center described along a single radial line (figure 4.10). Where Recorde shows how the postulated circles can be produced, it is enough for Billingsley to assert that they can be. Like Recorde (figure 4.8 above), Billingsley shows two straight lines failing to make a surface (figure 4.10), but does not show the feat being accomplished when lines are allowed to bend. As before, it is a simple fact in Billingsley's presentation that straight lines cannot enclose a surface; he does not strive like Recorde to graphically detail the scope and import of the claim as he makes it.

The same disanalogy applies to the theorems of the two works. Billingsley's and Recorde's illustrations for Euclid's theorem that a pair of circles can cross at most twice[43] both depict a circle crossed four times by an eye shape,

---

[42] Fol.1^v. [43] Theorem *lv* (Sig.*i*2^r) in Recorde, book 3 proposition 10 (Fol.89^r) in Billingsley.

the standard figure (with very few exceptions) for this proof in the Euclidean canon. This non-example provides a starting point for a proof by contradiction which is spelled out in Billingsley's translation but only hinted at by Recorde. Recorde, however, also includes another circle of the same size to show how circles do indeed cross,[44] as well as an ovular 'tunne forme' to show, along with the eye form, that only 'irregulare formes' may violate the theorem. He thus adds a surplus of pedagogic detail to facilitate understanding of the range of the theorem's implications. Billingsley's figure is an accessory to the proof of the theorem, never seeking to show more than the relationships between different objects cited in the proof and providing a means of visualizing the series of letters and shapes to which the textual demonstration refers. A similar contrast in approaches is present throughout Recorde's and Billingsley's works, reaching all the way back to the definitions.

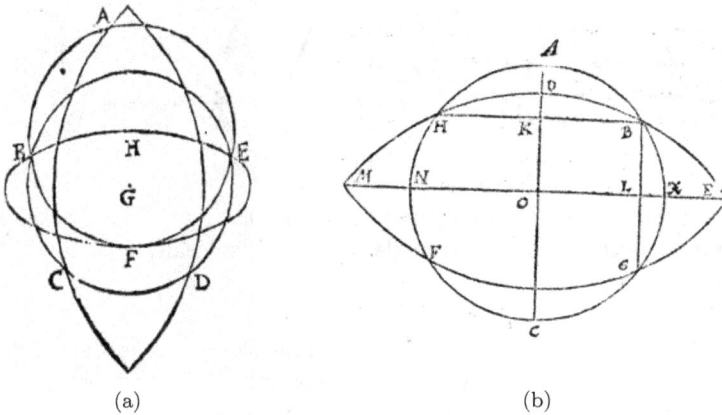

Fig. 4.11: Proofs about intersecting circles: (a) Recorde (1551, Sig.$i.2^r$), (b) Billingsley (1570, Fol.89^r–89^v)

This is not to say that Billingsley wholly disregards pedagogic considerations or indications of how geometry might look in practice. Both Recorde and Billingsley include figures demonstrating how compass marks might economically be produced in service of a construction (figure 4.12). In Recorde's case, the construction is a practical non-rigorous shortcut. For Billingsley, on the other hand, the figures showing compass marks indicate how only certain arcs of circles need be drawn in order 'readily' to produce triangles in good Euclidean form – they are not allowed to stand in for the thoroughly Euclidean constructions in later proofs.

---

[44] Pacioli (1523) shows only two circles crossing at two points for this theorem.

(a)

(b)

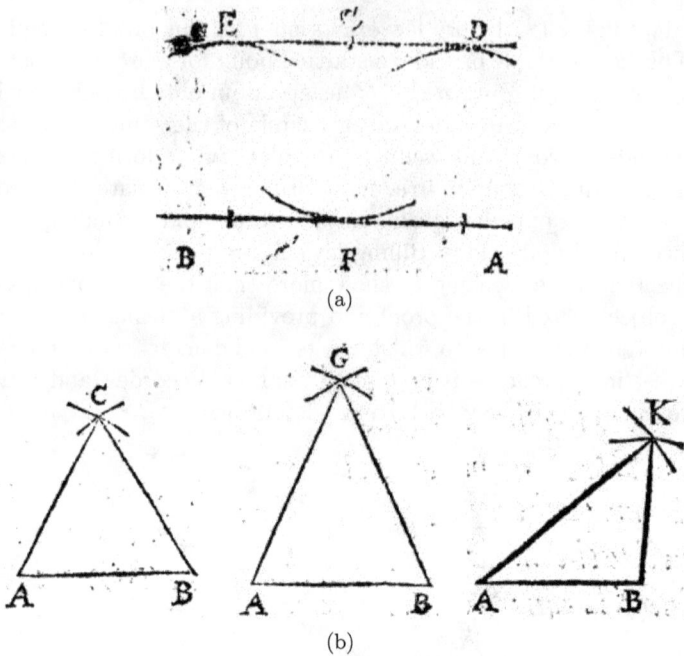

Fig. 4.12: Illustrations involving compass marks: (a) Recorde (1551, Sig.$D2^r$),
(b) Billingsley (1570, Fol.10^r, triangles arranged vertically in original)

The comparison between Recorde and Billingsley takes another dimension
in their diagrams for the Pythagorean theorem, relating the lengths of the
sides of a right triangle. Recorde builds his figure from a right triangle whose
sides are in the ratio of 3-4-5, dividing each side and its associated square
accordingly. He writes that "by the numbre of the divisions in eche of these
squares, may you perceave not onely what the square of any line is called,
but also that the theoreme is true, and expressed plainly bothe by lines and
numbre."[45] Because his aim is to illustrate the theorem in as comprehensible and multifarious a way as possible, Recorde depicts a right triangle with
the simplest combination of sides whose lengths are related by ordinary ratios of integers. This allows him to make the demonstration "bothe by lines
and numbre" in a readily graspable format, and his textual explanation is a
step-by-step description identifying the features of the figure with the general
claims of the proposition – first identifying the shapes, then showing how to
read their respective areas from the diagram, and finally affirming that the

---

[45] Sig.e4^r.

proposition is satisfied in the depicted case before describing how to use the proposition to find unknown sides for other right triangles.

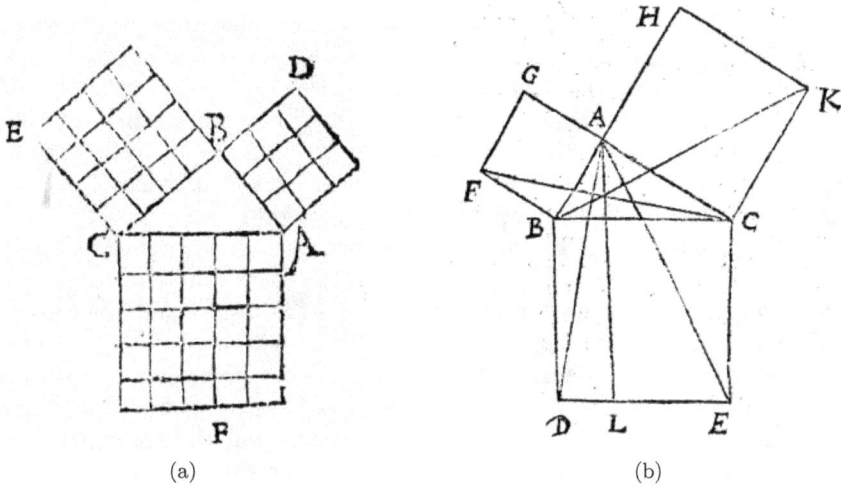

Fig. 4.13: Diagrams for the Pythagorean theorem: (a) Recorde (1551, Sig.e4^r), (b) Billingsley (1570, Fol.58^r)

This figure is a partial exception to Recorde's rule of adding extra details to his figures for those who would learn their conventional proofs. The standard Euclidean proof, corresponding to Billingsley's diagram, requires a number of auxiliary lines to allow the areas of the squares to be compared by a means other than counting unit squares – something which would not even be possible with the triangle in Billingsley's figure because his sides and hypotenuse do not appear to correspond to any simple Pythagorean triple of integers when measured. Recorde's diagram, however, manages to invoke its Euclidean counterpart. The orientation of the triangles and squares is an obvious parallel. Easier to miss, Recorde labels a point 'F' at the bottom of his diagram which corresponds not to any of the corners or crossings of his figure but to the point labeled as 'L' from the vertical auxiliary line in Billingsley's figure.

In the case of the Pythagorean theorem, as, indeed, with most of Euclid's propositions, Recorde's text can hardly be construed to provide even the outline of a conventionally rigorous argument. Recorde's gestures at this distant rigor – the 'F' label, the extra lines in other diagrams – point rather to an ideal of what geometry is and what it is about. *Pathway*'s readers did not learn to prove, but they saw what proofs looked like and, perhaps more importantly, they saw what proofs could show. Recorde's is thus manifestly a geometry of showing and, insofar as it was practical for his vernacular readers, of doing as well.

Billingsley's geometry, like Recorde's, aims to explain the meaning and value of both geometry's results and practices. Unlike Recorde, however, Billingsley insists on doing geometry even when it is not a simple matter of filling in a dotted line or hanging a ball of lead from an archway. In this sense, Billingsley's text appears to us as a work *of* geometry, while Recorde's seems more *about* geometry. This reflects, in large part, the different pedagogical approaches taken by the two translators. But it also reveals a bias I would like to suggest is distinctly posterior to these writings.

For both Recorde and Billingsley, proofs are essential to geometric knowledge. Recorde, however, presents a geometry in which man's senses and actions are prior even to the proofs. Focusing on the results and applications of Euclidean geometry as they are available to perception, *Pathway* need not be seen as deficient for lacking the sort of rigor later imagined as the heart of the geometric method. Rather, Recorde's geometry treats first things first: the sensible takes priority over the rational throughout the book, just as it did in Recorde's definitions. Billingsley, then, departs from Recorde only insofar as he gradually allows the rational to assert itself where the senses do not suffice. This is not to argue that this one contrast need overthrow our present received view of Early Modern Euclidean rigor, but to suggest that other readings are possible, and indeed may account for some features of texts that might otherwise pass without notice.

Before outlining the dramatically different illustrative strategies in Digges's *Pantometria*, a few words on those of Dee's preface are in order. Indeed, Dee's preface is striking for its lack of geometric illustrations, and contains only three small diagrams relating to geometry's applications and his famous taxonomic diagram of the mathematical arts and sciences, despite treating several concepts related to some which Billingsley finds necessary to illustrate extensively. Dee's text thus stands as an important corrective for the comparisons undertaken in this article, a role which will become more explicit when all three volumes are directly compared at the end. For Dee, the relationships between mathematical objects and sciences are to be understood logically and schematically, not diagrammatically. Even the diagrams he does use, including the fold-out diagram at the end of his preface, emphasize that it is the order of the mathematical sciences which is at the heart of his work, not the understanding of their constituents.

4.5 Digges's Geometry in Context

Like Dee's preface, Thomas Digges's appended discourse on geometric solids is only sparsely illustrated. His definitions include canonical projections of Platonic solids along with ornately lettered name labels (figure 4.14) and some other geometric projections and line diagrams for complicated constructions and calculations appear later. But there is little to indicate that he intends readers to understand his results with the help of in-text diagrams. As a compendium of new mathematical results, Digges's discourse is the least pedagogical of the texts here considered, and its transmission of Euclid is more by way of form and style than textual content or imagery. Thus, he begins with definitions and presents results in the form of Euclidean propositions, using Euclidean terminology and rhetoric throughout.

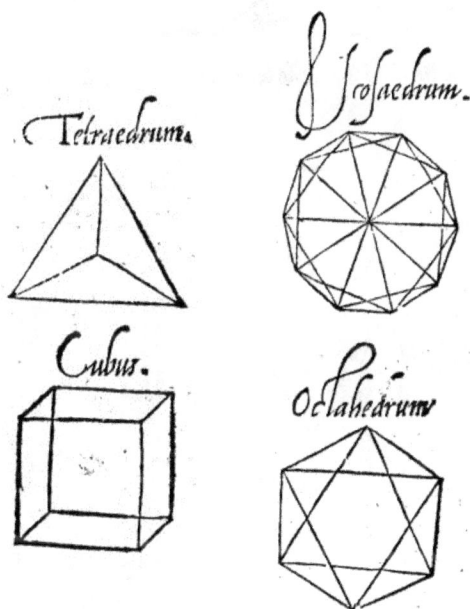

Fig. 4.14: Images of polyhedra from Thomas Digges's (1571, discourse, Sig.$T2^r$–$T2^v$)

The *Pantometria*, on the other hand, is richly illustrated with conventional geometric figures and examples, plans for surveying instruments, and, above all, detailed scenes of geometry in practice. Leonard Digges's definitions, like Recorde's, are illustrated and labeled in a way that allows them to stand on their own without textual explanations. Indeed, the language of Digges's definitions is so spare and unelaborated in comparison with the rest of the text and with Billingsley and Recorde's renderings that it can easily be seen as secondary to the diagrams, a perfunctory nod to the norms of Euclidean exposition. Without much aid from their surrounding text, the diagrams of

the *Pantometria* systematically illustrate geometric concepts and their relationships. They do not, like Recorde's, aim to show the wide variety of objects
under consideration, but rather depict each concept in a single case in order
to establish a working vocabulary.

The B;ode o; Obtuse Angle is greater than the Orthogonal.

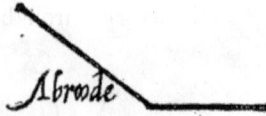

A brode

The Acute o; fharpe, is leffer than the right angle.

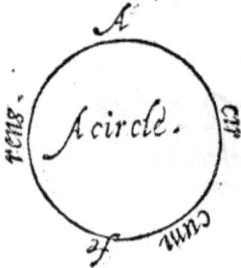

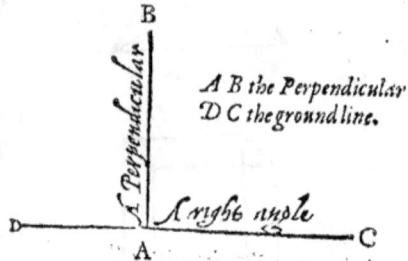

a sharpe

A circle.

B

A Perpendicular

A B the Perpendicular
D C the groundline.

A right angle

D ——————— A ——————————— C

Fig. 4.15: Definitional figures from Digges (1571, Sig.$B1^v$–$B2^r$)

These single cases, however, are not portrayed in isolation. Even Digges's
simplest definitional images are arranged in what might be called 'conceptual
scenes' which show how his concepts are related. Thus, a point appears alongside two types of lines, and terms related to angles, circles, or perpendicular
lines are joined in single composite images (figure 4.15 and figure 4.28 below).
The diagrams establish a touchpoint for a new geometric vocabulary, and
help the reader to systematize the large variety of new definitions by visually
associating related terms and images.

The geometric concepts of the definitions are then given contextual meaning within elaborate scenes of surveying and warfare. These scenes impose geometric lines, measures, and instruments on landscapes and in settings where
they might be used. They often contain additional buildings, people, stat-
ues, and decorations, in many cases significantly more stylized than the ob-

jects most closely implicated in the geometry under consideration, in order to establish the setting.[46] In some cases, geometric measurements are made by surveyors (figure 4.16 shows a surveyor making three measurements over time), but in other scenes geometric features are superimosed on the landscape without the benefit of an instrument, observer, or either of the two (e.g. figure 4.17). Additional people in the scenes help stage the measurement either directly, as where a finely attired woman gestures at the surveyor (figure 4.18), or indirectly, as where armies stand and wait for the geometer to finish his work (figure 4.19) or hunting parties chase game which is perhaps to be served in the hall being measured (figure 4.20). The latter includes hunting parties in both the foreground and background, corresponding to surveyors at either end of the hall.

Longimetra.

Fig. 4.16: Scene with time-lapse measurements from Digges (1571, Sig.$D1^r$)

The difficult work of bridging representational conventions in landscape art and geometry often creates striking oddities in the scenes. The figures from the text use a standard repertoire of techniques to establish depth and perspective, but these techniques are not applied to the geometric figures overlaying the landscapes. The result is that where the geometric figure itself has depth (that is, when it is not in the plane perpendicular to the viewer's line of sight) there is a visible incongruity with between the geometry and its scene. Attempting

---

[46] Many of these visual features can be found in, for instance, Pacioli 1523, Fine 1544, or Frisius 1557.

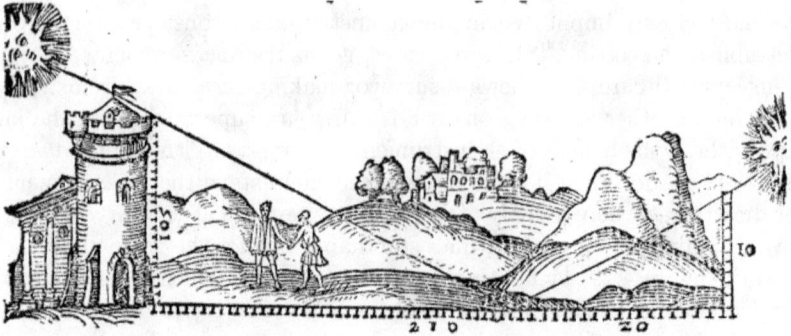

Fig. 4.17: Measuring a tower using the sun from Digges (1571, Sig.$D1^v$)

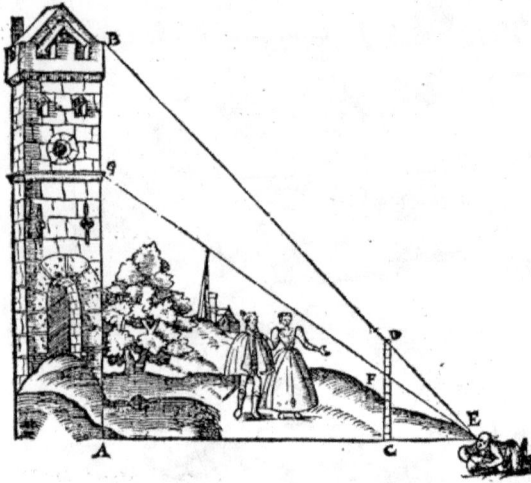

Fig. 4.18: Geometric scene with well-attired onlookers from Digges (1571, Sig.$E1^v$)

to establish an identity between geometric and landscape drawing, Digges's figures do not quite succeed in either. One can see, for instance, that the lines in figure 4.21 describe a right-triangular section of a pasture (not least because a draftsman's square is drawn in at one corner), but the pasture's nearby square corner appears obtuse from its perspectival rendering. Digges means to show, as in all of his situated figures, how geometry can be made manifest in otherwise familiar scenes, but shows somewhat inadvertently just how much work this manifestation entails.

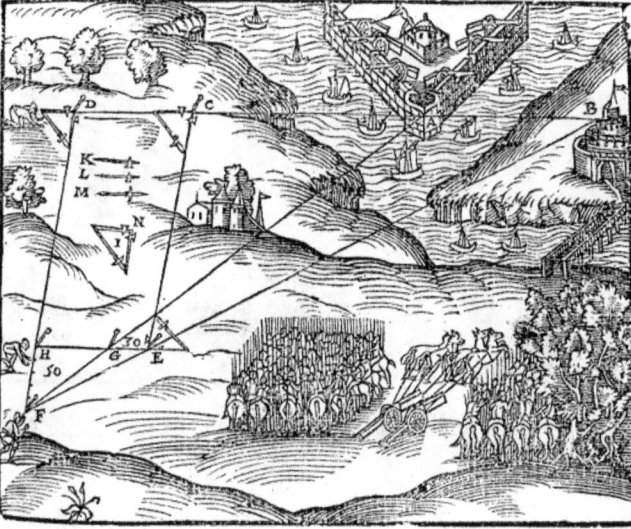

Fig. 4.19: A military scene from Digges (1571, Sig.$F2^r$)

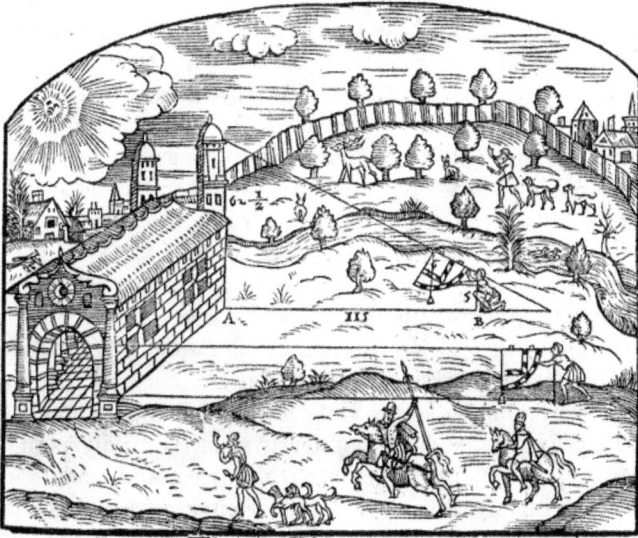

Fig. 4.20: Scene with hunters and a hall from Digges (1571, Sig.$D4^r$)

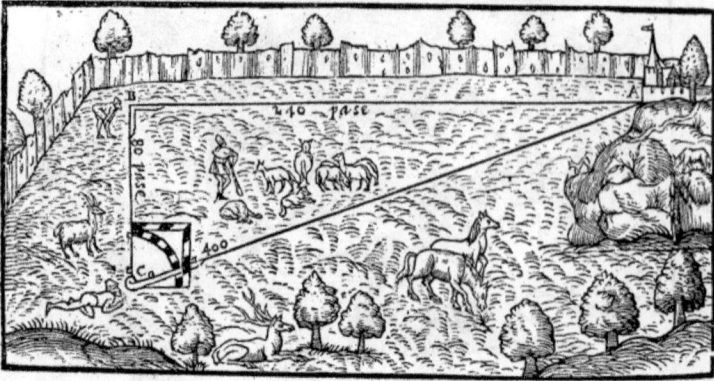

Fig. 4.21: Surveying a pasture, from Digges (1571, Sig.$G4^v$)

A similar tension emerges where depth is created by other means than perspective. Figure 4.22 depicts the determination of the "true water levell from a fountaine."[47] Digges makes a fountain's height plausible with a winding path, but the depth associated with the path vanishes in the fountain's geometrization. In scenes such as this, geometry is not necessarily made visually realistic, but rather is given a situational context where the geometer's figures and the viewer's scenes can comfortably (if not always naturally) coincide. More broadly, distortions of scale and other visual simplifications or embellishments in the *Pantometria* give rise to scenes which do not precisely depict actual users of geometry in their past or anticipated work. Rather, they conjure a constellation of images which appeal to geometry as a practical, worldly, and even glamorous endeavor and reinforce the plausibility of both the geometric methods themselves and their purported applications.

Nothing better represents the vexed nexus of geometry and familiar experience than the appearance of instruments in the *Pantometria*. Geometry, after all, was wholly alien to the work's vernacular readers, and was made less so by association with familiar scenes and contexts. Measuring and surveying instruments, on the other hand, are not nearly so otherworldly as the geometric entities they help to produce. They are real objects and readers may indeed have seen them without knowing their full role in geometry, but the work and its illustrations are also premised on the presumption that such instruments be also unfamiliar and outside of the normal experience, in both form and use, of Digges's audience.

Within the *Pantometria*, geometric instruments play a number of roles. In the work's many scenes, the instruments work both to establish geomet-

---

[47] Digges (1571, Sig.$K1^v$).

Fig. 4.22: Geometric scene with a fountain on a hill from Digges (1571, Sig.$K2^r$). (Here, 'fountain' is synonymous with 'well')

ric properties and show their manners of measurement. Thus, a draftsman's square produces right angles in two related ways: it shows which angles in a scene are right angles by virtue of their association with the instrument and it shows how such angles can be produced by the surveyor who would apply the lessons of the construction or calculation from the scene. Embedded quadrants have a similar function for non-right angles, and lengths are shown with regularly spaced marks along lines or with labels indicating a certain number of paces.

In this way, scenes show more than just contexts for the geometry of the *Pantometria*. They also use depictions of instruments within those contexts to bridge the scenes of the work and the sites of the work's potential application. Showing instruments in use, Digges also shows how they are to be used. This principle holds in the scenes discussed above, which depict instruments alongside their users, but it also applies where instruments are shown free of the surveyors or geometers who might use them. In figure 4.19, for instance, Digges shows how to produce an angle using "three staves, halberdes, billes, or any such like things, K L M"[48] and depicts several such arrangements, both on their own and being used by a surveyor. In figure 4.23, the geometric tools are given full-name labels within the scene and are shown performing a simple geometric measurement on their own, independent of the geometer's interventions.

---

[48] Sig.$F1^v$.

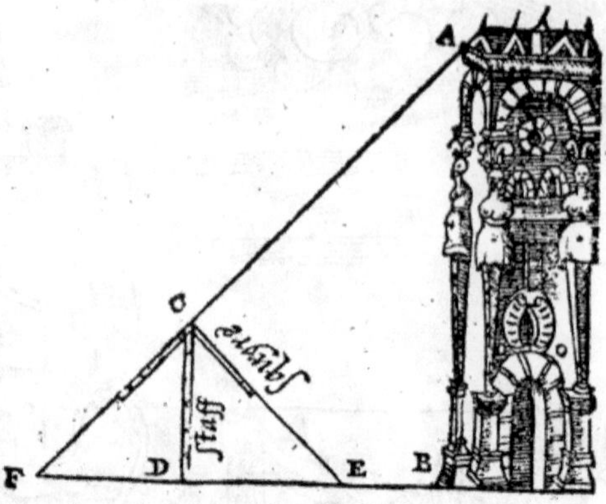

Fig. 4.23: Geometric scene with labeled instruments from Digges (1571, Sig.$E2^r$)

In some places, the line between geometric instrument and geometric object is blurred in Digges's diagrams. He refers, for instance, to smaller triangles which one could construct and measure as a replacement for cumbersome trigonometric calculations. These triangles are depicted within the larger scene in order to show all the geometer's resources in one and the same image, but appear out of scale so that their construction and dimensions can be more legibly rendered. Figure 4.24 has one such triangle with whose aid the geometer in the example measures a much larger similar triangle. Here, the auxiliary triangle floats somewhat apart from the scene, away from its users and in an otherwise unfilled part of the landscape. It is a necessary part of the calculation, but it could, implies Digges, be anywhere, at any scale. Artistic convenience here coincides with mathematical principles about similar triangles and their use in calculations across large scales. If the auxiliary figure is particularly complex or its associated context particularly difficult to depict, it might even be shown alone as a stand-in for a more detailed geometric scene, as is the case for the construction in figure 4.25 for an example involving the determination of distances between landmarks.

Measuring instruments themselves also appear in isolation and with considerable detail in the *Pantometria*. Illustrations such as those of figure 4.26 would have taught readers how to imagine the details of the coarsely schematized instruments in Digges's scenes as well as how to build such instruments for themselves. From the details of the images, one can discern something of the instrument's materials, features, and even assembly. The images accompany written instructions which guide the reader through each instrument's production, along with some indications regarding its use. *Pantometria* was, after all, a text of practical and applied geometry. In order to make Euclidean

Fig. 4.24: Scene with an auxiliary triangle from Digges (1571, Sig.$G3^v$)

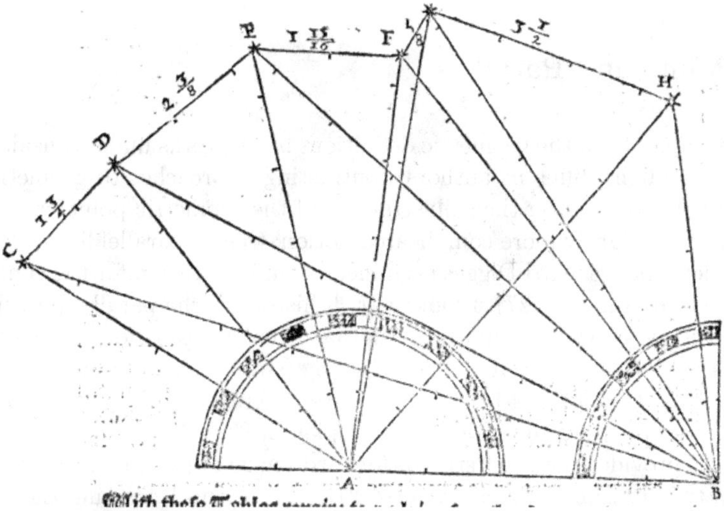

Fig. 4.25: Geometric figure for an extended computation from Digges (1571, Sig.$K4^v$)

geometry relevant to the *Pantometria*'s readers, the work had to fill in the gaps between Euclidean ideals and geometric experience. Instrument-making was, for these users of geometry, one of the most essential mechanisms for this gap-filling.

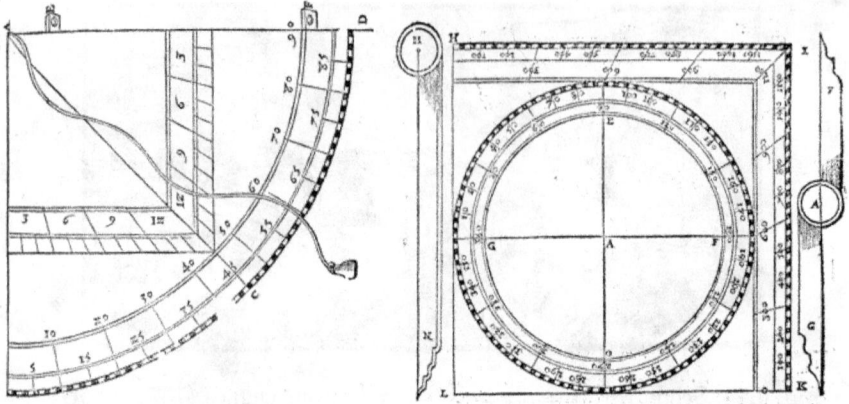

Fig. 4.26: A quadrant and a theodolite from Digges (1571, Sig.$C4^r$, $I2^r$)

4.6 Points and Parallels

Two examples from the geometric definitions in the works under consideration help to highlight different authors' contrasting approaches to geometric figures. Before considering their illustrations of the geometric point, it will help to examine a slightly more complicated notion: that of parallel lines. Recorde, Billingsley, and Leonard Digges each include an illustration for the concept of a parallel line (figure 4.27). From the rich history of the parallel postulate in Euclidean geometry it should be clear that what it means for two lines to be parallel is by no means self-evident. Staging parallel lines with different sorts of diagrams, the authors bring to the fore different aspects of the 'parallel' concept and different roles parallel lines play in Euclidean geometry.

Digges provides the most straightforward image of parallel lines. His figure depicts two horizontal line segments of the same length framing the caption 'Paralleles.' As with Digges's other definitional figures, this one serves to establish an operating vocabulary. The two depicted lines are parallels, and act as a point of reference for future invocations of the parallel concept without being exhaustive of all possibilities for its manifestation. Digges defines parallel lines as ones "so equedistantly placed"[49] that they never meet, and underscores this notion by making his example lines not just equidistantly placed but also of the same length. Parallel lines, according to Digges, are characterized by their levelness and their equalness, a message reinforced by the spare details of his image.

---

[49] Sig.$B4^r$. Equidistance seems an unusual criterion to modern readers, but is not hard to find in Early Modern texts. In addition to Digges and Recorde, their contemporary Petrus

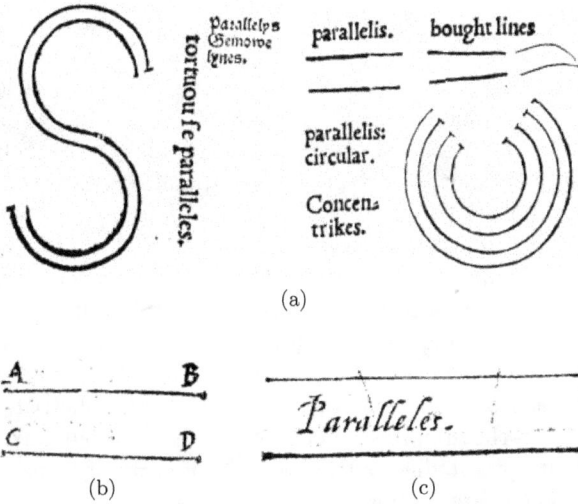

Fig. 4.27: Figures for parallel lines: (a) Recorde (1551, Sig.$A4^r$–$A4^v$), (b) Billingsley (1570, Fol.5^v), (c) Digges (1571, Sig.$B4^r$)

Parallel lines for Billingsley have a very similar image as for Digges: he too uses evenly spaced horizontal segments of the same length. But where Digges labels his with a full word, Billingsley marks his parallel lines with four letters placed at the endpoints of the two segments. In Billlingsley's text, the system of reference established by the letters – that of referring to lines by pairs of points therein – is more important than Digges's crude taxonomic nomenclature. That the lines are parallel is something to be proved or stipulated in the text with the help of labels and not something to be observed from the ostensible properties of their appearance on the page. Billingsley thus uses the labeling scheme for his figure to displace the property of parallelism into the written text, even as his image reproduces the same conceptual shorthands – evenness, levelness, equalness – as does Digges's.

Recorde also has an image of two horizontal segments with the label 'parallelis', but this is just the second of four labeled examples used to illustrate the concept. On the page where he introduces the parallel concept, the image is of two parallel S-shaped curves. In Recorde's textual definition, equal spacing is the paramount feature of parallel lines, and the 'tortuouse paralleles' of his first figure emphasize this point by showing that no matter where a line turns its parallel must turn with it in order to stay evenly spaced. As suggested above, Recorde's definitions deal far more in curved lines than do

Ramus (1580, 14) has equidistance as his explicit definition for parallel lines.

the subsequent parts of his text, and his definition for parallel lines is no exception. Curves allow Recorde economically to embed geometric multiplicity in relatively few figures. A single S-curve shows how parallel lines work in every direction and under any transformation.

Also in keeping with other definitions, Recorde illustrates the parallel concept contrastively by showing two pairs of non-parallel lines next to his canonical horizontal parallels. This pairing has two functions. First, it shows just how parallel lines differ from non-parallels. Even though Recorde's non-examples do not cross, one can see very clearly where they will meet by following the courses of the paired lines from left to right. Recorde thus provides a visual gloss to aid in recognizing parallelism and non-parallelism in figures where these properties might be ambiguous. But second, Recorde establishes the image of even horizontal lines as the canonical one for the parallel concept. Both by giving it the simplest label – 'parallelis' without any adjectives – and by setting it opposite his contrastive examples, Recorde makes horizontal parallels a default reference point for the concept.

Recorde's parallel concentric circles complete his suite of examples. No longer equal in length, these continue to show the 'equality' aspect of parallelism by covering the same angular scope. Moreover, these have marked endpoints in order to identify the appropriate corresponding points for judging even spacing, suggesting how such points might be used to gauge parallelism more generally. They show yet another way to judge parallelness, and emphasize that equal length is not the only possible clue indicating the parallelness of two lines. Recorde thus closes his illustration of the parallel concept by insisting, as he does throughout his definitions, on the concept's multiplicity and wide scope.

Unlike parallel lines, the simple geometric point appears at first to be an unproblematic concept whose properties are largely self-evident from the common experience of any potential reader. Perhaps because it is the first defined object in the Euclidean corpus, the point receives a level of attention seemingly out of proportion to its obvious simplicity. All but Thomas Digges offer definitions. Of those, all but the elder Digges add explanatory notes and all but Dee offer illustrations (figure 4.28). The challenge, for our authors, was to establish a relationship between the points of common knowledge and the geometric points of the Euclidean texts. For our authors' readership, points were recognizable commodities whose manifestation in geometry was nonetheless completely alien. Textual and visual cues conspire to transfigure common points into Euclidean points, and thereby to set each work on a suitably rigorous Euclidean foundation. There is not much to distinguish the geometric points' visual manifestation in the works under consideration, but the subtle differences that do exist become quite stark and significant upon consideration of their textual context.

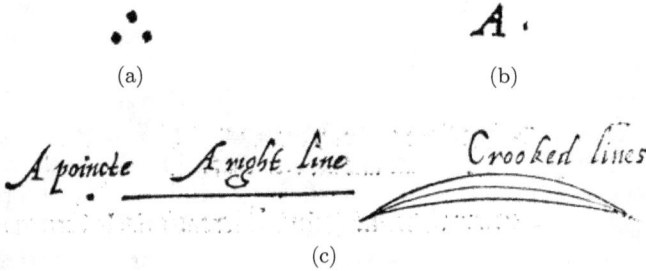

Fig. 4.28: Figures for the geometric point: (a) Recorde (1551, Sig.$A1^r$), (b) Billingsley (1570, Fol.1^r), (c) Digges (1571, Sig.$B1^r$)

Billingsley illustrates the geometric point with a small dot in the outside margin to the right of a label 'A', which is printed slightly larger than the labels in Billingsley's other geometric diagrams. Digges shows a large dot labeled 'A pointe' arrayed horizontally next to a similarly labeled right line and collection of crooked lines in a figure below his third definition. Recorde's three exemplary points, placed at the end of his paragraph, are about the size of his punctuation marks and are arranged in a small upright equilateral triangle. Although Dee does not include a drawing of a point, he does connect Euclidean points to visual experience by explaining that "by visible formes, we are holpen to imagine, what our Line Mathematicall, is. What our Point, is."[50]

Indeed, for both Dee and Billingsley there is an explicit call for the reader's imagination to make the final leap from visual to geometric points. Billingsley calls a point "the least thing that by minde and understanding can be imagined and conceived: then which, there can be nothing lesse, as the point A in the margent."[51] The printed mark in the margin is small, but it is surely not the least thing imaginable. Rather, Billingsley's figure indicates the relevant features of points for his exposition, including their smallness and amenity to labeling by a single letter. Billingsley's explanation instructs the reader to regard his textual model as the least thing imaginable, with the letter next to it understood as the point's name.

Dee's and Digges's use of point-figures sit at two possible extremes. Dee, on the one hand, does not illustrate points at all. Not wishing to build his preface on Euclidean diagrams, he has no need for the visual geometric literacy so necessary for following the others' expositions. Points are pointedly not illustrated. It is the point's (textually manifested) philosophical relation to other

---

[50] Sig.$a2^r$.   [51] Fol.1^r.

objects and ideas which is important to Dee, not its operational centrality in geometric proofs and figures. Dee's words are rich and elaborate, placing points in a broader schema for all of mathematics.

Digges, by contrast, bluntly and without elaboration states that "A *Point* I call whiche cannot be divided, whose parte is nothing."[52] His definition would have been of little use to *Pantometria*'s readers. Instead, their information about points came primarily from his diagram, which shows a representative point and indicates its relation to lines. The definitions for lines and right lines, which explicitly describe lines as having points for their extremities, combine with the point's definition and depiction to guide the reader to imagine points as the indivisible bounds of short segments of lines. This operational understanding covers Digges's uses of points, for which the 'whose parte is nothing' aspect plays no formal role.

Recorde, finally, places his points within a geometric arrangement. Even when they are the sole subject of the illustration, his points participate in a larger geometric context. Moreover, Recorde's triangle of points, like the text that accompanies them, emphasizes more than any of the other images how truly common the geometric point should be to his readers. He stages his points amidst punctuation marks and descriptions of pen pricks so as to establish their meaning in the familiar contexts of writing and reading. As his first geometric illustration, the points in Recorde's figure begin the difficult work of bridging everyday experience and geometry by showing how geometric texts produce meaning through arrangements of familiar forms.

4.7 Conclusion

The first heralds of English geometry – Recorde, Billingsley, Dee, and Leonard and Thomas Digges – produced, over a twenty year period, three starkly different geometric texts. Their attempts to translate Euclid's *Elements* into vernacular English brought with them an opportunity to reimagine the whole of geometry for a new audience. As the comparisons in this essay indicate, visual images figured centrally in this reimagining. Our authors used geometric figures and diagrams to show geometry vividly to their readers, and their different strategies of illustration place different emphases and establish different priorities in the English geometries they aimed to create.

All translations involve the attempt to convey meaning from one idiom and context to another. It is axiomatic that a translator is faced with a wide range of textual considerations which can dramatically affect the meaning of the resulting work. Nor is it surprising that similar considerations play out in the

---

[52] Sig. $B1^r$.

non-textual elements of the translation. More than just decorations or elaborations, diagrams in geometric texts are crucial vehicles for both meaning and context. The diagrams in the first English geometries irrefutably participated in those geometries' construction.

For all the ink that has been spilled in the analysis of the *Elements* and for all the comparatively few analyses of the neologisms and other non-diagrammatic features of its first English translations, much remains to be learned from the role diagrams play in Euclid's re-renderings. Studying diagrams under translation, like the corresponding study of the translation of words and phrases, can say a great deal about the work of interest. It can show how the work was received, what it meant to its translators and users, and also what is possible in the work's interpretation and transmission.

Studies such as this one open the way for a richer discussion of the purpose and function of Euclidean diagrams in general. They show, for instance, what features of Euclidean diagrams were considered important, by whom and for whom. They show how the relationships between diagrams and their textual context change over time and between audiences. They show, moreover, how a simple geometry and its associated visual tradition maintained, at least for the Early Moderns, a remarkable level of interpretive and representational flexibility.

As Euclid crossed into a new tongue, his translators each refracted different features of his geometry. By contrasting their diagrams, I argue, we may better glimpse the contrasts between the Euclidean translators, and ultimately gain a better insight into what it means to translate Euclid.

Acknowledgements I would like to thank Jackie Stedall, Jonathan Harris, Maarten Van Dyck and Albrecht Heeffer, this chapter's two anonymous referees, Stephanie Kelly, the University of Cambridge Scientific Images seminar, and the staff and curators of the Cambridge and Cornell University Rare Books departments. I received many helpful comments and suggestions at the conference from which this volume is derived. This paper is part of a research project begun as an MPhil candidate at St. John's College, Cambridge, and continued as an MSc candidate in the University of Edinburgh's School of Social and Political Science. The images for the figures in this paper are provided by permission of the Master and Fellows of St. John's College, Cambridge (Billingsley's and Digges's texts), and the Huntington Library (Recorde's text).

References

Abbreviations: CamUL = Cambridge Universtiy Library; CUKL = Cornell University Kroch Library, Rare Books & Manuscripts; CULAC = Cornell University Library Adelmann Collection; CULHC = Cornell University Library Hollister Collection; CULHSC = Cornell University Library History of

Science Collection; DNB = Oxford Dictionary of National Biography, online, Oxford University Press; HL = Huntington Library, online access via Early English Books Online; OBL = Bodleian Library, online access via Early English Books Online; SJCL = St. John's College Library, Cambridge; STC = Short Title Catalogue, 2nd edition; Yale University Library, online access via Early English Books Online.

1. Alexander, Amir, 1995. "The Imperialist Space of Elizabethan Mathematics". *Studies in the History and Philosophy of Science* 26(4): 559–591.

2. Archibald, Raymond C., 1950. "The First Translation of Euclid's Elements into English and its Source". *American Mathematical Monthly* 57(7): 443–452.

3. Barrow-Green, June, 2006. "'Much necessary for all sortes of men': 450 years of Euclid's *Elements* in English". *BSHM Bulletin* 21:2–25.

4. Benedetti, Giovanni Battista, 1553. *Resolvitio omnium Euclidis problematum aliorumq[ue] ad hoc necessario inventorum una tantummodo circini data apertura.* Venice: B Caesanum. CULHSC.

5. Bennett, James A., 1986. "The Mechanics' Philosophy and the Mechanical Philosophy". *History of Science* 24: 1–28.

6. Billingsley, Henry, 1570. *The Elements of Geometrie of the most auncient Philosopher Euclide of Megara. Fathfully (now first) translated into the Englishe toung, by H. Billingsley, Citizen of London. Whereunto are annexed certaine Scholies, Annotations, and Inventions, of the best Mathematiciens, both of time past, and in this our age. With a very fruitfull Prӕface made by M. J. Dee, specifying the chiefe Mathematicall Sciences, what they are, and wherunto commodious: where, also, are disclosed certaine new Secrets Mathematicall and Mechanicall, untill these our daies, greatly missed. Printed by John Daye, dwelling over Aldersgate beneath Saint Martins. 1570.* London. STC 10560. CamUL (Adams.4.57.1), HL.

7. Cardano, Girolamo, 1554. *De subtilitate libri XXI. nunc demum recogniti atq[ue] perfecti. [Hieronymi Cardani ... In Cl. Ptolemaei Pelusiensis IIII de astrorum iudiciis, aut, ut uulgò uocant, quadripartitae constructionis, libros commentaria, quae non solum astronomis & astrologis, sed etiam omnibus philosophiae studiosis plurimum adiumenti adferre poterunt. Nunc primùm in lucem aedita. Praeterea, eiusdem Hier. Cardani Geniturarum XII, et auditu mirabilia et notatu digna, & ad hanc scientiam recte exercendam observatu utilia, exempla. Atque alia multa quae interrogationibus & electionibus praeclarè serviunt, uanáque à veris rectè secernunt. Ac denique eclipseos, quam gravissima pestis subsecuta est, exemplum.]* Basel: Ludovicum Lucium. CULAC.

8. De Young, Gregg, 2005. "Diagrams in the Arabic Euclidean tradition: a preliminary assessment". *Historia Mathematica* 32(2):129–179.

9. De Young, Gregg, 2009, "Diagrams in ancient Egyptian geometry: Survey and assessment". *Historia Mathematica* 36(4):321–373.

10. Digges, Leonard, 1556. *A boke named Tectonicon briefelye shewynge the exacte measurynge, and speady reckenynge all maner Lande, squared Tymber, Stone, Steaples, Pyllers, Globes. &c. Further, declarynge the perfecte makynge and large use of the Carpenters Ruler, conteynynge a Quadrant Geometricall: comprehendynge also the rare use of the Squire. And in thende a little treatise adjoyned, openinge the composition and appliancie of an Instrument called the profitable Staffe. With other thinges pleasant & necessary, most conducible for Surveyers, Landmeters, Joyners, Carpenters, and Masons. Published by Leonarde Digges Gentleman, in the yere of our Lorde. 1556.* STC 6849.5. OBL.

11. Digges, Leonard, 1571. *A Geometrical Practise, named Pantometria, divided into three Bookes, Longimetra, Planimetra and Stereometria, containing Rules manifolde for mensuration of all lines, Superficies and Solides: with sundry straunge conclusions both by instrument and without, and also by Perspective glasses, to set forth the true description or exact plat of an whole Region: framed by Leonard Digges Gentleman, lately finished by Thomas Digges his sonne. Who hath also thereunto adjoyned a Mathematicall treatise of the five regulare Platonicall bodies, and their Metamorphosis or transformation into five other equilater uniforme solides Geometricall, of his owne invention, hitherto not mentioned of by any Geometricians. Imprinted at London by Henrie Bynneman Anno. 1571.* London. STC 6858. CamUL (Peterborough.B.3.15), YUL.

12. Digges, Leonard, 1591. *A Geometrical Practical Treatize Named Pantometria, divided into three Bookes, Longimetra, Planimetra, and Stereometra, Containing rules manifolde for mensuration of all Lines, Superficies and Solides: with sundrie strange conclusions both by Instrument and without, and also by Glasses to set forth the true Description or exact Platte of an whole Region. First published by Thomas Digges Esquire, and Dedicated to the Grave, Wise, and Honourable, Sir Nicholas Bacon Knight, Lord Keeper of the great Seale of England. With a Mathematicall discourse of the five regular Platonicall Solides, and their Metamorphosis into other five compound rare Geometricall Bodyes, conteyning an hundred newe Theroremes at least of his owne Invention, never mentioned before by anye other Geometrician. Lately Reviewed by the Author himselfe, and augmented with sundrie Additions, Diffinitions, Problemes, and rare Theoremes, to open the passage, and prepare away to the understanding of his Treatize of Martiall Pyrotechnie and great Artillerie, hereafter to be published. At London, Printed by Abell Jeffes. Anno. 1591.* London. STC 6859. CamUL (M.15.2), HL.

13. Digges, Thomas, 1573. *Alæ Seu Scalæ Mathematicæ, quibus visibilium remotissima Cœlorum Theatra conscendi, & Planetarum omnium itinera novis & inauditis Methodis explorari: tùm huius portentosi Syderis in Mundi Boreali plaga insolito fulgore coruscantis, Distantia, & Magnitudo immensa, Situsq[ue]; protinùs tremendus indagari, Deiq[ue]; stupendum ostentum, Terricolis expositum, cognosci liquidissimè possit.* London. STC 6871. HL.

14. Drake, Stillman, 1970. "Early Science and the Printed Book: The Spread of Science Beyond the Universities". *Renaissance and Reformation* 6(3): 43–52.

15. Easton, Joy B., 1966. "A Tudor Euclid." *Scripta Mathematica* 27(4): 339–355.

16. Easton, Joy B., 1967. T"he Early Editions of Robert Recorde's Ground of Artes". *Isis* 58(4): 515–532.

17. Feingold, Mordechai, 1984. *The mathematicians' apprenticeship: Science, universities and society in Englannd, 1560–1640.* Cambridge: Cambridge University Press.

18. Fine, Oronce, 1544. *Orontii Finei Delphinatis, Lutetiae Liberalium disciplinarum Professoris regii Liber de Geometria practica sive de practicis longitudinum, planorum & solidorum: hoc est, linearum, superficierum, & corporum mensionibus aliisq[ue] mechanicis, ex demonstratis Euclidis elementis corollarius. Ubi et de Quadrato Geometrico, et virgis seu baculis mensoriis. Nunc primumm apud Germanos in lucem emissus.* Strassburg: Ex officina Knoblochiana, per Georgium Machaeropoeum. CULHC.

19. Frisius, Gemma, 1557. *Gemmæ Frisii, Medici et Mathematici, de radio astronomico & Geometrico liber. In quo multa quæ ad Geographiam, Opticam, Geometriam & Astronomiam utiliss.sunt, demonstrantur. Illustriss. Comiti de Feria dicatus. Adjunximus brevem tractationem Ioannis Spangebergii & Sebastiani Munsteri de Simpliciore Radio, quem Baculu[m] Iacob vulgus nominat.* Paris: Apud Guilielmum Cavellat, in pingui gallina, ex adverso Collegii Cameracensis. CULHC.

20. Gaskell, Philip, 1972. *A New Introduction to Bibliography.* Oxford: Clarendon Press.

21. Grynaeus, Simon, 1533. *Eukleidou Stoicheion: Bibl. ie' ek ton theonos sunousion; Eis tou autou to proton exegematon Proklou biblios; Adiecta præfatiuncula in qua de disciplinis mathematicis nonnihil.* Basel: Apud Joan Hervagium. CULHSC, SJCL (Aa.1.45).
22. Heath, Thomas L., 1956 [1926]. *The Thirteen Books of Euclid's Elements, translated from the text of Heiberg, with introduction and commentary. Second edition, revised with additions. Volume I: Introduction and Books I, II.* New York: Dover.
23. Heninger Jr., S.K., 1969. "Tudor Literature of the Physical Sciences". *The Huntington Library Quarterly* 32(2): 101–133.
24. Hill, Katherine, 1998. "'Juglers or Schollers?': Negotiating the Role of a Mathematical Practitioner". *The British Journal for the History of Science* 31(3): 253–274.
25. Hirschvogel, Augustin, 1543. *Ein aigentliche und grundtliche Anweysung in die Geometria, sonderlich aber, wie alle Regulierte und Unreglierte Corpora in den grundt gelegt und in das Perspectiff gebracht, auch mit iren Linien auffzogen sollen werden.* Nuremburg. CULHSC.
26. Howson, Geoffrey, 1982. *A history of mathematics education in England.* Cambridge: Cambridge University Press.
27. Johnson, Francis R., 1936. "The Influence of Thomas Digges on the Progress of Modern Astronomy in Sixteenth-Century England." *Osiris* 1: 390–410.
28. Johnson, Francis R., 1944. "Latin versus English: The Sixteenth-Century Debate over Scientific Terminology". *Studies in Philology* 41(2): 109–135.
29. Johnson, Francis R. and Sanford V. Larkey, 1935. "Robert Recorde's Mathematical Teaching and the Anti-Aristotelian Movement". *The Huntington Library Quarterly* 7: 59–87.
30. Johnston, Stephen, 2004a. Digges, Leonard (c.1515–c.1559). DNB.
31. Johnston, Stephen, 2004b. Digges, Thomas (c.1546–1595). DNB.
32. Johnston, Stephen, 2004c. Recorde, Robert (c.1512–1558). DNB.
33. Keller, Agathe, 2005. "Making diagrams speak, in Bhāskara I's commentary on the Āryabhatīya". *Historia Mathematica* 32(3): 275–302.
34. McConnell, Anita, 2004. Billingsley, Sir Henry (d. 1606). DNB.
35. McRae, Andrew, 1993. "To Know One's Own: Estate Surveying and the Representation of the Land in Early Modern England". *The Huntington Library Quarterly* 56(4): 333–357.
36. Nemorarius, Jordanus, 1533. *Liber Iordani Nemorarii viri clarissimi De ponderibus propositiones XIII & earundem demonstrationes: multarum [que] rerum rationes sanè pulcherrimas complectens, nunc in lucem editus.* Nuremburg: Johannes Petreius. CULHSC.
37. Pacioli, Luca, 1523. *Summa de arithmetica geometria, proportioni, et proportionalita.* Toscolano: Paganino Paganini. CULHC.
38. Ramus, Petrus, 1580. *Arithmeticae libri duo: Geometriae septem et viginti.* Basel: per Eusebium Episcopium, & Nicolai fratris hæredes. CULHSC.
39. Ratdolt, Erhard, 1482. *Elementa geometriae.* Venice. CULHSC, SJCL (Aa.1.43).
40. Recorde, Robert, 1551. *The pathway to knowledg, containing the first principles of Geometrie, as they may moste aptly be applied unto practise, bothe for use of instrumentes Geometricall, and astronomicall and also for projection of plattes in everye kinde, and therfore muche necessary for all sortes of men. and The Second Booke of the Principles of Geometry, containing certaine Theoremes, whiche may be called Approved truthes. And be as it were the moste certaine groundes, wheron the practike conclusions of Geometry ar founded. Wherunto are annexed certaine declarations by examples, for the right understanding of the same, to the ende that the simple reader might not justly complain of hardnes or obscuritee, and for the same cause ar the demonstrations and just profes omitted, untill a more convient time. 1551.* Imprinted at London in Poules churcheyarde, at the signe of the Brasenserpent, by Reynold Wolfe. London. STC 20812. CamUL (Peterborough.G.4.15, Syn.7.55.45), HL.

41. Roberts, R. Julian, 2004. Dee, John (1527–1609). DNB.
42. Rose, Paul Lawrence, 1977. "Erasmians and Mathematicians at Cambridge in the Early Sixteenth Century". *The Sixteenth Century Journal* 8(2): 47–59.
43. Taylor, E.G.R., 1954. *The Mathematical Practitioners of Tudor & Stuart England.* Cambridge: Cambridge University Press.
44. Xylander, Wilhelm, 1562. *Die sechs erste Bücher Euclidis, vom Anfang oder Grund der Geometri: ... Auss griechischer Sprach in die Teütsch gebracht, aigentlich erklärt, auch mit verstentlichen Exampeln, gründlichen Figurn, und allerlai den nutz für Augen stellenden Anhängen geziert, dermassen vormals in Teütscher Sprach nie esehen worden. ; Alles zu Lieb und Gebrauch den kunstliebenden Teütscher, so sych der Geometri und Rechenkunst anmassen, mit vilfältiger Mühe und Arbait zum trewlichsten erarnet, und in Truckh gegeben, durch Wilhelm Holtzman, genant Xylander, von Augspurg.* Basel: Jacob Kündig, Joanns Oporini Kosten. CUKL.
45. Zetterberg, J. Peter, 1980. "The Mistaking of 'the Mathematicks' for Magic in Tudor and Stuart England". *The Sixteenth Century Journal* 11(1): 83–97.

Chapter 5
The symbolic treatment of Euclid's *Elements* in Hérigone's *Cursus mathematicus* (1634, 1637, 1642)

Maria Rosa Massa Esteve

Abstract The publication in 1591 of *In artem analyticem isagoge* by François Viète (1540–1603) constituted an important step forward in the development of a symbolic language. This work was diffused through many other algebra texts, such as the section entitled *Algebra* in the *Cursus mathematicus, nova, brevi et clara methodo demonstratus, per notas reales & universales, citra usum cuiuscunque idiomatis, intellectu faciles* (Paris, 1634/1637/1642) by Pierre Hérigone (1580–1643). In fact, Hérigone's aim in his *Cursus* was to introduce a symbolic language as a universal language for dealing with both pure and mixed mathematics using new symbols, abbreviations and margin notes. In this article we focus on the symbolic treatment of Euclid's *Elements* in the first volume of the *Cursus* in which Hérigone replaced the rhetorical language of Euclid's *Elements* by symbolic language in an original way. Since Hérigone stated that he had followed Clavius's *Elements* (1589) in the writing of this first volume, we compare some demonstrations found in both authors' works as regards the style and the use of other propositions from Euclid's *Elements*, with the aim of clarifying the significance and the usefulness of Hérigone's new method of demonstration for a better understanding of mathematics.

Key words: Pierre Hérigone; Symbolic language; *Cursus mathematicus*; Euclid's *Elements*; Seventeenth century; Clavius's *Elements*.

Centre de Recerca per a la Història de la Tècnica, Departament de Matemàtica Aplicada I. Universitat Politècnica de Catalunya

5.1 Introduction

Pierre Hérigone[1] (1580–1643) published his *Cursus Mathematicus* (1634/ 1637/1642) in a period when the algebraization of mathematics was taking place. One of the fundamental characteristics of this process was the introduction of algebraic procedures to solve geometric problems.[2] In this process, the creation of a formal symbolic language to represent algebraic equations and geometric constructions and curves became one of algebra's essential features.[3] For this reason, the publication in 1591 of *In Artem Analyticen Isagoge* by François Viète (1540–1603) constituted an important step forward in the development of a symbolic language for mathematics.[4]

Viète's work was transmitted through various texts on algebra, such as the *Algebra* section of Hérigone's *Cursus Mathematicus* (hereafter referred to as the *Cursus*). We have analyzed this section in a recently published article (Massa, 2008), in which we show that while Hérigone used Viète's statements to deal with equations and their solutions, his notation, presentation and procedures were indeed quite different. Furthermore, we analyzed some of Hérigone's improvements that derived from a generalization of Viète's examples.

We now focus our research on the symbolic treatment of Euclid's *Elements* in the first volume of Hérigone's *Cursus* and its usefulness for rendering Math-

---

[1] Very little is known about Hérigone's life. Per Stromholm claims that he was from the Basque Country and that he taught mathematics in Paris. For more information see Stromholm (1972, p.6) and Knobloch (2001, p.13–14).

[2] Therefore, two new developments occurred in mathematics: first, the creation of what is now named analytic geometry, and second, the emergence of infinitesimal calculus. The two new disciplines achieve their ends through connections between algebraic expressions and geometric curves, on the one hand, and between algebraic operations and geometric constructions on the other. There are many useful studies on this subject, including Mahoney (1980, p.141–156), Mancosu (1996, p.84–86) and Panza (2005).

[3] In fact, the notation is not present in algebraic works in Arabic. Abbreviations are first used to represent the unknown quantities in the arithmetic works of the Renaissance period and algebraic procedures were expressed in syncopate form. The widespread use of symbolic notation began in the middle of the sixteenth century. There are many useful studies on the evolution of symbolic language, including Wallis (1685), Cajori (1928–29), Pycior (1997) and Stedall (2002).

[4] Viète used symbols to represent both known and unknown quantities, and was thus able to investigate polynomial equations in a completely general form. He conceived of equations in terms of Euclidean ideas of proportion. The equation $x^2 + bx = d^2$, for example, can be written as $x(x + b) = d^2$ and therefore as a proportion $x : d = d : (x + b)$. Solving the equation is therefore equivalent to finding three lines in continued proportion. Viète showed the usefulness of algebraic procedures for analysing and solving problems in arithmetic, geometry and trigonometry. The purpose of Viète's analytical art, in his own words, was to solve all kinds of problems. For more information see Viète (1646), Giusti (1992) and Bos (2001).

ematics more comprehensive. The aim of this paper is to show how Hérigone replaces the rhetorical language of Euclid's *Elements* with a symbolic language, as well as to analyze some examples of this procedure as a useful means of obtaining new results.

Since Hérigone stated in the *Prolegomena* to his *Elements* that he had followed the *Elements* (1589) of Christoph Clavius (1538–1612) for the writing of this first volume, we compare some demonstrations found in the texts by both authors, examining their style and the order and use of other propositions from Euclid's *Elements* in order to clarify the significance and the usefulness of reformulating rhetorical text into symbolic language.

We divide the article into three sections: the first section deals with the features of Hérigone's "new method" in the *Cursus*, the second describes his procedure of symbolically treating Euclid's *Elements* to make demonstrations, and our final section analyzes some examples of geometrical propositions in Hérigone's *Elements*, which facilitated the production of new demonstrations in the *Cursus*.

5.2 Hérigone's new method

In order to understand the reasoning used by Hérigone in his work, we must analyze the principal features of Hérigone's new method of demonstration described in the *Cursus*: the original system of notation, the axiomatic-deductive reasoning and the presentation of the propositions.

Hérigone wrote an encyclopaedic textbook consisting of five volumes known as the *Cursus Mathematicus*.[5] The first four volumes were published in 1634. The first and second volumes of the *Cursus* deal with pure mathematics. The first volume deals with geometry and the second volume is devoted to arithmetic and algebra. The third and fourth volumes deal with mixed mathematics, that is to say, with the mathematics required for practical geometry, military or mechanical uses, geography, and navigation. The fifth and last volume of the first edition, published in 1637, includes spherical trigonometry and music. Later, in the second edition (1642), Hérigone added the sixth and final volume, which contains two parts dealing with algebra; it also deals with perspective and astronomy.

Published in parallel Latin and French columns on the same page, the first edition, whose full title is *Cursus mathematicus, nova, brevi et clara methodo*

---

[5] Hérigone published an edition of the first six books of Euclid in 1639 (Hérigone, 1639), but Stromholm (1972, p.299) claims that these are "little more than the French portion of Volume 1 of the *Cursus*." For more information on the parts of the *Cursus* see Massa (2008, p.287).

*demonstratus, per notas reales & universales, citra usum cuiuscunque idioma-
tis, intellectu faciles* ["Course of Mathematics demonstrated by a brief and
clear new method through real and universal symbols,[6] which are easily understood without the use of any language"],[7] states that Hérigone devised a
new method of demonstration to understand Mathematics in a straightforward manner.

Hérigone also claimed that he had invented a new method for making
demonstrations briefer and more intelligible that did not require the use of
any language. In the preface to the first volume, which bore the dedication
"Au lecteur" [To the reader] he explains,

> There is no doubt at all that the best method for teaching the sciences is that in which
> brevity is combined with ease. But it is not always easy to attain both, particularly
> in mathematics, which, as Cicero pointed out, is highly obscure. Having considered
> this myself, and seeing that the greatest difficulties arise from an understanding of
> the demonstrations, on which the knowledge of all parts of mathematics depend, I
> have devised a new method, brief and clear, of making demonstrations, without the
> use of any language.[8]

Indeed, Hérigone's stated aim in the *Cursus* was to introduce a symbolic
language as a universal language for dealing with both pure and mixed mathematics. Moreover, Hérigone stressed the importance of knowing the symbols
and understanding the demonstrations performed with this notation. His way
of reasoning through the steps of the demonstration is axiomatic-deductive,
as we explain below.

Thus, the first feature of Hérigone's new method is his system of notation;
he uses many new symbols and abbreviations (which he calls "notes") and

---

[6] We have translated the expression "notes" as "symbols;" however, in Hérigone's view
"notes" include symbols and abbreviations.

[7] The title in French is "Cours Mathematique demonstré d'une nouvelle briefve et Claire
methode. Par notes reelles & universelles, qui peuvent estre entendues sans l'usage d'aucune
langue." In writing this article the author has referred to the copy held in the Bibliothèque
Nationale de France.

[8] Car on ne doute point, que la meilleure methode d'enseigner les sciences est celle, en
laquelle la briefveté se trouve conjoincte avec la facilité : mais il n'est pas aisé de pouvoir
obtenir l'une & l'autre, principalement aux Mathematiques, lesquelles comme temoigne Ci-
ceron, sont grandement obscures. Ce que considerant en moy-mesme, & voyant que les plus
grandes difficultez estoient aux demonstrations, de l'intelligence desquelles dépend la cog-
noissance de toutes les parties des Mathematiques : i'ay inventé une nouvelle methode de
faire les demonstrations, briefve & intelligible sans l'usage d'aucune langue. /Nam extra
controversiam est, optimam methodum tradendi scientias, esse eam, in qua brevitas per-
spicuitati coniungitur, sed utramque assequi hoc opus hic labor est, praesertim in Mathematicis disciplinis, quae teste Cicerone, in maxima versantur difficultate. Quae cum animo
perpenderem, perspectumque haberem, difficultates quae in erudito Mathematicorum pul-
vere plus negotij facessunt, consistere in demonstrationibus, ex quarum intelligentia Mathematicarum disciplinarum omnis omnino pendet cognitio : excogitavi novam methodum

margin notes (which he calls "citations"). We may claim that his notation is entirely original; indeed, most of the symbols had not appeared in any previous book. For example, in *Algebra*, Hérigone, like Viète, uses vowels to represent unknown quantities and consonants to represent known or given quantities. To represent powers, Hérigone writes the exponents on the right side of the letter (so the square is represented by a 2, the cube by a 3 and so on). See table 5.1 below.

Signs	Viète (1590s)	Harriot (1631)	Hérigone (1634)	Descartes (1637)
Equality	*Aequalis*	$=$	2|2	$\propto$
Greater than	*Maior est*	$>$	3|2	Plus grande
Less than	*Minus est*	$<$	2|3	Plus petite
Product of a and b	A and B	ab	ab	ab
Addition	*plus*	$+$	$+$	$+$
Subtraction	*minus*	$-$	$\sim$	$-$
Ratio	*ad*		$\prod$	à
Square root	$VQ.$	$\sqrt{}$	$V2$	$\sqrt{}$
Cubic root	$VC.$	$\sqrt{c}$	$V3$	$\sqrt{c}$
Squares	*Aquadratus, Aquad*	aa	$a2$	a^2, aa
Cubes	*Acubus, Acub*	aaa	$a3$	a^3

Table 5.1: Table of notations from Massa (2008, p.289).

Furthermore, Hérigone provides alphabetically ordered explanatory tables of abbreviations and symbols (which he calls "explicatio notarum"). For example, there is a mark for the side of the square, a sign meaning 'perpendicular', and a symbol for representing ratios. (See figure 5.1.)

Hérigone also gives explanatory tables for the citations (which he calls "explicatio citationum") at the beginning of each of the volumes of which the *Cursus* is composed. The citations always refer either to propositions in Euclid's *Elements* or to the *Cursus* itself. In the margin of the demonstrations of propositions, Hérigone cites, line by line, the numbers corresponding to the theorems he has used.[9]

---

demonstrandi brevem & citra ullius idiomatis usum intellectu facilem. (Hérigone, 1634, I, *Ad Lectorem*). All translations are the author's own.

[9] In the Ancient copies of Greek editions of Euclid there are no references in the margin to the theorems he used. However, these references are introduced in Renaissance editions of Euclid, particularly in Clavius, which was evidently Hérigone's model, as he himself points out. We would like to draw attention to Hérigone's elucidation of Clavius, in which it is not just Clavius's works that are mentioned; Hérigone explained that he had used Clavius's order and text for Euclid's *Elements*, as well as for the three books of Theodosius's *Spherics* and for the fourth book up to the eighteenth proposition. See Hérigone (1642, VI, p. 241).

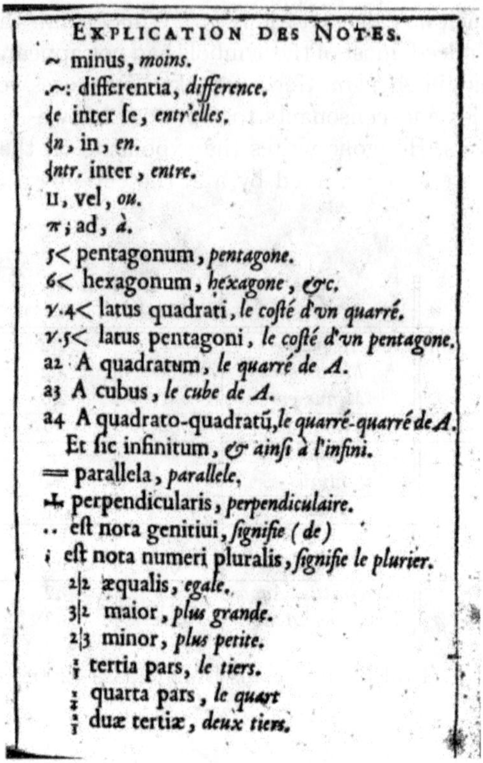

EXPLICATION DES NOTES.
~ minus, *moins.*
.~: differentia, *difference.*
⟨e inter se, *entr'elles.*
⟨n, in, *en.*
⟨ntr. inter, *entre.*
II, vel, *ou.*
π; ad, *à.*
5< pentagonum, *pentagone.*
6< hexagonum, *hexagone, &c.*
ν.4< latus quadrati, *le costé d'vn quarré.*
ν.5< latus pentagoni, *le costé d'vn pentagone.*
a2 A quadratum, *le quarré de A.*
a3 A cubus, *le cube de A.*
a4 A quadrato-quadratū, *le quarré-quarré de A.*
Et sic infinitum, *& ainsi à l'infini.*
= parallela, *parallele.*
⊥ perpendicularis, *perpendiculaire.*
.. est nota genitiui, *signifie (de)*
; est nota numeri pluralis, *signifie le plurier.*
2|2 æqualis, *egale.*
3|2 maior, *plus grande.*
2|3 minor, *plus petite.*
$\frac{1}{3}$ tertia pars, *le tiers.*
$\frac{1}{4}$ quarta pars, *le quart*
$\frac{2}{3}$ duæ tertiæ, *deux tiers.*

Fig. 5.1: Hérigone's table of abbreviations (Hérigone, 1634, I, f. bv^r)

Thus, for example, "c.l.60.10" means "Corollary of the lemma of the proposition X.60" (See figure 5.2).[10]

The second feature of Hérigone's method is the axiomatic-deductive reasoning explicitly described by him. In the preface to the reader, Hérigone emphasizes that the introduction of margin notes is key for following the steps of the demonstration and this trait is used in this method, unlike in the "vulgar and common" or ordinary method. He criticizes other authors who use the "vulgar and common" method. We do not know the exact meaning of this expression, but since it was Hérigone's belief that it was difficult to understand the demonstrations, this expression acquires its significance for

On Clavius and his influence on other seventeenth-century authors, see Knobloch (1988) and Rommevaux (2006).

[10] Corollaire du lemme de la soixantième du dixième./ Corollarium lemmatis sexagesimae decimi (Hérigone, 1634, I, unpaginated).

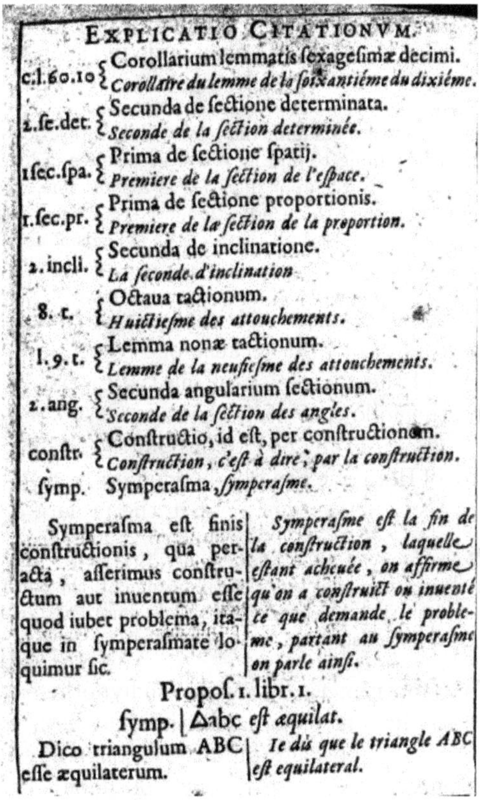

EXPLICATIO CITATIONVM.

c.l.60.10	Corollarium lemmatis sexagesimæ decimi. *Corollaire du lemme de la soixantiéme du dixiéme.*
2.sc.det.	Secunda de sectione determinata. *Seconde de la section determinée.*
1sec.spa.	Prima de sectione spatij. *Premiere de la section de l'espace.*
1.sec.pr.	Prima de sectione proportionis. *Premiere de la section de la proportion.*
2.incli.	Secunda de inclinatione. *La seconde d'inclination*
8.t.	Octaua tactionum. *Huictiesme des attouchements.*
l.9.t.	Lemma nonæ tactionum. *Lemme de la neufiesme des attouchements.*
2.ang.	Secunda angularium sectionum. *Seconde de la section des angles.*
constr.	Constructio, id est, per constructionem. *Construction, c'est à dire, par la construction.*
symp.	Symperasma, *symperasme.*

Symperasma est finis constructionis, qua peracta, asserimus constructum aut inuentum esse quod iubet problema, itaque in symperasmate loquimur sic. | *Symperasme est la fin de la construction, laquelle estant acheuée, on affirme qu'on a construict ou inuenté ce que demande le probleme, partant au symperasme on parle ainsi.*

Propos. 1. libr. 1.

symp. | △abc *est æquilat.*

Dico triangulum ABC esse æquilaterum. | *Ie dis que le triangle ABC est equilateral.*

Fig. 5.2: Hérigone's explanatory table of citations (Hérigone, 1634, I, f. bvii^{v})

designing the methods used by other authors in contrast to the new method he is introducing. In Hérigone's own words:

> I also stress that in the ordinary method many words and axioms are used without prior explanation, but in this method there is nothing that has not already been explained and conceded in the premises; even in the demonstrations, which are somewhat longer, all that was proved in the sequence of the demonstration are cited with Greek letters.[11]

[11] Soient aussi qu'en la methode ordinaire on se sert beaucoup de mots & d'axiomes sans les avoir premierement expliquez, mais en cette methode on ne dit rien qui n'aye esté expliqué & concedé aux premises ; mesme aux demonstrations, qui sont quelque peu longues, on cite par lettres Grecques, ce qui a esté demonstrée en la suite de la démonstration. /Huc etiam accedit, quòd in vulgari & communi docendi ratione, plurima proferantur vocabula,& axiomata absque ulla illorum in praemisis explicatione : sed in hac methodo nihil adfertur, nisi fuerit in praemissis explicatum & concessum. Quum etiam longiores occurrunt demonstra-

Hérigone goes on to describe his axiomatic-deductive reasoning for the demonstrations, and adds that he will give an example in the first proposition of the first book. In Hérigone's own words,

> And as each consequence depends immediately on the proposition cited, the demonstration follows from beginning to end by a continue series of legitimate, necessary and immediate consequences, each one included in a short line, which can be solved easily by syllogisms, because in the proposition cited as well as in that which corresponds to the citation one can find all parts of the syllogism, as one may see in the first demonstration of first book, which has been reduced by syllogisms.[12]

Hérigone's originality resides not only in the explicit explanation of axiomatic-deductive reasoning, but also because one can find in one symbolic line the major premise and the conclusion, using the former symbolic line as the minor premise. In the following section we analyse the syllogism and the identification of the premises in the demonstration.

The third feature of Hérigone's method of demonstration is the presentation of propositions. He also stresses this point in the preface to the reader,

> The distinction of the proposition in its members, that is, the part in which the hypothesis is advanced, the explanation of the requirement, the construction or preparation and the demonstration, likewise relieves the memory and makes it very helpful for understanding the demonstration.[13]

Indeed, Hérigone's propositions are proved from hypotheses and well-established properties. Sometimes he states the equalities that he needs for the demonstration in a "Praeparatio" paragraph after the hypothesis. He also divides his demonstrations into separate sections: hypothesis (known and unknown

---

tiones, quae iam in serie demonstrationis sunt probata, litteris Graecis citantur (Hérigone, 1634, I, *Ad Lectorem*).

[12] Et parce que chaque consequence depend immediatement de la proposition citée, la demonstration s'entretien depuis son commencement jusques à la conclusion, par une suite continue de consequences legitimes, necessaires & immediates, contenues chacune en une petite ligne, lesquelles se peuvent resoudre facilement en syllogismes, à cause qu'en la proposition citée, & en celle qui correspond à la citation, se trouvent toutes les parties du syllogisme: comme on peut voir en la premiere demonstration du premier livre, qui a esté reduite en syllogismes. /Et quoniam singulae consequentiae ex propositionibus allegatis immediate pendent, demonstratio ab initio ad finem, serie continua, legitimarum, necessariarumque consecutionum immediatarum, singulis lineolis comprensarum aptè cohaeret: quarum unaquaeque nullo negotio in syllogismum potest converti, quòd in propositione citata, & in ea quae citationi respondet, omnes syllogismi partes reperiatur: ut videre est in prima libri primi demonstratione, quae in syllogismos est conversa (Hérigone, 1634, I, *Ad Lectorem*).

[13] La distinction de la proposition en ses membres, savoir en l'hypothese, l'explication du requis, la construction, ou preparation, & la demonstration, soulage aussi la memoire, & sert grandement à l'intelligence de la demonstration. /Praeterea distinctio propositionis in sua membra, scilicet in hypothesin, explicationem quaesiti, constructionem, vel praeparationem, & demonstrationem non parum iuvat quoque memoriam, & ad intelligendam demonstrationem multùm prodest. (Hérigone, 1634, I, *Ad Lectorem*).

quantities); explanation or requirement; demonstration, and conclusion. In the margin he writes the number of propositions of Euclid's *Elements* that he is using. He occasionally gives the numerical solution (for example in an equation) in a section headed "Determinatio". In geometric constructions, he provides the instructions needed to make the drawing in a paragraph referred to as "Constructio".[14]

Let us see how Hérigone works when proving an algebraic identity in the *Algebra* (see Figure 5.3). He proves the algebraic identity, which in modern

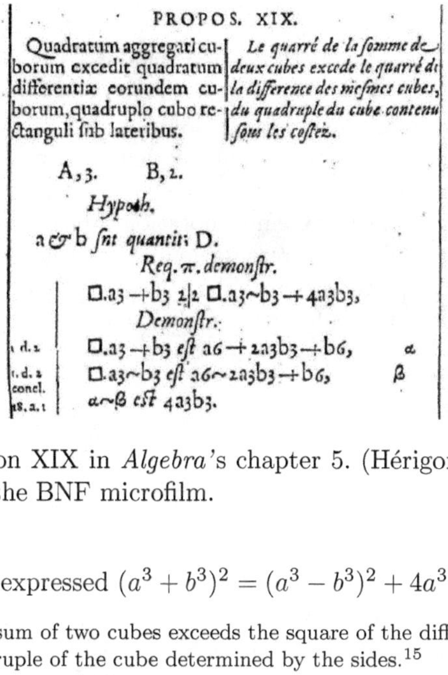
PROPOS. XIX.

Quadratum aggregati cuborum excedit quadratum differentiæ eorundem cuborum, quadruplo cubo rectanguli sub lateribus. | *Le quarré de la somme de deux cubes excede le quarré de la difference des mesmes cubes, du quadruple du cube contenu sous les costez.*

A,3. B,2.

Hypoth.

a & b *snt quantit.* D.

Req. π. *demonstr.*

□.a3 + b3 2|2 □.a3 ~ b3 + 4a3b3,

Demonstr.

4 d.2 □.a3 + b3 *est* a6 + 2a3b3 + b6, α

5 d.2 □.a3 ~ b3 *est* a6 ~ 2a3b3 + b6, β

concl. 18.a.1 α ~ β *est* 4a3b3.

Fig. 5.3: Proposition XIX in *Algebra*'s chapter 5. (Hérigone, 1634, II, p. 46) Reproduced from the BNF microfilm.

notation would be expressed $(a^3 + b^3)^2 = (a^3 - b^3)^2 + 4a^3b^3$, as follows:

> The square of the sum of two cubes exceeds the square of the difference of the same cubes by the quadruple of the cube determined by the sides.[15]

[14] We would also like to point out that Pietro Mengoli (1626-1686), Hérigone's follower, writes all his demonstrations in Hérigonean style by dividing them into a "Hypothesis," "Demonstratio," "Praeparatio" and "Constructio." Furthermore, in the margin he cites line by line all the propositions and properties he has used according to an axiomatic-deductive reasoning. Thus, under the influence of Hérigone, who considered Euclid's *Elements* the point of reference par excellence, Mengoli brings together, as he says, a "conjuntis perfectionibus" [perfect conjunction] of classical mathematics and modern mathematics to obtain new theories and new results. See Massa (1997, 2003, 2006a, 2006b, 2009).

[15] Le quarré de la somme de deux cubes excede le quarré de la difference des mesmes cubes, du quadruple du cube contenu sous les costez. /Quadratum aggregati cuborum excedit quadratum differentiae eorundem cuborum, quadruplo cubo rectanguli sub lateribus (Hérigone, 1634, II, p. 46).

Hérigone's Notation	Modern Notation
Hypoth.	Hypothesis
a & b snt quantit; D.	*a* and *b* are given quantities.
Req. Π. Demonstr.	It is required to prove that:
$.\Box a3 + b32\|2.\Box a3 \sim b3 + 4a3b3$, *Demonstr.*	$(a^3 + b^3)^2 = (a^3 - b^3) + 4a^3b^3$, Demonstration.
$1.d.2^{16}.\Box.a3 + b3\ est\ a6 + 2a3b3 + b6, \alpha$	II.def.1 $(a^3 + b^3)^2$ is $a^6 + 2a^3b^3 + b^6, (\alpha)$
$1.d.2.\Box.a3 \sim b3\ est\ a6 \sim 2a3b3 + b6, \beta$	II.def.1 $(a^3 - b^3)^2$ is $a^6 - 2a^3b^3 + b^6, (\beta)$
Concl. $18.a.1.^{17}\ \alpha \sim \beta\ est\ 4a3b3.$	Conclusion. I. axiom.18 $\alpha - \beta$ is $4a^3b^3$.

Table 5.2: Modern translations of Hérigone's notations

It is worth pointing out that Hérigone formulates the identity to prove and even the definitions and axiom used in symbols, without rhetorical explanations or verbal descriptions. He also divides his demonstration into separate sections: Hypothesis, requirement to prove, demonstration and conclusion.

We may conclude that Hérigone was convinced that this new method of demonstration with his new system of notation, his axiomatic-deductive reasoning and his new manner of presentation is the clearest, most concise and most suitable for rendering the mathematics more comprehensively. In the preface, after analyzing the features of his new method Hérigone affirms: "These are the principal commodities to be found in our new method of demonstration".[18]

5.3 The reformulation of Euclid's *Elements* in symbolic language

The first volume of the *Cursus* contains Euclid's *Elements* and *Data*, Apollonius's *Conics*[19] and an exposition of Viète's *Doctrine of angular sections* (see Figure 5.4). Hérigone presents the fifteen[20] books of Euclid's *Elements*, which is also one of the first translations of Euclid's *Elements* into a symbolic language. In fact, Isaac Barrow (1630–1677) in the letter *Ad lectorem* in his own edition of the *Elements* (1659), mentioned Hérigone as an example to

---

[18] Voila les principales commoditez qui se trouvent en notre nouvelle méthode de demonstrer. /Atque haec sunt commoda, quae in hac nova methodo demonstrandi reperiuntur (Hérigone, 1634, I, unpaginated).

[19] At the end of Euclid's *Data*, Hérigone's stated aim was to introduce his new method of demonstration into the five texts on Apollonius's *Conics* restored by Snell (3 texts), Ghetaldi and Viète as well as into the section of angles invented by Viète. (Hérigone 1634, I, p.889–935).

[20] Hérigone, like Clavius, mentions that only the first thirteen books are attributed to Euclid and that the other two are attributed to Hypsicles Alexandrinus (Hérigone, 1634, I,

follow both for reducing Euclid's *Elements* to one volume and for turning it into a symbolic language (Barrow, 1659, unpaginated).[21]

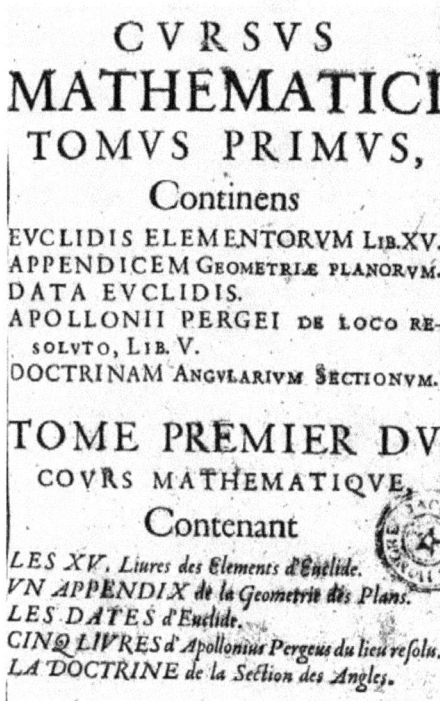

Fig. 5.4: Hérigone's frontispiece to Volume I of the *Cursus*. (Hérigone, 1634, I, f. aii^v).

Although Hérigone uses the Latin version of Clavius's 1589 edition of the *Elements* only the statements and some figures for the propositions match

---

Prolegomena).

[21] Barrow for his part explained that Hérigone's reformulation is for the gratification of those readers who prefer symbolical to verbal reasoning. In his introduction, Heath also explained this circumstance when he described the principal translations and editions of the Elements. "The first six books 'demonstrated by symbols, by a method very brief and intelligible' by Pierre Hérigone, mentioned by Barrow as the only editor before him who had used symbols for the exposition of Euclid" (Heath, 1956, p.108). However, Barrow was partially mistaken, since Oughtred, in 1631, in the first edition of the *Clavis Mathematicae* had also rewritten some propositions of Euclid's *Elements* in symbolic language. Harriot had also done this even earlier but his version was never published and remains in manuscript form. See Stedall (2007, p.386). On the influence of Hérigone's *Cursus*, see Cifoletti (1990)

in both texts. The style of Hérigone's propositions, dividing his demonstrations into separate sections, is not found in Clavius's *Elements*. Moreover, Clavius, unlike Hérigone, describes the demonstration and the corresponding construction for each of his propositions and problems rhetorically.

Like Clavius in his *Prolegomena*, Hérigone's *Prolegomena* to the *Elements* discusses the classification of Mathematics; however, Hérigone did not follow Clavius's classifications. In Clavius' *Prolegomena* the order of the parts (Arithmetic, Music, Geometry and Astronomy) and the division into pure and mixed mathematics are the same as those by Proclus in his commentary.[22] In contrast, Hérigone ordered the four parts as Arithmetic, Geometry, Astronomy and Music and while like Clavius he considered mathematics to be divided into pure and mixed mathematics, Hérigone only mentioned Optic, Mechanics, Astronomy and Music as mixed.[23]

Hérigone, in accordance with Clavius, divides the fifteen books of Euclid's *Elements* into four parts[24] and this paragraph in both *Prolegomena* is identical word for word. There is a further part in Hérigone's *Prolegomena* called "The principles of Mathematics,"[25] which is also very similar to the corresponding part in Clavius. Both considered the principles of Mathematics as being divided into three types: the definitions, the postulates and the axioms or common notions.[26]

However, Hérigone goes further to add new "scholia" to Clavius's propositions, which he later uses to justify his demonstrations, and an appendix to

---

and Massa (2008, p.298–299).

[22] Hérigone explained that the Pythagoreans divided mathematics into four categories: arithmetic, geometry, astronomy and music. He said others divided mathematics into pure and mixed mathematics, specifying that in pure mathematics quantity was recognized as being separate from matter. He considered that pure mathematics should be divided according to the kind of quantity (either continuous or discrete) into geometry and arithmetic, and that mixed mathematics should be divided into optics, mechanics, astronomy and music. See Hérigone (1634, I, *Prolegomena*). Clavius also divided mathematics into pure and mixed Mathematics, pure Mathematics includes Arithmetic and Geometry and mixed Mathematics includes Astrology, Perspective, Geodesy, Canonical or Music, Calculation and Mechanics. See Clavius (1589, section II, *Prolegomena*).

[23] On the status of the mathematical disciplines in sixteenth century, see Axworthy (2004, p.62–80).

[24] The first part contains the first six books, which deal with planes. The second includes the subsequent three books, which deal with numbers. The third part contains only Book X, which deals with commensurable and incommensurable lines, while the last part is composed of the last five books, which treat the science of solids. See Hérigone (1634, I, *Prolegomena*). Like Clavius, Hérigone specifies the part corresponding to each book in the titles, for example, Book XI reads "The first book on the science of solids."

[25] Des principes des Mathematiques. /De principiis Mathematicis (Hérigone, 1634, I, *Prolegomena*).

[26] Hérigone claims that he added new axioms to the principles of Mathematics whenever he considered them necessary for the demonstrations. He specifies that he included a letter

Book VI, where Hérigone explains sums and products of lines, justifying them by propositions from his own *Elements*. In addition, Hérigone introduces this appendix with the claim that its problems and theorems are necessary for understanding Algebra and Astronomy.[27]

In fact, throughout the *Cursus*, Hérigone insists on the fundamental role of Euclid's *Elements* for understanding mathematics. Hérigone deals with geometry and arithmetic in the first and second volumes, respectively, and in the preface to the second volume he justifies treating geometry before arithmetic by claiming that geometry enables a better understanding of arithmetic:

> On the one hand, it is certain that knowledge of numbers is absolutely necessary for considering symmetry and incommensurability of a continuous quantity, of which Geometry constitutes one of the principal objects. On the other hand, there are some demonstrations in our arithmetic that cannot be understood without the help of the first books of Euclid's *Elements*.[28]

Moreover, when Hérigone discusses the importance of algebra in Volume VI (1642), he again stresses that the only requirement for solving the equations is an understanding of Euclid's *Elements*.[29]

We may assume that Hérigone believed that an understanding of Euclid's *Elements* also served a propaedeutic function in his *Cursus*.[30]

---

to distinguish his new axioms from Clavius's and Euclid's axioms.

[27] A ces six livres des Elements d'Euclide, j'adiousteray un appendix de divers problèmes & theoremes, dont les uns sont necessaires à l'Algebre, les autres à l'Astronomie ; /His sex elementorum Euclidis libris, annectam variorum problematum atque theorematum appendicem; quorum alia ad Algebram, alia ad Astronomiam. [To these six books of Euclid's *Elements*, I add an appendix with some problems and theorems, some of which are necessary for Algebra and others for Astronomy.] (Hérigone, 1634, I, p.302).

[28] Car d'un coté il est constant que la connaissance des nombres est absolument requise à la considération de la symétrie et incommensurabilité de la quantité continue, desquelles la Géométrie fait un de ses principaux objets ; et d'autre part, il y a des démonstrations en notre Arithmétique qui ne peuvent être entendues sans le secours des premiers livres des Eléments d'Euclide. /Quantitatis enim continuae symmetriam & incommensurabilitatem, quas praecipue inquirit Geometra nusquam intelliget imparatus à numeris : Neque ex adverso percipi possunt Aritmeticae nostrae quaedam demonstrationes, sine previa cognitione priorum elementorum Euclidis. (Hérigone, 1634, II, unpaginated)

[29] Supplément de l'Algèbre . Les équations d'Algèbre sont d'autant plus difficiles à expliquer qu'elles sont hautes en l'ordre de l'échelle. Et n'est pas besoin d'autres préceptes particuliers, que de l'intelligence des éléments d'Euclide pour trouver la valeur d'une racine constituée en sa base. /Omnis algebrae aequatio quo altiorem scalae tenet locum, eo difficiliorem habet explicationem. Nec ullo praecepto particulari, praeter Euclidis elementorum notitiam, opus est, ad exhibendum radicis in sua base existentis valorem. [Supplement on Algebra. The higher the degree of equations in algebra, the more difficult it is to solve them. There is no need for particular rules other than an understanding of Euclid's *Elements* to find the value of a root that constitutes the base [of the equation]]. (Hérigone, 1642, VI, p.1)

[30] On the propaedeutic function in Euclid's *Elements*, Tartaglia and Clavius, see Axworthy (2004, p.13–38).

Volume I of the *Cursus* includes a translation into French of Euclid's *Elements*, which Hérigone reformulates in his new symbolic language in an original way. So all Euclidean propositions are expressed using symbolic expressions; for example, Pythagoras's theorem in Proposition I.47 from Euclid's *Elements* is expressed as "$\square.bc\ 2|2\ \square.ab + \square.ac$."[31]

However, it is of the utmost importance to analyze how Hérigone replaces rhetorical language in the *Cursus* using his own *Elements* expressed in symbolic language. He introduces original symbols and abbreviations ("notes") and margin notes ("citations") to represent axioms, postulates and definitions. In fact, Hérigone classifies the citations used in the demonstrations as follows:

> There are seven types of citations in mathematical demonstrations, that is to say, the postulates, the problems, the definitions, the axioms, the theorems, the hypotheses and the constructions: of which the two first pertain to the construction or to the preparation and the other five to the demonstrations.[32]

His procedure for the citations is as follows: first, he writes the statement of the axiom, postulate or definition in rhetorical language similar to Clavius's *Elements*; second, he writes the symbol or abbreviation deduced from this axiom, postulate or definition, and finally, he offers an explanation of this abbreviation (*Explicatio notarum*). For example, the note "3.p.1." refers to Euclid's Postulate I. 3: "To describe a circle with any centre and distance" (see figure 5.5). Then Hérigone replaces Clavius's rhetorical language by these symbolic expressions and abbreviations defined previously. For example, where Clavius has "Centro *A*, & intervalo rectae *AB*, describatur circulus *CBD*," Hérigone writes "*abcd* est O" and notes in the margin "3.p.1.," referring to the sentence deduced from Euclid's Postulate I.3. Similarly, throughout Clavius's text Hérigone replaces rhetorical explanations by symbolic language. Let us take one example, the first proposition in Book I, where Hérigone uses this abbreviation and other similar ones in the construction and in the demonstration (see Figure 5.6). Hérigone's statement is expressed as follows: "On a finished straight line, to make an equilateral triangle."[33]

---

[31] In these demonstrations, Hérigone writes a paragraph headed "praeparatio" in which he expresses parallel lines using the symbol "==," angles using the symbol "<" and a right angle using the symbol "$\square$" (Hérigone, 1634, I, p.55–56).

[32] Aux demonstrations Mathematiques il y a sept genres de citations, à savoir, les postulats, les problèmes, les definitions, les axiomes, les theoremes, les hypotheses,& les constructions : desquels les deux premiers appartiennent à la construction, ou preparation, & les cinq autres à la demonstration. /In demonstrationibus Mathematicis sunt septem citationum genera, scilicet, postulata, problemata, definitiones, axiomata, theoremata, hypotheses, & constructiones: quorum duo priora, ad constructionem, aut praeparationem, reliqua quinque ad demonstrationem pertinent (Hérigone, 1634, I, Rrr iiij). This clarification is found at the end of volume 1 under the title: "Annotations on the first volume of *Cursus Mathematicus*".

[33] Sur une ligne droite donnée & terminée, descrire un triangle equilateral. /Super data

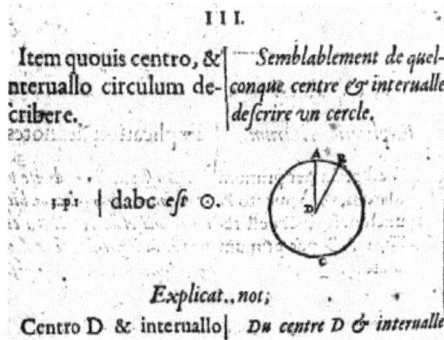

Fig. 5.5: Postulate I.3 of Hérigone's *Elements* (Hérigone, 1634, I, f. dvii^v).

In this proposition, both describe how to construct an equilateral triangle, Clavius using rhetorical explanations and Hérigone using his symbolic language with repeated references to Euclid's *Elements*. The connecting thread is the geometric construction of the solution. Hérigone replaces Clavius's rhetorical explanations and instructions with symbolic language. He proceeds by replacing each of Clavius's rhetorical sentences by his own corresponding abbreviation, and in the margin he makes a note referring to Hérigone-Euclid's propositions, postulates or axioms used and defined previously. For instance, when Clavius has, "Ex quarum utrovis, nempe ex C, ducantur duae rectae lineae CA, CB, ad puncta A & B," Hérigone writes the abbreviations "ac & bc, snt —" and makes a note in the margin "1.p.1.," thus referring to Postulate 1 of Book I of Hérigone's first volume. Similarly, for the demonstration, where Clavius has: "Quoniam rectae AB, AC, ducuntur ex centro A, ad circumferentiam circuli CBD, erit recta AC, recta AB, aequalis," Hérigone writes, "ac 2|2 ab" and makes a note in the margin "15.d.1.," referring to Euclid I. definition 15.

Like Clavius, Hérigone makes a new demonstration by syllogisms; however, the procedure is not exactly the same. Clavius makes the demonstration in a scholium and begins the sequence by the last syllogism of the demonstration. See for example, the order in Clavius's demonstration; he begins,

All triangles that have three equal sides[35] are equilateral.
The triangle ABC has three equal sides.
Therefore the triangle ABC is equilateral.

---

recta linea terminata, triangulum aequilaterum constituere. (Hérigone, 1634, I, p.1). The statement and the figure are identical to those of Clavius.

[35] Here Clavius makes a small letter "d" and in the margin he writes "d. 23.def." (Clavius, 1589, p.28).

ELEM.. EVCLID. LI. I.

PROBL. I. PROPOS. I.

SVPER data recta linea terminata, triangulum æquilaterum constituere.

Sur une ligne droicte donnée & terminée, descrire un triangle equilateral.

Hérigone's Notation	Modern Notation
Hypoth.	Hypothesis.
ab est—D	*AB* is a given straight line.
Req. π. fa.	It is required to make:
△*abc æquilat.*	*ABC* equilateral triangle.
Constr.	Construction.
3.*p*.1. *abcd est* ⊙,	I.postulate.3. *ABCD* is a circle of center *A* and distance *AB*,
3.*p*.1. *bace est* ⊙,	I.postulate.3. *BACE* is a circle of center *B* and distance *BA*,
1.*p*.1. *ac & bc, snt—,*	I.postulate.1 *AC* and *BC* are straight lines,
Symp.[34] △*abc est aequilat.*	Symperasma. I say that the triangle *ABC* is equilateral
Demonstr.	Demonstration.
Constr. abcd & bace snt ⊙,	Construction. *ABCD* and *BACE* are circles,
15.*d*.1. *ac* 2|2 *ab*,	I.definition.15. *AC* = *AB*,
15.*d*.1. *bc* 2|2 *ba*,	I.definition.15. *BC* = *BA*,
1.*a*.1. *ac* 2|2 *bc*,	I.axiom.1. *AC* = *BC*,
Concl.	Conclusion.
23.*d*.1. △*abc est aequilat.*	I.definition.23. *ABC* is an equilateral triangle.

Fig. 5.6: Proposition I.1 (Hérigone, 1634, I, p.1) and modern translations of Hérigone's notations.

The minor will be confirmed by this other syllogism:[36]

Clavius continues the demonstration by syllogisms until the first sentence, which is the construction of the circumference.

Hérigone also makes the same demonstration in a scholium, but explains the four syllogisms beginning with the first line of demonstration[37], and the major and minor premise as well as the conclusion can easily be identified in each syllogism. See Hérigone's demonstration by syllogisms:

> This demonstration is made by four syllogisms, as one can perceive from the number of citations.
>
> I SYLLOGISM.
>
> The straight lines traced from the centre to the circumference are equal to each other.
>
> But the straight lines AC & AB are traced from the centre to the circumference.
>
> Therefore the straight lines AC & AB are equal to each other.[38]

If we consider the citation: "I. definition. 15. $AC = AB$," we can see that the major premise is "I. definition. 15.," and that the minor premise is the line immediately preceding it: "$ABCD$ and $BACE$ are circles," and that the conclusion is: "$AC = AB$." For the second syllogism, Hérigone explains that it is the same as the first. The conclusions of the two first syllogisms serve for the minor premise in the third syllogism. Let us now consider the third syllogism:

> III SYLLOGISM.
>
> Things those are equal to the same are equal to each other.

---

[36] Omne triangulum habent tria latera aequalia, est equilaterum. Triangulum ABC, tria habet aequalia latera. Triangulum igitur ABC, est aequilaterum. Minorem confirmabit hoc alio syllogismo. (Clavius, 1589, p.28). The same sequence by syllogisms is found in the appendix by Alessandro Piccolomini entitled *Commentarium de Certitudine Mathematicarum Disciplinarum* (Roma, 1547). (Piccolomini, 1547, p.99 r–99v). According to Rommevaux (2005, p.52), Clavius makes no reference to Piccolomini, although he probably knew this work, which forms part of the debate on the certainty of mathematics in the sixteenth century. There are many useful works on this *quaestio*, including Mancosu (1996) and Romano (2007).

[37] This same order of syllogisms is found in the demonstration by syllogisms of Dasypodio's work on Euclid's *Elements*. (Dasypodio, 1566, A ij).

[38] Cette demonstration se fait par quatre syllogismes, comme il appert du nombre des citations. I. SYLLOGISME. Les lignes droites menées du centre à la circonference, sont égales entre elles. Mais les lignes droites AC & AB sont menées du centre à la circonference. Donc les lignes droites AC & AB sont égales entr'elles. /Haec demonstratio sit quatuor syllogismis, ut perspicuum est ex numero citationum. I SYLLOGISMUS. Rectae lineae quae ducuntur à centro ad circunferentiam, sunt inter se aequales. Sed rectae AC & AB ducuntur à centro ad circunferentiam. Igitur rectae AC & AB sunt inter se aequales. (Hérigone, 1634, p.1–2)

But the straight lines AC & CB are equal to the same straight line.
Therefore the straight lines AC & CB are equal to each other.[39]

In this case, "I. axiom. 1.$AC = BC$," the major premise is the first axiom, while the minor premise is deduced from the conclusions of the first and second syllogisms: $AC = AB$ and $BC = BA$, and the conclusion of the third syllogism is $AC = BC$. These conclusions enable the minor premise in the last syllogism to be deduced.

IV SYLLOGISM.
All triangles that have three equal sides are equilateral.
But the triangle ABC has three equal sides.
Therefore the triangle ABC is equilateral.[40]

In this case, "I. definition. 23. ABC is an equilateral triangle," the major premise is I.d.23, while the minor premise is deduced from the former conclusions $AC = AB$, $BC = BA$ and $AC = BC$, and the conclusion of the third syllogism is that "the triangle ABC is equilateral," which concludes the demonstration.

Hérigone makes no other demonstration by syllogisms and neither does he make any identification between symbolic lines of the demonstration and the premises of these syllogisms, although this may be deduced from his explanation in the preface: "The demonstration... included each one on a short line, which can be solved easily by syllogisms, because in the proposition cited as well as in that which corresponds to the citation one can find all parts of the syllogism."

Hérigone's originality resides not in demonstrating by syllogisms, but rather in recognizing that it is possible to identify all parts of the syllogism in symbolic lines, which transforms the demonstration by syllogisms into another one that is shorter and easier. Indeed, it is important to point out that Hérigone sought to introduce a new, briefer and more intelligible method for making demonstrations. Although the excess of abbreviations and new symbols may have caused his attempt to fail, there is no doubting the intelligence of Hérigone's approach.

---

[39] III. SYLLOGISME. Les choses égales à une mesme, sont égales entr'elles. Mais les lignes droites AC & CB sont égales à une mesme ligne droite. Donc les lignes droites AC & BC sont égales entr'elles. /III. SYLLOGISMUS. Quae eidem aequalia sunt, inter se sunt aequalia. Sed rectae AC & BC sunt eidem rectae aequales. Igitur rectae AC & BC sunt inter se aequales. (Hérigone, 1634, I, p.2).

[40] IV. SYLLOGISME. Tout triangle qui a trois costez égaux, est equilateral. Mais le triangle ABC a trois costez égaux. Donc le triangle ABC est equilateral. /IV. SYLLOGISMUS. Omne triangulum habens tria latera aequalia, est aequilaterum. Sed triangulum ABC tria habet aequalia latera. Igitur triangulum ABC est aequilaterum. (Hérigone, 1634, I, p.2).

5.4 The usefulness of Hérigone's new method

In this section we analyze two examples in order to show how Hérigone sometimes makes improvements on Viète's examples and Clavius's *Elements*, using his new method.

5.4.1 Equations in the Algebra

The first example refers to the treatment of equations in the *Algebra*. According to Hérigone, an understanding of Euclid's *Elements* is the basis for understanding arithmetic and solving equations. In *Algebra*, Hérigone used propositions from Euclid's *Elements* to justify algebraic demonstrations. Furthermore, all instructions, procedures and rhetorical explanations for geometrical constructions are replaced by Euclid's propositions and postulates, expressed or formulated in Hérigone's symbolic language. Thus, Euclid's *Elements* are deeply entrenched in the development of Hérigone's *Algebra*.

Algebra is a section in Volume 2, which consists of 20 chapters. Hérigone accepts Viète's view that the symbols of analytic art (or algebra) can be used to represent not just numbers but also values of any abstract magnitude.[41] Indeed, Hérigone explicitly distinguishes vulgar algebra, which deals with problems expressed in terms of numbers, from specious algebra, which deals with problems expressed in more general terms, by means of species or letters. This idea is very important because it is from Viète's algebra that mathematicians began to consider objects of algebra, the letters of which represent numbers and also figures, angles and lines.

Hérigone's Notation	Modern Notation
$ab - a2 \ 2|2 \ d2$	$xb - x^2 = d^2$
$b - a \prod d \prod a$	$\frac{(b-x)}{d} = \frac{d}{x}$
$a \ 2|2 \ \frac{1}{2}b + \sqrt{\begin{cases} b2\frac{1}{4} \\ -d2 \end{cases}}$	$x = \frac{1}{2}b + \sqrt{\frac{b^2}{4} - d^2}.$

Fig. 5.7: Modern translation of notations from Hérigone's *Algebra*.

As regards the treatment of equations, Hérigone, like Viète, transforms equations into a relationship between three proportional quantities. The key is the identification of the terms of an equation, both known and unknown quantities, as terms of a proportion. However, Hérigone always specifies whether

---

[41] On the comparison between Viète's and Hérigone's algebra, see Massa (2008).

the required quantity is a number or a line, and in the latter case he justifies the results by using propositions from Book VI of Euclid's *Elements*. Solving equations consists of three steps: equation, proportion and solution by rule (see Figure 5.7).

He again justifies the passage between these steps with the propositions of Euclid's *Elements*. In this equation d is the mean proportional and b is the sum of the extremes, both of which are given. He emphasizes that with these data that this equation can be solved geometrically by the scholium of Euclid VI.28. In fact, in the case that b is the difference of the extremes, he claims that he uses the scholium of Euclid VI.29, and in the other cases of equations he mentions propositions in the appendix to the six first books.

Moreover, if we analyze this scholium of Euclid VI.28 (see Figure 5.8), which is used to justify the rules of the quadratic equations and compare it with those of Clavius, we can say that the statement, the presentation and the procedure are different; in other words, this scholium is not present in Clavius's *Elements*. So this scholium was added by Hérigone, as he states:

> Given the mean of three proportional [straight lines] and the addition of the extremes, find the extremes.[42]

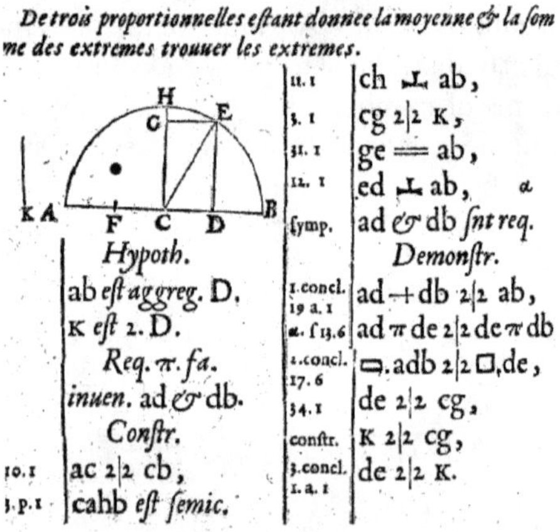

Fig. 5.8: "Scholium" of Euclid VI.28 (Hérigone, 1634, I, p.293)

The usual or classic figure in Euclid's *Elements*, with squares and rectangles or a semicircle with an inscribed triangle delimiting the diameter, is not found

---

[42] De trois proportionnelles estant donnee la moyenne & la somme des extremes trouver les extremes./ E serie trium proportionalium, data media & summa extremarum invenire extremas. (Hérigone, 1634, I, p.293–294).

in this construction. We can consider this new figure as Hérigone's canonical diagram for geometric constructions for quadratic equations. Hérigone states two rules for finding the value of the unknown in an equation with three terms where the degree of the highest power is double that of the lower power, and both rules are also illustrated by this same figure. In fact, one finds this figure and the reference to this scholium throughout the *Cursus*.[43] For example, when Hérigone deals with irrational numbers in *Algebra* he again explicitly cites and uses scholium VI.28 for justifying geometrically his method for finding the root of a binomial. Using the same figure as that used in the previous scholium, in the demonstration, Hérigone states:

> To find the square root of a given binomial. Let us assume that the bigger number of the binomial is the sum of sides and the smaller number (of the binomial) is four times the rectangle determined by the sides, therefore one will find the root by the scholium 28.6, as follows.[44]

In contrast, Clavius in his *Algebra* (1608, p.150) explains three rules for finding the square root of a binomial in rhetorical language; then he gives an example for every rule with an explanation consisting of two pages. Hérigone makes the demonstration in half a page, does not use the same numerical example as Clavius, and the mathematical procedure and presentation are also very different.

This new scholium and its figure allow Hérigone to make some new demonstrations and to illustrate new rules. We show this scholium as an example of Hérigone's new threads achieved with his new method of demonstration.[45]

5.4.2 Book X Definitions

Another example focusses on Hérigone's treatment of the first definitions in Book X of Euclid's *Elements*. Book X introduces the Euclidean theory of irrationality; it is difficult and full of definitions in a geometric context, but Hérigone, unlike Clavius in his Book X, always provides examples referring to

---

[43] In Hérigone's *Elements* there are many examples: in propositions II.6 and II.29, in scholium of proposition II.5, in propositions VI.29, X.16, X.18 and X.19.

[44] Extraire la racine quarrée d'un binôme donné. Soit supposé que le plus grande nombre du binôme est l'aggregé des costez, & le moindre nombre le quadruple du rectangle contenu sous les costez, puis on trouvera la racine par le scholie de la 28 du 6, comme s'ensuit. /Ex dato binomio extrahere radicem quadratam. Finge maius nomen binomij dati esse aggregatum laterum, minus nomen quadruplum rectanguli sub lateribus comprehensi, deinde invenietur quaesita radix per scholium 28.6 sic (Hérigone, 1634, II, p.254).

[45] For more detailed examples and improvements see Massa (2008).

numbers and translates the demonstrations in symbolic language.[46] Moreover, he adds some scholia and a new classification to clarify these ideas.

Both Clavius and Hérigone present all eleven of Campanus's definitions; however, Hérigone makes an addition with four scholies to clarify these ideas. Definitions X.1 and X.2 define magnitudes to be commensurable when measured by the same common measure and otherwise incommensurable. Definitions X.3 and X.4 concern commensurable straight lines in square; Hérigone states that they are commensurable in square when the squares on them are measured by the same area and otherwise incommensurable in square. Let us take the first definitions: commensurability and incommensurability.

> Commensurable magnitudes are those that are measured by the same common measure.[47] But incommensurable magnitudes are those that do not have any common measure.[48]

After the fourth definition, Hérigone introduces examples in numbers. He states that the lines $a = 7$ and $b = 5$ are commensurable in length (as always, first in symbols and then with his explanation of abbreviations). The lines $b = 5$, $c = \sqrt{10}$ and $d = \sqrt{8}$ are commensurable in square because the squares 25, 10 and 8 are commensurable in length. The lines $e = \sqrt{\sqrt{10}}$ and $f = \sqrt{\sqrt{8}}$ are incommensurable in square because the squares $\sqrt{10}$ and $\sqrt{8}$ are incommensurable in length.

In definitions X.5–X.7 Hérigone describes the rational straight line. Taking a rational straight line as a reference, the other lines commensurable in length and in square are called rational. And the lines incommensurable with respect to this line are termed irrational. In definitions X.8, X.9, X.10, and X.11, Hérigone describes rational and irrational figures.

Hérigone, unlike Clavius, adds four scholia to clarify the concepts. In the first scholium, Hérigone clarifies that incommensurable magnitudes cannot become commensurable, while irrational magnitudes can become rational ones by changing the rational that one takes as a reference. In fact, for Hérigone the notions of commensurable and rational are not parallel at all.

In the scholia II and III Hérigone specifies the relation of incommensurability in numbers by taking unity as the reference. In fact, the incommensurable numbers with respect to the unit are called irrationals or "surds." However,

---

[46] On the treatment of Book X, there are many interesting works including Fowler (1992) and Rommevaux (2001).

[47] Commensurables grandeurs sont celles-là lesquelles sont mesurees par une mesme commune mesure. /Commensurabiles magnitudines dicuntur, quas eadem mensura metitur (Hérigone, 1634, I, p.486). The text of this statement in Latin is identical in Clavius' s *Elements*.

[48] Mais les grandeurs incommensurables sont celles-là, lesquelles n'ont aucune commune mesure. /Incommensurabiles autem sunt, quarum nullam communem mensuram contingit reperiri (Hérigone, 1634, I, p.486).

and more importantly, he specifies the idea of a rational line expressed by an irrational number that can be rational if its square is expressed by a rational number.

In scholium IV Hérigone, unlike Clavius, introduces a "new" classification of the rational lines commensurable in length to each other according to three types illustrated by a geometric construction (see figure 5.9).

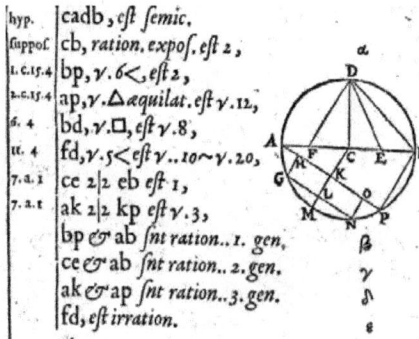

Fig. 5.9: Classification of the rational lines in Book X (Hérigone, 1634, I, p.491).

Having constructed an example, he takes the semi-diameter of the value 2 as the unit of measure; he then constructs the other lines using as major premises both corollaries of Euclid IV.15, which states: "In a given circle to inscribe an equilateral and equiangular hexagon,"[49] Euclid IV.6 states: "In a given circle to inscribe a square."[50], Euclid IV.11 states: "In a given circle to inscribe an equilateral and equiangular pentagon,"[51] and the common notion Euclid I. axiom. 7 states: "And things which are half of another thing or equal things are also all equal to each other."[52] He concludes that if $cb = 2$ is the rational line of reference; $bp = 2$ and $ab = 4$ are rational lines of the first type, because they are commensurable in length and one of them is equal to the rational reference; $ce = 1$ and $ab = 4$ are rational lines of the second type because they are commensurable in length and none of them is equal to the

---

[49] En un cercle donné, inscrire un hexagone equilateral & equiangle. /In dato circulo, hexagonum & equilaterum & equiangulum inscribere (Hérigone, 1634, I, p.169).

[50] Dans un cercle donné, inscrire un carré. /In dato circulo quadratum describere (Hérigone, 1634, I, p.156).

[51] En un cercle donné, inscrire un pentagone equilateral & equiangle. /In dato circulo, pentagonum & equilaterum & equiangulum inscribere (Hérigone, 1634, I, p.161).

[52] Et les choses qui sont moitiés d'une mesme, ou des choses egales, sont aussi egales entre elles. /Et quae eiusdem, vel aequalium sunt dimidia inter se sunt aequalia (Hérigone, 1634, I, p. c.iiij).

rational reference. For the third type $ak = \sqrt{3}$ and $ap = \sqrt{12}$ are rational because they are only commensurable in square with the rational reference and $fd = \sqrt{10 - \sqrt{20}}$ is irrational because it is neither commensurable in square nor in length.

Hérigone later uses this new classification of the rational lines and this figure to solve a problem with irrational numbers in the *Algebra*. The question consists in finding the side of a regular pent decagon inscribed in a circle. Hérigone states: "To find the side of a regular pent decagon inscribed in a given circle."[53]

5.5 Some final remarks

The first remark to be made is that, since Hérigone mentioned that he used the Latin version of Clavius (1589) to write his *Elements*, we have verified the statements of definitions in Latin and they turn out to be mostly identical. However, unlike Clavius, after every statement Hérigone gives no rhetorical explanations. Moreover, Hérigone adds some scholia and an appendix in order to explain the mathematics better.

He reformulates Clavius's *Elements* by using his symbolic language in an original way. Thus, Hérigone in the *Cursus* avoids rhetorical explanations and seeks to express all phrases symbolically. The steps are justified by citations referring to the propositions, axioms, postulates and definitions from Euclid's *Elements*, which are formulated in symbolic language in Volume 1 as well. We may surmise that Hérigone's presentation of this justification is once more a reflection of the great significance that Euclid's *Elements* held for him.

After showing Hérigone's examples in our Section 3, his procedure of replacing the rhetorical language of Euclid's *Elements* by symbolic language to make demonstrations in the *Cursus* becomes clear. This new method of demonstration using a universal language and logical sentences through an axiomatic-deductive reasoning is absolutely original and offers us the logical and clear structure of his thinking. Moreover, Hérigone emphasizes that his method is useful for both pure and mixed mathematics and he applies it in all parts of the *Cursus*. Perhaps the idea of extending his method to all mathematics arose from his reading of Clavius's *Elements*. Indeed, Clavius in the first demonstration of the first book after the demonstration by syllogisms claims that in this manner one can solve all Euclid's propositions as well as those of all other mathematicians.

---

[53] Quaestion III. Cap. XIX. Trouver le coté d'un quindecagonregulier inscrit dans un cercle donné. /Invenire latus quintidecagoni ordinati dato circuli inscripti (Hérigone, 1634, II, p.261).

However, in order to highlight the highly unusual relationship between the classical mathematics that Euclid's work represents and new algebra as it appears in the work of Hérigone, we would like to discuss whether the symbolic language introduced in Hérigone's *Elements* is useful to obtain new results or whether it is a different way of arriving at the same results. In other words, did Hérigone actually perform an algebraization of the *Elements*? I believe that he did not, at least not completely; Hérigone indeed translates the different notions, interprets the statement and the demonstrations in terms of symbolic notation, and at the same time gives numerical examples, but without overlooking the geometric context. He uses figures, abbreviations and symbols to establish and reinforce these meanings as well as to make them meaningful for his readers.

The strategies employed by Hérigone in order to render Euclid's geometry and all his mathematics intelligible to his audience also enable him to make some improvements. In my previous article on Hérigone's *Algebra* and in the examples from Section 4, it was shown that this different writing in logical statements allows him to obtain some new demonstrations, some new rules, some new paths and some new classifications. In fact, in the preface "to the reader," in the first volume of the *Cursus*, Hérigone lays claim to his contributions by stating:

> Those who love these divine sciences [Mathematics] may judge what I have contributed on my own behalf in every part of this *Cursus*, which I trust will prove to be of use and profit to them.[54]

References

1. Axworthy, Angela, 2004. *Le statut des disciplines mathématiques au XVIe siècle au regard des préfaces aux Éléments d'Euclide de Niccolò Tartaglia et de Christophe Clavius.* "mémoire de maîtrise." Tours: Centre d'Études Supérieures de la Renaissance.
2. Barrow, Isaac, 1659. *Euclidis elementorum libri xv brevi demonstrati.* London.
3. Bos, Henk J. M., 2001. *Redefining geometrical exactness: Descartes' Transformation of the Early Modern Concept of Construction.* New York: Springer-Verlag.
4. Cajori, Florian, 1928. *A History of Mathematical Notations.* I. *Notations in Elementary mathematics.* II. *Notations Mainly in Higher Mathematics.* La Salle, Illinois: Open Court, 1928–29 (Dover edition, 1993).
5. Cifoletti, Giovanna, 1990. "La méthode de Fermat: son statut et sa diffusion". *Cahiers d'histoire et de philosophie des sciences,* nouvelle série, vol. 33. Paris: Société française d'histoire des sciences et des techniques.

---

[54] Ceux qui aiment ces divines sciences iugeront ce que j'ay apporté du mien en chacune partie de ce Cours, que je souhaite qu'il leur soit utile & profitable. /Quid autem in singulis huius Cursus partibus praestiterim, iudicabunt studiosi, quibus opto hunc meum laborem utilem esse (Hérigone, 1634, I, *Ad Lectorem*).

6. Clavius, Christopher, 1589. *Euclidis Elementorum libri XV, accessit XVI de solidorum regularium [...] nunc iterum editi, ac multorum rerum accessione locupletati*, Rome: apud Sanctium et Socios.

7. Clavius, Christopher, 1608. *Algebra*. Rome: B. Zanetti.

8. Dasypodius, Konrad ; Herlinus, Christiano, 1566. *Analyseis geometricae sex librorum Euclidis*, Strasbourg.

9. Fowler, D. H., 1992. "An Invitation to Read Book X of Euclid's *Elements*". *Historia Mathematica* **19**, 233–264.

10. Heath, Thomas L. (Ed.), 1956. Euclid. The Thirteen Books of *The Elements*. New York: Dover.

11. Hérigone, Pierre, 1634/1637/1642. *Cursus Mathematicus nova, brevi et clara methodo demonstratus, Per NOTAS reales & universales, citra usum cuiuscumque idiomatis, intellectu, faciles/ Cours Mathematique demonstré d'une nouvelle briefve et Claire methode. Par notes reelles & universelles, qui peuvent estre entendues sans l'usage d'aucune langue*. 5 vols. (1634, 1637) plus a supplement (1642). Paris: For the author and Henry Le Gras.

12. Hérigone, Pierre, 1639. *Les six premiers livres des Éléments d'Euclide, démonstrez par notes d'une méthode très brière et intelligible*. Paris: For the author and Henry Le Gras (reprinted in 1644 Paris: for Simeon Piget).

13. Hérigone, Pierre, 1644. *Cursus Mathematicus nova, brevi et clara methodo demonstratus, Per NOTAS reales & universales, citra usum cuiuscumque idiomatis, intellectu, faciles*, 6 vols., second edition. Paris: For Simeon Piget.

14. Knobloch, Eberhard, 1988. "Sur la vie et l'oeuvre de Christophore Clavius (1538–1612)". *Revue d'histoire des sciences* XLI-3/4, 331–356.

15. Knobloch, Eberhard, 2001. "Klassifikationen". In Menso Folkerts, Eberhard Knobloch, Karin Reich (eds.) *Maß, Zahl und Gewicht, Mathematik als Schlüssel zu Weltverständnis und Weltbeherrschung*, 2nd edition Wolfenbüttel: Herzog August Bibliotheck. (First edition 1989), 5-14.

16. Mahoney, Michael S., 1980. "The beginnings of algebraic thought in the seventeenth century". In: Gaukroger, S. (Ed.), *Descartes: Philosophy, Mathematics and Physics*. Barnes and Noble/ Harvester, Totowa/ Brighton, 141–156.

17. Mancosu, Paolo, 1996. *Philosophy of Mathematics and Mathematical Practice in the Seventeenth Century*. Oxford: Oxford University Press.

18. Massa Esteve, Maria Rosa, 1997. "Mengoli on 'Quasi Proportions'". *Historia Mathematica* **24**, 257–280.

19. Massa Esteve, Maria Rosa, 2003. "La théorie euclidienne des proportions dans les 'Geometriae Speciosae Elementa' (1659) de Pietro Mengoli". *Revue d'histoire des sciences* **56**, 457–474.

20. Massa Esteve, Maria Rosa, 2006a. "Algebra and Geometry in Pietro Mengoli (1625–1686)". *Historia Mathematica* **33**, 82–112.

21. Massa Esteve, Maria Rosa, 2006b. "L'algebrització de les matemàtiques. Pietro Mengoli (1625–1686)". Barcelona: Societat Catalana d'Història de la Ciència i de la Tècnica, Filial de l'Institut d'Estudis Catalans.

22. Massa Esteve, Maria Rosa, 2008. "Symbolic language in early modern mathematics: The *Algebra* of Pierre Hérigone (1580–1643)". *Historia Mathematica* **35**, 285–301.

23. Massa Esteve, Maria Rosa; Delshams, Amadeu, 2009. "Euler's Beta integral in Pietro Mengoli's works". *Archive for History of Exact Sciences*, **63**, no. 3, 325–356.

24. Panza, Marco, 2005. *Newton et les origines de l'analyse, 1664–1666*. Paris: Blanchard.

25. Piccolomini, Alessandro, 1547. *In mechanicas quaestiones Aristotelis Paraphrasis paulo quidem plenior... Eiusdem commentarium de certitudine Mathematicarum Disciplinarum....* Rome: Bladum Asulanum.

26. Pycior, Helena M., 1997. *Symbols, Impossible Numbers, and Geometric Entanglements: British Algebra through the Commentaries on Newton's Universal Arithmetic.* Cambridge: Cambridge University Press.
27. Romano, Antonella, 2007. "El estatuto de las matemáticas hacia 1600". In *Los Orígenes de la Ciencia Moderna.* Actas Años XI y XII. Canarias: Fundación Orotava de Historia de la Ciencia, 277–308.
28. Rommevaux, Sabine, 2001. "Rationalité, exprimabilité: une relecture médiévale du livre X des *Éléments* d'Euclide". *Revue d'Histoire des mathématiques,* **7**, 91–119.
29. Rommevaux, Sabine, 2006. *Clavius une clé pour Euclide au XVIe siècle.* Mathesis, Paris: Vrin.
30. Stedall, Jackie A., 2002. *A discourse concerning algebra: English algebra to 1685.* Oxford: Oxford University Press.
31. Stedall, Jackie A., 2007. "Symbolism, combinations, and visual imagery in the mathematics of Thomas Harriot". *Historia Mathematica,* **34**, 380–401.
32. Stromholm, Per, 1972. Hérigone. In: Gillispie, C.C.(ed.) *Dictionary of Scientific Biography,* Scribner, 6, New York, 299.
33. Viète, François, 1591. *In artem analyticen isagoge. Seorsim excussa ab Opere restituaeæ mathematicaeæ analyseos, seu algebra nova.* Tournon: apud Iametium Mettayer typographum regium.
34. Viète, François, 1646. *Opera Mathematica.* Edition by Frans Van Schooten. Leyden. (Reprint Hildesheim: Olms, 1970).
35. Wallis, John, 1685. *A treatise of Algebra both Historical and Practical showing The Original, Progress, and Advancement thereof, from time to time; and by what Steps it hath attained to the Height at which now it is.* London: J. Playford for R. Davis.

Chapter 6
What more there is in early modern algebra than its literal Formalism

Marco Panza

Abstract It is often agued that early-modern algebra essentially resulted from the adoption of a new literal formalism (mainly thanks to the works of Viète, Harriot, Descartes, etc.). I will try to question this claim. I will not deny that this formalism plays a crucial role and I will even wonder which is this role. But I am suggesting that its introduction supervenes on a change in the conception concerning geometrical magnitudes, problems and theorems that has quite old roots. The paper is devoted to describing this change, by considering a single example.

Key words: literal formalism, Early Modern Algebra, Euclid, al-Khwārizmī, Thābit ibn Qurra

6.1 Introduction

There are two views about early-modern algebra very often endorsed (either explicitly or implicitly). The former is that in early-modern age, algebra and geometry were different branches of mathematics and provided alternative solutions for many problems. The latter is that early-modern algebra essentially resulted from the adoption of a new literal formalism. My present purpose

IHPST (CNRS and University of Paris 1, ENS Paris).

is to question the latter. In doing that, I shall also implicitly undermine the former[1].

I shall do it by considering a single example. This is the example of a classical problem. More in particular, I shall consider and compare different ways of understanding and solving this problem. Under the first understanding I shall consider, this problem appears as that of cutting a given segment in extreme and mean ratio. This is what proposition VI.30 of the *Elements* requires. In section 6.2, I shall expound and discuss Euclid's solution of this proposition[2]. Then, in section 6.3, I shall consider other propositions of the same *Elements*, and argue that they suggest another, quite different understanding of the same problem. Under this other understanding, this is viewed as the problem of constructing a segment meeting a certain condition relative to another given segment. One way to express this condition is by stating the first of the three trinomial equations studied in al-Khwārizmī's *Algebra*, by supposing that this equation is geometrically understood and a particular case of it is considered[3]. In section 6.4, I shall consider this option, by focusing in particular, on Thābit ibn Qurra's interpretation and solution of this equation.

---

[1] For a more articulated argumentation against this former view, cf. my [(Panza, 2007)]. The term 'early-modern algebra' is quite vague and open to many different understandings. What I mean by it is essentially the practice of dealing both with arithmetical and geometrical problems through a common approach mainly originated by Viète's and Descartes's achievements. Hence, early-modern algebra has to be conceived, in my parlance, as being essentially about geometrical concerns.

[2] According to its common use in geometry, the term 'solution' is equivocal: it can respectively denote the way a problem is solved, the argument that one relies on to solve a problem, the action of solving a problem, or even the object (or objects) whose construction is required by a problem . To avoid unfamiliar phrases, in what follows I shall have no other option than using this term in these different senses, on different occasions. I hope the context will be enough to avoid any misunderstanding. In particular, when I shall claim that the problem I shall consider admits a unique solution, if it is conceived as Euclid does, I shall not mean that there is only one way to solve it, or only one argument allowing to solve it, within Euclid's geometry. I shall rather mean that, supposing that a certain segment be given, in order to solve this problem so conceived, one has to construct a certain point (on this segment) which is univocally determined by the conditions of the problem. In other terms, my point will be that, this segment being given, each way of solving this problem so conceived, or each argument through which it is solved, have to result in the construction of the same point (on this segment).

[3] In my [(Panza, 2007)], I preferred avoiding to use the term 'equation' to refer to conditions like those that al-Khwārizmī's and, after it, al-Khayyām's *Algebra* ([Woepcke, 1851)]; [(Rashed, 1999)]), are about. I used instead the term 'equations-like problem' to refer to the problems associated to these conditions. My reason was that, for al-Khwārizmī and al-Khayyām, stating such a condition was not the same as presenting a mathematical object, as it is the case when a polynomial equation is presented in the context of the formalism of early-modern algebra ([(Panza, 2007)], p. 124). I still think that this reason is good. But, for short and simplicity, I do not follow here my previous convention and I call al-Khwārizmī's conditions 'equations', as it is usually done. I hope the previous remark to be enough for

The transition from the former understanding to the latter hinges on the transition from a way of conceiving geometrical magnitudes and the relative problems and theorems, to another way of doing the same[4]. Whereas the former conception is proper to Euclid's geometry, the latter is proper to early-modern algebra. My main point will be that this latter conception is independent of the appeal to any literal formalism, as is shown by the fact that it is already at work in Medieval Arabic geometry, in which there is no trace of such a formalism. Early modern algebra actually resulted from combining this conception with an appropriate use of a symbolic notation, which gave raise to the new literal formalism. In my view, this use was made possible by the adoption of this conception, but it was not merely a natural outcome of it. Hence, though early-modern algebra could have not development without the adoption of this conception, the latter cannot be reduced to the former. The purpose of section 6.5, the last one of my paper, will be that of discussing this connection.

6.2 Proposition VI.30 of the *Elements*

Proposition VI.30 of the *Elements* is a problem. It requires "to cut a given segment in extreme and mean ratio"[5]. Let AB be the given segment (Figure 6.1). The problem requires, in other terms, to construct a point E on it, such that

$$AB : AE = AE : EB. \tag{6.1}$$

Euclid's solution relies on that of proposition VI.29. This is a problem in turn, and requires "to apply to a given segment a parallelogram equal to a given rectilineal figure with excess of [another] parallelogram similar to a [third] given one". If the given segment is CA, the given rectilineal figure is

---

avoiding any misunderstanding: when this term is used to refer to these conditions, it has not to be taken to have the same sense as when it is used in the context of early-modern algebra.

[4] A quite similar transition (perhaps the same one, even if conceptualized in a partially different way) is described by R. Netz in his [(Netz, 2004)]. About the convergences and divergences of Netz's and my views on this matter, cf. [(Panza, 2007)], p. 117 (footnote 46). Also Netz's discussion is focused on a single problem. This is significantly more complex that the one I shall consider here. Namely, differently from this latter one, it is not solvable within Euclid's geometry, that is, through a construction by ruler and compass (or elementary construction, according to the parlance I adopted in [(Panza, 2010)]). Far from undermining the effectiveness of my example, the relative straightforwardness of the problem I shall consider is rather intended to make clear that the shift I want to describe does not depend on the adoption of any mathematical tools exceeding the simple ones proper to Euclid's geometry, but was just a shift in conception.

[5] For my quotations from the *Elements*, I base myself on Heath's translation ([(Heath, 1926)]), though I slightly modify it somewhere.

A ———————————— E ———————— B

Fig. 6.1: Proposition VI.30 of the *Elements*

Γ, and the given parallelogram is Δ (Figure 6.2),[6] the problem consists in producing CA up to a point G such that the parallelograms CFDG and AEDG be respectively equal to Γ and similar to Δ.

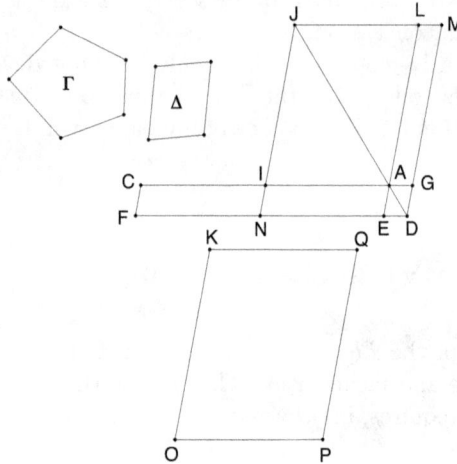

Fig. 6.2: Proposition VI.29 of the *Elements*

To solve this last problem, Euclid begins by taking the middle point I of CA, and constructing on IA the parallelogram IALJ similar to Δ. Then, he constructs a parallelogram OPQK similar to IALJ and Δ, and equal to Γ and the same IALJ taken together. At this point, it is enough to produce IJ up to a point N such that NJ = OK, and then to complete the parallelogram NDMJ in such a way that its diagonal DJ be collineal with the diagonal AJ of IALJ, and to produce CA up to the point, called G, where it cuts the side DM of this parallelogram. The point G will be the searched after one[7].

---

[6] For the purpose of denoting with the same letters the points with play an analogues role in the following constructions related to the solution of proposition VI.30, I change some of the letters in Euclid's diagrams as they appear in [(Heath, 1926)].

[7] The proof is simple. By construction, the equalities OPQK = Γ + IALJ = NDMJ hold. Hence Γ is equal to the gnomon NDMLAI. But, as I is the middle point of CA, and NDMJ is similar to IALJ, NEAI is equal both to FNIC and to AGML. Hence FNIC is equal to AGML, and then FDGC is equal to the gnomon NDMLAI, and consequently to Γ. On the other side,

To construct the middle point of a given segment is easy. According to the solution of propositions I.1 and I.9-10, it is enough to describe two circles having this segment as a common radius and its extremities as the respective centers, and to join their intersection points (Figure 6.3).

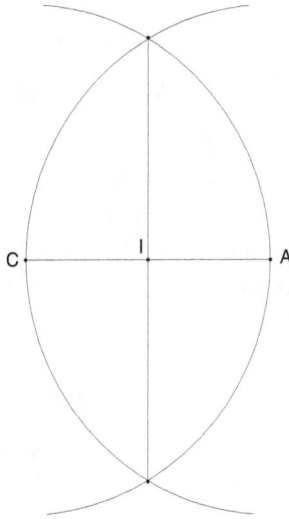

Fig. 6.3: A construction used in propositions I.9-10 of the *Elements*

It is also easy to construct a parallelogram similar to another given parallelogram on a given segment. Consider the case under examination. The segment IA being given (Figure 6.4), it is enough: to take on IA a point R such that RA be equal to the side ra of Δ; to construct on RA (according to the solution of proposition I.22) a triangle RAS having its sides equal to the sides of the triangle ras formed by tracing the diagonal rs of Δ (so that $\widehat{RAS} = \widehat{ras}$, according to proposition I.8 and the solution of proposition I.23); to trace the parallel IT to RS through I; to produce AS up to meet this parallel in L; and to complete the parallelogram IALJ.

Hence, the only critical step in the solution of proposition I.29 is the construction of the parallelogram OPQK (Figure 6.2). Euclid does not detail this construction, but it is easy to see that, in order to perform it, one has first to construct a rectilineal figure equal to Γ and IALJ taken together, and then apply the solution of proposition IV.25.

IALJ is similar to Δ by construction, with as consequence that also EDGA is so.

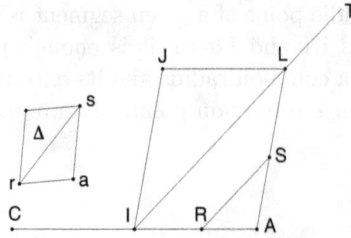

Fig. 6.4: The construction of a parallelogram similar to another given parallelogram on a given segment

This can be done in different ways, depending on the nature of such a rectilineal figure. An obvious possibility is to take this figure to be the parallelogram VALU (Figure 6.5) which is got by constructing on IJ and in the angle $\widehat{CIJ}$ the parallelogram VIJU equal to Γ.

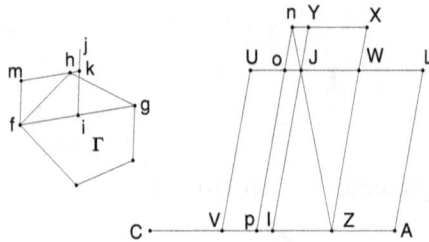

Fig. 6.5: Applying the solution of propositions I.42, I.44-45

This construction goes as follows, according to propositions I.42, and I.44-45. If Γ is not a triangle, divide it into several (non-overlapping) triangles by tracing appropriate segments joining two non-consecutive vertexes of it. Construct the middle point i, of a side fg of one of these triangles fgh. With i as vertex, construct the angle $\widehat{fij}$ equal to $\widehat{CIJ}$ (which can be done as explained above, according to the solution of propositions I.23) Through h, trace mk parallel to fg, respectively meeting in m and k the parallel to ij through f and ij itself). The parallelogram fikm is equal to the triangle fgh, and its internal angle $\widehat{fik}$ is, by construction, equal to the angle $\widehat{CIJ}$. It is then easy to construct another parallelogram JWXY equal to fikm, and consequently to fgh, and having the sides JY and JW collineal to IJ and JL, respectively (it is enough to produce IJ up to a point Y such that JY = fm and take a point W on JL such that JW = fi). Produce then XW up to meet CA in Z and complete the parallelograms ZXnp and IJop in such a way that the diagonal of the former be collineal to the diagonal ZJ of the other parallelogram IZWJ. The parallelogram plJo is equal to JWXY, and then to fgh. By repeating the same construction for all the triangles composing Γ, one then gets, step by step, the sought after parallelogram VIJU.

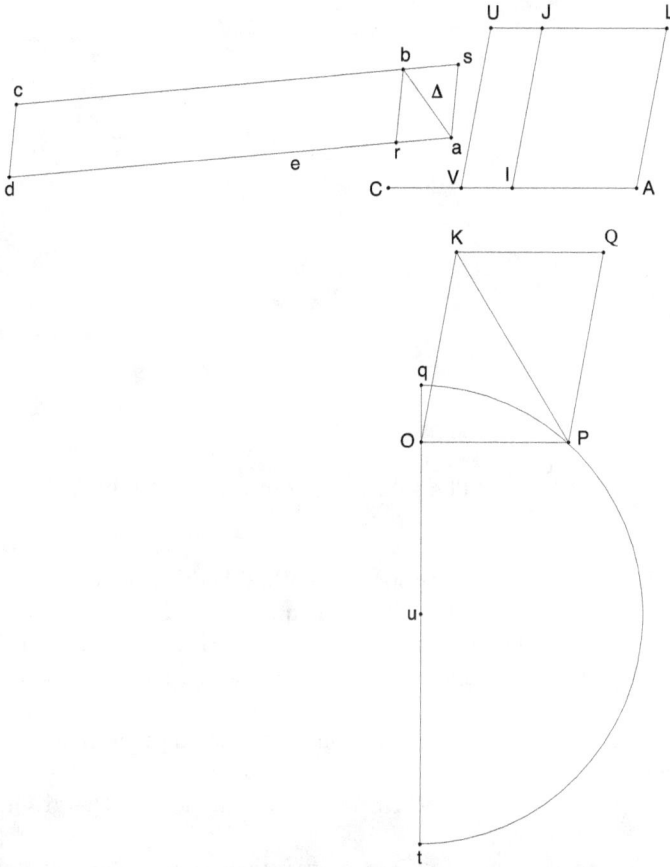

Fig. 6.6: Applying the solution of proposition IV.25

Once this last parallelogram has been constructed, the construction of the parallelogram OPQK (Figure 6.6) goes as follows, according to the solution of proposition VI.25. Proceeding in the same way as in the construction of the parallelogram VIJU, construct, on the side rb of Δ and in the angle $\widehat{erb}$ equal to $\widehat{ras}$, the parallelogram drbc equal to VALU. According to the solution of proposition IV.13, construct a segment OP mean proportional between dr and ra (supposing that tO = dr and Oq = ra, and that u is the middle point of tq, OP is the perpendicular through O to tq up the the circle of centre u and diameter tq). Then construct the two angles $\widehat{\mathsf{KOP}}$ and $\widehat{\mathsf{OPK}}$ equal to $\widehat{\mathsf{bra}}$ and $\widehat{\mathsf{rab}}$, respectively, and complete the triangle OPK and the parallelogram OPQK.

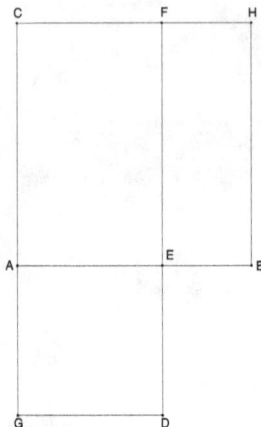

Fig. 6.7: The solution of proposition VI.30

At this point, the solution of proposition VI.30 is easy. It includes two steps. The former consists in constructing the square ABHC on the given segment AB (Figure 6.7). The latter consists in applying to the side AC of this square a rectangle equal to this same square with excess of a parallelogram similar to such a square, *i. e.* with excess of a square. This reduces to the construction, according to the solution of proposition VI.29, of the rectangle GDFC equal to the square ABHC and such that AG = GD. The sought after point is the intersection point E of the side DF of this rectangle and the given segment AB[8].

When applied to the case considered in proposition VI.30, the solution of proposition VI.29 is simplified with respect to the general case expounded above. This is because the role of both the rectilineal figure Γ and the parallelogram Δ is played by the same figure, and this figure is a square, namely the square ABHC. Let us see how this solution applies in this case.

Let the square ABHC be given (Figure 6.8). Construct the middle point I of AC, and, on AI, the square LAIJ. According to the general procedure expounded above, one should then construct the rectangle JIVU equal to ABHC.

---

[8] The proof is easy. As GDFC = ABHC and AEFC is a common part of both, GDEA = EBHF.

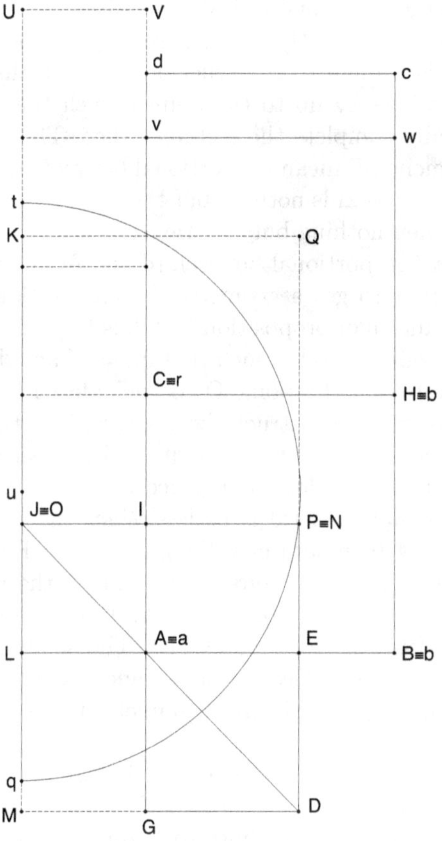

Fig. 6.8: Resuming Euclid's whole argument for solving proposition VI.30

There would be a quite simple way to do it without following the general step-by-step procedure expounded above: insofar as ABHC can be divided in four squares equal to LAIJ, it would be enough to produce LJ up to a point U such that JU be equal to 4 times LJ, and then complete the rectangle JIVU (which would be the same as constructing four squares equal to LAIJ on each other). But this is useless, in fact. Insofar as the role of the parallelogram Δ is played by the square ABHC, one can take, indeed, the side CH of this square to coincide with the side rb of Δ (compare Figure 6.8 with Figure 6.6) to the effect that, after having constructed the rectangle JIVU, one should construct on CH ≡ rb the rectangle rbcd equal to LAVU. Now, as this last rectangle

_____

Hence, according to proposition I.14: GA : EF = EB : AE, that is: AB : AE = AE : EB, as required.

would be, by construction, equal to LAIJ and ABHC taken together, this step can easily be performed directly, without relying on the construction of the rectangle JIVU. It is enough to construct on CH ≡ rb the square rbwv equal to ABHC, then produce Cv up to the point d such that vd be equal to the half of AI, and finally complete the rectangle rbcd. The next step consists in constructing a segment OP mean proportional between rd and the homologous side of △. But, insofar as △ is nothing but the square ABHC, a coincides with A and this side is then nothing but AC. Hence, what has to be constructed is a segment OP mean proportional between rd and AC ≡ ar. The more natural way to do it, in order to get a geometrical configuration analogous to that which enters the solution of proposition VI.29, is to perform this construction on the segment qt collineal to LJ, such that qL = LJ and Jt = rd (to the effect that qt = Ad), by taking the point O to coincide with the point J. If u is the middle point of qt, to construct the point P, it is then enough to trace the semicircle of center u and diameter qt, and the perpendicular OP to qt through J up to this semicircle. If one proceeds this way, the construction of the square OPQK equal to CHcd is useless, since the point P directly coincide with the point N (compare now Figure 6.8 with Figure 6.2 and Figure 6.7), the point D is got as the intersection point of the perpendicular to JN through N and the diagonal AJ of the square LAIJ produced, and the point G is got by completing the rectangle GDNI. The sough after point E is the intersection point of the side DN of this rectangle and the given segment AB (the construction of the point M which completes the rectangle MDNJ is then useless).

Insofar as ABHC is a square, AI is equal to the half of the given segment AB, and LAIJ is equal to a square having the half of such a given segment as side. Hence, the rectangle rbcd is equal both to a square having the given segment as side taken together with a square having the half of such a given segment as side, and to a rectangle having as sides the given segment and a segment equal to this same given segment taken together with another segment equal to a fourth of it. This is also the case of the square OPKQ. Its side OP is then the side of a square equal to such a last rectangle. It follows that the segment AE results from cutting off the segment LA, equal to a half of the given segment, from the segment LE equal to a side of a square equal to this same rectangle.

This description is cumbersome. Still, the adoption of an appropriate notation allows rephrasing it in a quite simple way. Call the given segment 'a'. For any segment α, denote then with '$\frac{\alpha}{n}$' (where 'n' denotes, in turn, a natural number greater than 1) any segment equal to one n-th of α, and with '$S(\alpha)$' any square having as side a segment equal to α. For any other segment β, denote then with '$R(\alpha, \beta)$' any rectangle having as sides two segments equal to α and β, respectively. Use moreover the sign '$+$' and '$-$' to indicate the

requirements of taking two geometric objects together and of cutting off one
of them from the other, respectively

Using this notation, one can then write the following equalities:

$$AB = a;$$
$$AI = \frac{a}{2};$$
$$LAIJ = S\left(\frac{a}{2}\right);$$
$$rbcd = S(a) + S\left(\frac{a}{2}\right) = R\left(a, a + \frac{a}{4}\right) = OPQK.$$

Let now a^* be a segment such that

$$R\left(a, a + \frac{a}{4}\right) = S(a^*).$$

Then:

$$OPQK = S(a^*) \quad ; \quad OP = a^* \quad ; \quad AE = a^* - \frac{a}{2}.$$

Euclid's construction as a whole can then be rephrased through three quite
simple equalities:

$$AB = a \quad ; \quad R\left(a, a + \frac{a}{4}\right) = S(a^*) \quad ; \quad AE = a^* - \frac{a}{2}. \qquad (6.2)$$

Mathematically, the possibility of this rephrasing leaves no doubt. Still,
this is not the same as admitting that this rephrasing is appropriate for the
purpose of describing Euclid's solution as it is, in fact. If this were so, it
would be natural to wonder why Euclid expounds such a solution in such a
relatively complex way, when he could have described it through such three
simple equalities or, at least, through a vernacular reformulation of them,
for example as follows: construct any segment equal to the given one and a
fourth of it taken together; construct any square equal to the rectangle having
segments equal to this last segment and to the given one as sides (which is the
same as finding a mean proportional between two sides of this last rectangle);
cut off a segment equal to a half of the given segment from a segment equal
to a side of this square; this results in a segment meeting the condition of the
problem.

This question is ill-stated, however. Since behind such a rephrasing of Euclid's solution – and, *a fortiori*, behind that consisting of equalities (6.2) –
there is an understanding of this solution that is quite far from Euclid's.

I call 'purely quantitative' a geometric problem that asks for the construction of whatever segments (or whatever geometric objects of a certain sort,
which are determined if appropriate segments are so), which are supposed

to meet some conditions depending only on the relative size of these segments and of other given ones, regardless to their respective positions. In other terms, such a problem asks for the construction of whatever segments belonging to the equivalence classes of segments which are specified by stating these conditions (the relevant equivalence relation being equality). Let a be any given segment. Hence, one would be stating purely quantitative problems by requiring, for example, to construct a segment x such that

$$a : x = x : a - x, \qquad (6.3)$$

or that

$$R(x, a + x) = S(a), \qquad (6.4)$$

whatever the given segment a and the positional relations of x and a might be. But this is not what propositions VI.30 and VI.29 require. They are rather, I would way, positional problems, each of which have a unique possible solution (or a finite number of possible symmetric solutions)[9].

Proposition VI.30 is a problem of partition of a given segment. It requires constructing a certain point on such a given segment, and is then solved if and only if this very point is constructed. According to a classical terminology, proposition VI.29 is a problem of application of an area with excess of a parallelogram similar to a given one. Though classical, this denomination is potentially misleading, however. Strictly speaking, Euclid is not asking to construct any parallelogram having a certain area (whatever an area might be for him). He is rather requiring to construct a certain parallelogram having a base collineal with a given segment and a vertex coincident with an extremity of such a segment.

To be more precise, in proposition VI.30, the segment AB being taken as given (Figure 6.1), Euclid does not require to construct any segment x smaller then AB and equal to a mean proportional between a segment equal to AB itself and another segment equal to that which have to be joined to x itself for getting a segment equal to AB. He rather requires to construct a certain determinate point on AB, namely the point E (which is a much more natural requirement to be advanced in his language). Analogously, in proposition VI.29, the segment CA being taken as given (Figure 6.2), Euclid does not require to construct any parallelogram equal to Γ suitable for being divided into two parallelograms, one of which has a base equal to CA and the other is similar to Δ[10]. He rather requires to construct a certain determinate

---

[9] Cf. footnote (2), above.

[10] Notice that if Euclid's proposal, in stating proposition VI.29, had been that of advancing a purely quantitative problem, it would have been pointless to take Γ and Δ to be any polygon and any parallelogram, respectively. This would have involved no gain of generality with respect to taking Γ and Δ to be a square and a rectangle, respectively.

parallelogram, having a vertex in C and a base collineal to CA, namely the parallelogram FDGC (or at least one of the four symmetric parallelograms FDGC, CGD'F', D"EAG', and G'AE'D''': Figure 6.9).

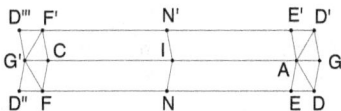

Fig. 6.9: The four symmetric parallelograms solving Proposition VI.30

Now, the previous rephrasing of the solution of proposition VI.30 respects its being concerned with the construction of a certain determinate point on the given segment, namely with the point E on AB (Figure 6.1). But, within the context of this rephrasing, the fact that the object to be constructed be a point on the given segment, rather than any segment equal to $a^* - \frac{a}{2}$, appears as an immaterial and extrinsic constraint. Such a rephrasing is based, indeed, on the understanding of the successive steps of the solution as responses to purely quantitative sub-problems, which is clearly not Euclid's understanding.

It is true that the segment OP (Figure 6.6) entering the solution of proposition VI.29 is supposed to be any mean proportional between any two segments equal to dr and ra, respectively. Still, this appears to be more a trick for avoiding a too intricate diagram than an intrinsic feature of this solution. Once the segment OP and the associate parallelogram OPQK are constructed, Euclid immediately constructs the segment JN, equal to the side OK of this parallelogram in its appropriate position (Figure 6.2). In other terms, he is certainly concerned with a purely quantitative condition, but he understands it as an ingredient of the positional problems he is interested to.

The previous rephrasing of Euclid's own solution of proposition VI.30 is then unfaithful to such a solution. But is it also, merely, a pointless modernization of Euclid's argument? In the following part of my paper I shall show that this is not the case, in fact.

6.3 Comparing Proposition VI.30 with Other Proposition of the *Elements*

If proposition VI.30 is compared with proposition I.3 of the same *Elements*, it becomes natural to understand the former as requiring to divide the given segments into two segments, one of which results from cutting off the other from the given one. Using, as above, the sign '−' to indicate the requirement of cutting off a segment from another, one could then rephrase the condition (6.1) as follows

$$AB : AE = AE : AB - AE. \qquad (6.5)$$

This rephrasing is not innocent as it displaces the focus of the problem from the construction of the point E into the construction of a segment AE having a certain relation with the given segment AB. Once this change of focus is admitted, the requirement that the sought after segment be taken on AB, that is, that the segment AE, its extremity E being on AB appears as an extrinsic constraint: as a pointless specification of a purely quantitive condition like

$$AB : x = x : AB - x, \qquad (6.6)$$

which only differs from condition (6.3) in the way in which the given segment is denoted. The passage from (6.5) to (6.6) or (6.3) is not anodyne, however, since it results in transforming the positional problem stated by proposition VI.30 into a purely quantitative one.

This is a radical change, in my mind. But it does not depend on the acquisition of some new mathematical resources with respect to Euclid's (the crucial identification of the segment EB with the result of cutting off AE from AB is openly licensed and also suggested by proposition I.3, as said). It merely depends on a change of focus, or, more generally, a shift in conception.

One could object that also the use of a literal notation and of an operational arithmetical sign like '−', i.e. of terms like 'a', 'x', and '$a - x$', is relevant. In other terms, one could object that condition (6.3) and the problem connected to it are significantly different from condition (6.1) and proposition VI.30 for a reason that hinges essentially on the fact that the former are stated by using a literal notation and an operational arithmetical sign like '−'. This is wrong, however, and depends on a confusion (which is not less misleading for its being recurrent). This is the confusion between the semantic function of these terms and their syntactical function[11].

The apparent plausibility of the objection depends on the fact that the terms 'a','x' and '$a - x$' are implicitly supposed to be involved in the whole formalism of early-modern algebra. If it were so, these terms would be also used for their syntactical function, to the effect, for example, that the condition (6.3) should be taken to be *ipso facto* equivalent to the condition

$$x^2 + ax = a^2. \qquad (6.7)$$

---

[11] My use of the two adjectives 'semantic' and 'syntactical' in order to account for the distinction I want to point out is far from mandatory. I could have used another terminology, instead. The advantage of my choice is that these adjectives immediately evoke some opposite features of language that this distinction is concerned with. Still, there are many other aspects of the meaning (or meanings) usually ascribed to these adjectives which are not relevant for my present purpose. By using these adjectives, I do not intend to evoke these. In section 6.5 below, I shall come back to this distinction, by trying to clarify it further.

But nothing like this is implied in the passage from (6.1) to (6.3), as I have described it above. In (6.3), the terms 'a','x' and 'a − x' are only intended to denote some segments (either given or sought after), that is, they are merely used for their semantic function[12].

Hence, this last condition (6.3) is not significantly distinct from the requirement that the segment to be constructed be a mean proportional between a segment equal to the given one and another segment equal to that resulting from cutting off a segment equal to it from a segment equal to the given one. On the other hand, this condition is not equivalent to the condition (6.7), since the passage from (6.3) to (6.7) depends on a formalism that goes not necessarily together with the use of the terms 'a','x' and 'a − x' according to their semantic function.

It follows that, according to their intended function (which is here only the semantic one), the use of terms like 'a','x' and 'a − x' is nothing more but a convenient linguistic trick for stating a purely quantitative problem. This use allows to avoid cumbersome vernacular expressions in favor of shorter, nimbler and unambiguous ones, and could thus help for going ahead in complex reasoning, or for expounding complex arguments in a simplified and clearer way. But it is in no way essential for the purpose of stating such a purely quantitative problem. Hence, the passage from proposition VI.30 to a purely quantitative problem (which is the passage I focus my attention on) does not depend on the use of such terms, that is, on the use of a literal notation and of an operational arithmetical sign like '−'.

This being said, consider now proposition II.2 of the same *Elements*. It is a theorem: "if a segment is cut at random, the rectangle contained by the whole

---

[12] Notice however that there is an essential difference between the symbol 'a' on one side, and the symbols 'x', and 'a − x', as well as the symbols '$\frac{a}{2}$', '$\frac{a}{4}$', '$a + \frac{a}{2}$', '$a + \frac{a}{4}$', 'a^*', '$a^* - \frac{a}{2}$', '$S(a)$', '$S\left(\frac{a}{2}\right)$', '$S\left(a + \frac{a}{2}\right)$', '$R\left(a, a + \frac{a}{2}\right)$', '$S(a^*)$', '$R(x, a + x)$' used above, on the other side. The symbol 'a' is intended to be a proper name of a certain segment, and is then, on this respect, analogous to terms like 'AB', 'AE', 'EB', or even 'ABHC' or 'CGDF' that are just used by Euclid as proper names of segments and polygons, respectively. The symbols 'x', 'a − x', '$\frac{a}{2}$', '$\frac{a}{4}$', etc. are intended to be, instead, non-definite descriptions denoting any element of a certain class of equivalence of geometrical objects of the appropriate sort (the relevant equivalence relation being equality). As a consequence, there is an essential difference between the meaning to be assigned to the sign '=' in formulas like 'AB = a' where this sign stands between two proper names and in formulas like 'AI = $\frac{a}{2}$', 'IJLA = $S\left(\frac{a}{2}\right)$' or '$S(a) + S\left(a + \frac{a}{2}\right) = R\left(a, a + \frac{a}{4}\right)$' where it stands between a proper name and a non-definite description, or between two non-definite descriptions. In the former case, this symbol indicates identity; in the latter it merely indicates equality. These differences reflect a distinctive feature of purely quantitative problems: the fact that they involve one or more principal or independent segments, in terms of which the segments to be constructed are characterized, and that these segments are taken to be given as such at the beginning of the construction that is intended to solve the problem, whereas the segments constructed during this construction are (for the very nature of the problem) whatever elements of certain classes of equivalence of segments defined in terms of the given ones.

A   C   B

D   F   E

Fig. 6.10: Proposition II.2

and each of the [two] parts is equal to the square on the whole". The expression
'the rectangle contained by the whole and each of the two parts' refers to the
rectangle contained by the whole and one of the two parts and the rectangle
contained by the whole and the other part taken together. Supposing that the
segment AB is cut at random in C[13] (Figure 6.10), Euclid takes his theorem
to state, indeed, that "the rectangle contained by AB, BC together with the
rectangle contained by BA, AC is equal to the square on AB". To prove that
this is so, he constructs the square DEBA, traces the perpendicular FC to AB
through C, and takes the rectangles DFCA and FEBC to be "the rectangle
contained by BA, AC" and "that contained by AB, BC", respectively, since
AD = AB = BE. In other terms, supposing that α, β, and γ be three distinct
segments, Euclid takes the rectangle "contained by" α, β to be the same as
(and not just equal to) the rectangle "contained by" α, γ, if $\beta = \gamma$ (which is
what allows him giving sense to the notion of rectangle "contained by" two
collineal segments). This suggests that he takes expressions like 'the rectangle
contained by α, β' and 'the square on α' to refer to any rectangle having
as sides two segments equal to α and β, and to any square having as side a
segment equal to α, respectively. These are purely quantitative notions: they
do not depend on the mutual positions of the relevant segments, but only on
the fact that some of them are equal to others.

Despite this, proposition II.2 appears as a positional theorem: a theorem
whose content depends on the mutual position of the objects it is concerned
with. It asserts something about whatever segment split up into two other

---

[13] I come back now to using the same letters as in Euclid's diagrams as they appear in
[(Heath, 1926)].

segments by any point on it, namely that any pair of rectangles one side of which is equal to the former segment, while the other side is respectively equal to the two parts of it, are, if taken together, equal to any square having this former segment as side.

This understanding is reinforced by Euclid's proof of this theorem, which is entirely diagrammatic. The segment AB being considered, the crucial step in in this proof is the claim that the rectangles DFCA and FEBC taken together are equal to the square DEBA. As this claim is not further justified, it is obvious to reconstruct Euclid's argument as follows: the rectangles DFCA and FEBC taken together coincide with the square DEBA; but, according to common notion I.4, "things which coincide with one another are equal to one another"; hence, the rectangles DFCA and FEBC taken together are equal to the square DEBA; still, these rectangles are those contained by BA, AC and by AB, BC, respectively; then these last rectangles taken together are equal to the square DEBA. This argument crucially depends on the mutual position of the rectangles DFCA and FEBC and the square DEBA. And this is just what reinforces the understanding of proposition II.2 as a positional theorem.

Still, there is no doubt that for Euclid, any rectangle is equal to any other rectangle having equal sides (this is a particular case of proposition I.35), and any pair of segments respectively equal to the two parts into which a third segment is split up are, if taken together, equal to this last segment and to any other segment equal to it. Hence, the same argument also proves the following implication: if two segments taken together are equal to a third one, then any pair of rectangles having as sides two segments equal to this third one and to the two others, respectively, are, if taken together, equal to any square having as side a segment equal to the third one.

This is no longer a positional theorem, however. Using the same notation as before, it can be rephrased as follows:

$$\text{if} \quad a = b + c \quad \text{then} \quad R(a, b) + R(a, c) = S(a). \tag{6.8}$$

Moreover, if three segments a, b and c are such that $b + c = a$, then if a segment equal to b is cut off from a segment equal to a, a segment equal to c results. Hence, from implication (6.8) it follows that

$$R(a, b) + R(a, a - b) = S(a). \tag{6.9}$$

Both implication (6.8) and equality (6.9) are purely quantitative theorems: they are about some geometric objects which are supposed to meet some conditions depending only on the relative size of these same objects, regardless to their respective positions (*i. e.*, they are about the equivalence classes of these geometric objects that are specified by stating these conditions, the relevant equivalence relation being equality).

One can resist, of course, to understanding proposition II.2 as being equivalent to these theorems, and, then, in ascribing these theorems to Euclid. Still, without admitting that, for Euclid, the rectangle contained by two segments α and β is the same as the rectangle contained by α and γ, if $\beta = \gamma$, the very statement and proof of proposition II.2 are hard to understand. And, once this is admitted, the understanding of this proposition as being equivalent to implication (6.8) and equality (6.9) becomes very natural. Nevertheless, I do not want to argue that this understanding is Euclid's. For my present purpose, it is enough to have determined such an understanding and to have shown that it is at least suggested by the very way Euclid expresses himself and reasons.

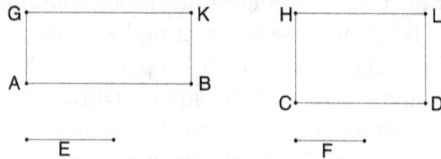

Fig. 6.11: Proposition VI.16 of the *Elements*

Consider now proposition VI.16 of the *Elements*, in particular its first part: "if four segments are proportional, the rectangle contained by the extremes is equal to the rectangle contained by the means". The terms 'the rectangle contained by the extremes' and 'the rectangle contained by the means' are subject to the same considerations relative to the analogous terms occurring in proposition II.2 and its proof. In proving proposition VI.16, Euclid considers four given segments, AB, CD, E, and F (Figure 6.11) such that

$$AB : CD = E : F$$

and constructs, on AB and CD, respectively, two rectangles ABKG and CDLH, such that AG = F and CH = E; then he reasons on these last rectangles. It is then clear, again, that he takes the rectangle contained by two segments α and β to be the same as the rectangle contained by α and γ, if $\beta = \gamma$. It is thus natural to understand proposition VI.16 as stating the following theorem: if four segments are proportional, then any rectangle having as sides two segments equal to the extremes is equal to any rectangle having as sides two segments equal to the means. In the same notation as before:

$$\text{if} \quad a : b = c : d \quad \text{then} \quad R(a, d) = R(b, c).$$

Now, by comparing this implication with (6.3), one gets

$$R(a, a - x) = S(x).$$

Hence, from (6.9), it follows:

$$S(x) + R(a, x) = S(a). \tag{6.10}$$

If the previous rephrasing of propositions II.2, VI.16 and VI.30 is admitted, this last proposition can be reduced to the problem of constructing a segment x such that this last equality holds, supposing that the segment a is given.

Compare now this equality with the construction that provides Euclid's solution of this proposition. If AB (Figure 6.7) is the given segment a, then AE satisfies the former: if one takes AE to be a value of x, then the rectangle AEFC and the square GDEA are values of $R(a, x)$ and $S(x)$, respectively, and, if they are taken together, they are equal to the square ABHC which is a value of $S(a)$.

The terminology I use requires an explanation. As I use them, the terms 'x', '$R(a, x)$', '$S(x)$', and '$S(a)$' are not singular, that is, they do not respectively refer to a unique object, but rather to whatever element of a certain equivalence class of objects. By saying that a certain segment is a value of x, I mean that this segment is an element of the equivalence class to whatever element of which 'x' refers (as the relevant equivalence relation is equality, it follows that, if α and β are two values of x, then $\alpha = \beta$). The same convention applies to '$R(a, x)$', '$S(x)$', and '$S(a)$' and to any other terms like those: by saying, for example that a certain square is a value of $S(a)$, I mean that it is a square having a segment equal to a as side.

This explains the relation between the equality (6.10) and the geometrical configuration that Euclid's solution of proposition VI.30 depends on: this configuration provides a positional model for this equality. Other models are possible, however. And this is not only because one can construct the point E – providing the unique solution of this proposition, according to Euclid's understanding – in many other ways. But also and overall because, even if the given segment is always the same – that is, a – there are infinite many segments other than AE which satisfy this equality as values of x^{14}. All of them provide a solution of the purely quantitative problem of constructing a segment x such that equality (6.10) holds.

All these remarks are not enough for solving such a problem. And, it is neither enough for this purpose to observe that, according to implication (6.8), from equality (6.10), equality (6.4) follows, and that this last equality is such that constructing a segment x that satisfies it is the same as solving a problem of application of an area, understood as a purely quantitative problem (namely the problem of application of an area whose solution is involved in Euclid's solution of proposition VI.30). What is still lacking is a construction

---

^{14} Cf. footnote (12), above.

of a segment suitable for providing a value of x. An appropriate formulation of the problem can help, however, in identifying a positional model suggesting a simple construction of such a segment. To say the same in another way: in order to look for a solution of the problem, one can try to reduce it to some other simpler problem, which can be done, in turn, by appropriately transforming the condition that this problem is concerned with.

6.4 Thābit ibn Qurra's interpretation of al-Khwārizmī's first trinomial equation

One such transformation is implied by Thābit ibn Qurra's solution of al-Khwārizmī's first trinomial equation. Expounding this treatment is a way of showing that the rephrasing of problems like proposition VI.30 as purely quantitative problems, and the appeal to appropriate other propositions of the *Elements*, understood in turn as purely quantitative theorems, for justifying such a rephrasing are not merely pleasant possibilities suggested to a modern reader by certain aspects of Euclid's text and arguments, but were already part of mathematical practice several centuries before early-modern algebra developed. This is the purpose of the present section.

Al-Khwārizmī's *Book of Algebra and al-Muābala*[15] deals with a combinatorial system including three basic elements, or modes: "Numbers", "Roots", and "Squares". These elements are combined so as to get six equations: three are binomial equations, the three other trinomial equations[16]. For short (using a notation which is not al-Khwārizmī's), one could write them as follows:

$$S = R \ ; \ \ S = N \ ; \ \ R = N \ ;$$
$$S + R = N \ ; \ \ S + N = R \ ; \ \ R + N = S$$

(where 'N' refers to a Number, 'R' to some Roots, and 'S' to some Squares; the meaning to assign to the sign '$+$' will be clarified later).

These equations can be firstly understood as (shortenings of) statements of general canonical problems about Numbers, Roots, and Squares. As these equations exhaust the possible combinations admitted by the system (suppos-

---

[15] Al-Khwārizmī's treatise was probably written in the first part of 9th century. I base my analysis on Rashed's edition and its French translation ([(Rashed, 2007)]). This is accompanied by a rich introduction which, among other things, stresses the connection between Al-Khwārizmī's algebra, Thābit's treatment of his trinomial equations, and the tradition of the *Elements*. Though my views are not always the same as Rashed's, much of what I shall say on these matters has been inspired by what he argues for in this introduction. For a much earlier edition of al-Khwārizmī's treatise, together with an English translation, cf. [(Rosen, 1831)].

[16] Cf. footnote (3), above.

ing that the commutativity of '+' is taken for granted, and repetitions are not allowed), these are all the problems of this sort that can be stated. Still, these same equations can also be understood as possible forms of different problems about what Numbers, Roots, and Squares are taken to stand for. In this case, the fact that these equations exhaust the possible combinations admitted by the system merely entails that these problems can have six different forms, that is, that al-Khwārizmī's algebra is concerned with all the problems that can take one of these forms, under an appropriate interpretation of Numbers, Roots, and Squares.

But what is it that Numbers, Roots, and Squares can stand for? In other terms, which are their possible interpretations? There are two: Numbers, Roots, and Squares can both stand for numbers, in the usual sense of this term, or for geometrical magnitudes, namely for rectangles whose sides are taken to be equal to certain segments which are either given or to be constructed.

This being said, let us stop with generalities about al-Khwārizmī's algebra. On one side, I do not want to say too much that I shall not have room to justify (through a detailed analysis of al-Khwārizmī's arguments). On the other side, the consideration of a single example is enough for my present purpose and will also allow to clarify some of the previous generalities.

This is the example of the general canonical problem corresponding to the first trinomial equation, that is, $S + R = N$, or, in al-Khwārizmī's own parlance: "The Squares and the Roots equal to a Number" ([(Rashed, 2007)], pp. 100-101).

Al-Khwārizmī describes its solution by expounding his well-known arithmetical algorithm in a particular case where one Square and ten Roots are equal to thirty-nine (dirhams). The consideration of this particular case is enough to make the arithmetic interpretation manifest. In this particular case, the problem consists in determining a number (which al-Khwārizmī takes to be a number of dirhams) such that by adding its square to the number resulting by multiplying it by ten one gets thirty-nine. In modern terms, one has then to solve the quadratic equation:

$$x^2 + nx = m$$

where $n = 10$, $m = 39$ and x is taken to be a number[17] (which implies that the Square is nothing but a product of numbers).

---

[17] I cannot enter here into the quite delicate question concerning the suppositions that one should make about the nature of this number, which is of course a crucial question to be addressed for clarifying the arithmetical interpretation of al-Khwārizmī's general canonical problems (but is not relevant for my present purpose). For a discussion of the notion of number in Arabic algebra, cf., for example, [(Oaks, 2010)], § 2, and [(Oaks, 2011)], § 5.

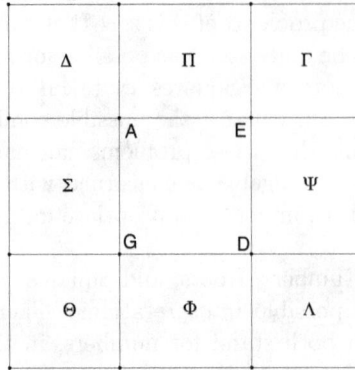

Fig. 6.12: The first geometrical demonstration by al-Khwārizmī

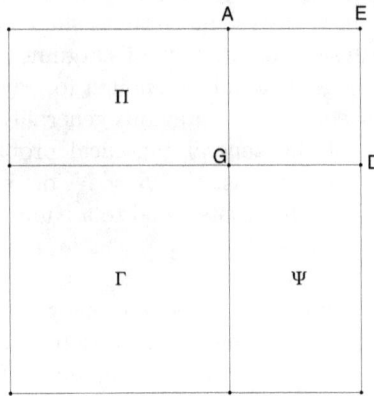

Fig. 6.13: The second geometrical demonstration by al-Khwārizmī

In order to justify this algorithm, he shifts to geometry ([(Rashed, 2007)], pp. 108–113). Namely, he offers two different positional models for the problem. In both of them, the Square is interpreted as being the square GDEA (Figures 6.12 and 6.13)[18] and its sides are interpreted as being equal to one Root. In the first model, the four small squares Γ, Δ, Θ, Λ (Figure 6.12) are then taken to have sides equal to a fourth of the number of Roots; the square GDEA and the four rectangles Π, Σ, Φ, Ψ taken together are taken to be

---

[18] For reasons of uniformity with respect to the letters used to denote points in the diagrams relative to Euclid's proposition VI.30, I change the letters in al-Khwārizmī's diagrams as they appear in the French translation included in [(Rashed, 2007)], pp. 110, 112.

equal to the Number; and the problem is conceived as that of constructing a side of the square GDEA, supposing that the sides of the squares Γ, Δ, Θ, Λ are equal to a fourth of a given segment, and the square GDEA and the four rectangles Π, Σ, Φ, Ψ taken together are equal to a given rectangle or square. In the second model, instead, the square Γ (Figure 6.13) is taken to have sides equal to a half of the number of Roots; the square GDEA and the two rectangles Π, Ψ taken together are taken to be equal to the Number; and the problem is conceived as that of constructing a side of the square GDEA, supposing that the sides of the square Γ are equal to half a given segment, and the square GDEA and the two rectangles Π, Ψ taken together are equal to a given rectangle or square.

Let a and b be two given segments. Using the previous notation, the condition to be met, as respectively interpreted in the former and in the latter model, could be written as follows:

$$S(x) + 4R(\frac{a}{4}, x) = S(b) \qquad \text{and} \qquad S(x) + 2R(\frac{a}{2}, x) = S(b). \qquad (6.11)$$

Insofar as it is natural to conceive the four rectangles Π, Σ, Φ, Ψ in the former model (Figure 6.12), and the two rectangles Π, Ψ in the latter (Figure 6.13), as being equal respectively to the four and the two equal parts in which is divided a rectangle having a side equal to one Root and another side equal to the number of Roots, these two conditions appear as appropriate rephrasings of the same condition:

$$S(x) + R(a, x) = S(b). \qquad (6.12)$$

Under the geometric interpretation, the problem appears then to consist in constructing a segment x such that this condition obtains, provided that a and b are given segments: in the particular example considered, a is supposed to measure 10 (unities of length), and b is supposed to be such that any square having a side equal to it measures 39 (unities of surface).

I do not want to argue that al-Khwārizmī actually understands his problem this way I will limit myself to remark that this understanding is compatible with his geometric arguments. These arguments go as follows.

In both models, the largest square has sides equal to one Root plus the half of the number of Roots. In the former model, it is also equal to the Number (which is equal to the square GDEA and the four rectangles Π, Σ, Φ, Ψ taken together: Figure 6.12) plus four squares whose sides are equal to a fourth of the number of Roots (the squares Γ, Δ, Θ, Λ). In the latter model, it is also equal to the Number (which is equal to the square GDEA and the two rectangles Π, Ψ taken together: Figure 6.13) plus a square whose sides are equal to a half of the number of Roots. Since, four squares whose sides are equal to a fourth of the number of Roots are equal, if taken together, to a square whose sides are equal to a half of the number of Roots, in both cases, the Root is found by subtracting a half of the number of Roots from the side of a square equal to the Number taken together with a square whose sides are equal to a half of the number of Roots. In the example considered, it is then equal to the side of a square equal to $39 + 25 = 64$ (unities of surface), i. e. 8 (unities of length), minus the side of a square equal to 25 (unities of surface), i. e. 5 (unities of length), which is the same as that which is prescribed by the arithmetic algorithm.

Under the previous understanding, this is the same as observing that condition (6.12) is equivalent to the conditions

$$S(b) + 4S\left(\frac{a}{4}\right) = S(x + \frac{a}{2}) \quad \text{and} \quad S(b) + S\left(\frac{a}{2}\right) = S(x + \frac{a}{2}). \quad (6.13)$$

The problem is thus reduced to two other problems whose solution is quite simple: a and b being given, construct a square equal to a square whose sides are equal to b taken together with a square whose sides are equal to a half of a; then cut off a segment also equal to a half of a from a side of this square.

Now, it is enough to suppose that a is the same segment as b, to reduce condition (6.12) to condition (6.10). Hence, al-Khwārizmī's handling of his first trinomial equation suggests reducing the problem stated in Euclid's proposition VI.30, understood as a purely quantitative problem, to the problem of constructing a segment x such that

$$S(a) + 4S\left(\frac{a}{4}\right) = S(x + \frac{a}{2}) \quad \text{or} \quad S(a) + S\left(\frac{a}{2}\right) = S(x + \frac{a}{2}). \quad (6.14)$$

This reduction is different in nature from that through which I passed above from proportion (6.3) to equality (6.10). Whereas this last reduction is justified by a mere inspection of the purely quantitative condition that the problem requires to meet, the former is justified by the inspection of two positional models of the original problem. In al-Khwārizmī's argument, condition (6.10) is firstly rephrased under the forms of conditions (6.11)[19]; these two

---

[19] This passage depends in fact on the replacement of $R(a, x)$ with $4R(\frac{a}{4}, x)$ and $2R(\frac{a}{2}, x)$, respectively, which is not justified by an inspection of the two positional models, but rather suggests them. A justification of this replacement could then depend on an appeal to proposition II.1 of the *Elements* understood as a purely quantitative theorem, in analogy with the understanding suggested above for proposition II.2. Al-Khwārizmī does not, however, make any explicit reference to the *Elements*, and it is far from sure that this reference is implicit for him. Proposition II.1 is the following: "If there be two segments and one of them be cut into any number of parts whatever, the rectangle contained by the two [whole] segments is equal to the rectangles contained by the uncut segment and each of the parts [in which that other segment is cut]." Understood as a purely quantitative theorem, this proposition can be rephrased through the following implicationIf $b = c + d + \ldots + e$ then $R(a, c) + R(a, d) + \ldots + R(a, e) = R(a, b)$, where a and b are any two given segments. Implication (6.8) can be clearly taken as a particular case of this last one.

conditions then suggest two positional models; the inspection of these models
suggests two new purely quantitative conditions, namely conditions (6.13);
finally conditions (6.14) are got by supposing that segment a be the same as
segment b[20].

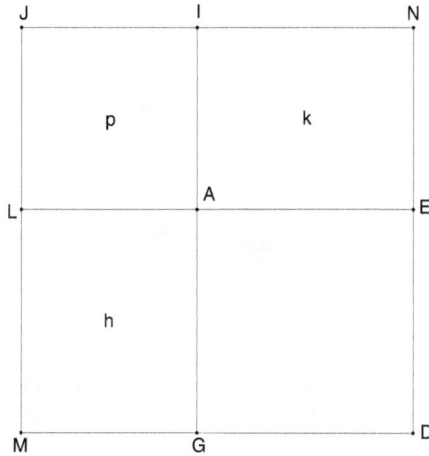

Fig. 6.14: The second geometrical demonstration by al-Khwārizmī of his first
trinomial equation: a symmetric configuration for $a = b$

In Thābit's solution of al-Khwārizmī's equation, this difference disappears.
The occasion for expounding this solution is a short treatise, composed around
the middle of 9th century: *Restoration of algebraic problems through geometrical demonstrations*[21]. The aim of this treatise is just that of making manifest
the geometric interpretation of al-Khwārizmī's trinomial equations: this is the

---

[20] Notice that, in the case where segment a is the same as segment b, it is enough to take
a configuration symmetric to that corresponding to the latter model (Figure 6.14) to get
back to the configuration constituted by the square **MDNJ** (Figure 6.8) entering into Euclid
solution of proposition VI.30. The argument leading to this configuration is however quite
different in the two cases.

[21] I base my analysis on Rashed's edition and French translation ([(Rashed, 2009)],
pp. 159-169), which is accompanied by an introduction stressing the connection between
Thābit's treatise and al-Khwārizmī's (*ibid.* pp. 153-158). The same translation is also inte-
grally quoted (with the exception of the two first introductory lines) in [(Rashed, 2007)],
pp. 33-41, where the same connection is also stressed, as I have already said in footnote
(15). For an earlier edition of Thābit's treatise, together with a German translation, cf.
[(Luckey, 1941)]. A quite similar argument is also advanced in Abū Kāmil's *Algebra* dating
back to the end of 9th century. An English translation of the relevant passage (together
with the appropriate references to the original text and its critical editions) in included in

restoration that its title refers to. I consider of course only Thābit's handling
of the first of these equations.

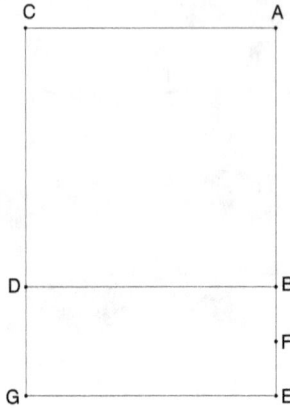

Fig. 6.15: Thābit ibn Qurra's solution of al-Khwārizmī's first trinomial equation

Thābit opens his argument by referring to a quite simple diagram (Figure
6.15). In it, the Square is interpreted as being the square DBAC[22]. The segment
EB and the rectangle GEBD are then taken to be equal, respectively, to the
number of Roots and to the Roots themselves. Hence, the whole rectangle
GEAC is equal to the Square and the Roots taken together, and then to the
Number.

In order to fix the interpretation of the segment EB and the rectangle GEBD
as being respectively equal to the number of Roots and to the Roots themselves, Thābit requires that EB be "how many times the unity by which lines
are measured as the supposed number of Roots", and remarks that "the product of AB and the unity by which lines are measured is the Root", to the effect
that "the product of AB and BE is equal to the Roots" ([(Rashed, 2009)],
p. 160-161). It seems then that Thābit intends his diagram as a geometric illustration of the problem arising from the arithmetical interpretation of
al-Khwārizmī's equation. Once having offered this illustration, he remarks,
however, that the problem associated to this equation reduces to "a known
geometric problem": supposing the segment EB to be "known", one "joins"

Appendix B of [(Oaks, 2011)].

[22] Apart from the addition of 'G', the letters are now those of Thābit's diagram, as it
appears in the French translation included in [(Rashed, 2009)], p. 162.

the segment BA to it, and one supposes that the "product of EA and BA is known"; the problem consists then in determining the segment BA.

Supposing that α and β are two collineal segments, like BA and EB, or BA and EA, Thābit's expression 'the product of α and β' seems to be intended as to have the same reference as Euclid's expression 'the rectangle contained by α, β', namely any rectangle having having as sides two segments equal to α and β, respectively. This suggests two things. The former is that Thābit conceives the "known geometric problem" (which is, of course, perfectly independent of the arithmetic interpretation of al-Khwārizmī's equation) as a purely quantitative problem: supposing that a is a given segment (to which BE is supposed to be equal), construct a segment x such that $R(x, a + x)$ is equal to a given rectangle or square. The latter thing is that Thābit is willing to solve this problem by appealing to appropriate propositions of book II of the *Elements*, understood as purely quantitative theorems, in turn. Both things are confirmed by the way Thābit continues.

Before coming to that, another remark is appropriate. The geometrical configuration represented by Thābit's diagram is the same as that which proposition II.3 of the *Elements* is concerned with. This is a theorem, again: "if a segment is cut at random, the rectangle contained by the whole and one of the [two] parts is equal to the rectangle contained by the [two] parts and the square on the aforesaid part". The same considerations made above about proposition II.2, also apply in this case[23], and suggest an analogous understanding of the former proposition as a purely quantitative theorem:

$$\text{if} \quad a = b + c \quad \text{then} \quad R(a, b) = R(b, c) + S(b),$$

from which it immediately follows that

$$S(b) + R(b, c) = R(b, b + c),$$

or, by appropriate replacements,

$$S(x) + R(a, x) = R(x, a + x). \tag{6.15}$$

Hence, it is enough to apply proposition II.3 so understood, in order to rephrase condition (6.12) as:

$$R(x, a + x) = S(b), \tag{6.16}$$

which just corresponds to the "known geometric problem" to which Thābit reduces al-Khwārizmī's equation, this problem being understood as a purely quantitative one, and is nothing but a generalization of condition (6.4).

---

[23] Euclid's proof of proposition II.3 is perfectly analogous to his proof of proposition II.2.

There is thus room for advancing the claim that Thābit's diagram is intended as an implicit reference to proposition II.3 used in order to reduce al-Khwārizmī's equation, interpreted under the form of condition (6.12), to such a known geometric problem. This is all the more plausible as this diagram does not play any further role in the continuation of Thābit's argument, which it is now time to consider.

He explicitly appeals to proposition II.6 of the *Elements* by claiming that, according to it, if F is the middle point of EB, then the "the product of EA and AB" and the "square of BF" taken together are equal to the "square of AF" ([(Rashed, 2009)], p. 162-163).

Proposition II.6 is also a theorem: "if a segment is cut in half and a segment is joined to it in a straight line, the rectangle contained by the whole [segment] with the joined segment and the joined segment, taken together with the square on the half, is equal to the square on the [segment] composed by the half and the joined segment". To prove it, Euclid considers a segment AB (Figure 6.16) cut in half at C and produced up to a point D. Then he takes: the "rectangle contained by the whole [segment] with the joined segment and the joined segment" to be the "rectangle contained by AD, DB", which he identifies, in turn, with the rectangle KMDA (supposing that MD = BD); the "square on the half" to be the "square on CB", which he identifies, in turn, with the square EGHL (supposing that EL = CB); the "square on the [segment] composed by the half and the joined segment" to be the "square on CD", which he identifies, in turn, with the square EFDC. What the proposition asserts is thus that the "rectangle contained by AD, DB", *i. e.* the rectangle KMDA, taken together with the "square on CB", *i. e.* the square EGHL, is equal to "square on CD", *i. e* the square EFDC. The proof is then very easy, since it is enough to notice that the rectangle KLCA is equal to the rectangle GFMH.

Thābit limits himself to applying Euclid's proposition to the configuration constituted by the segment EA (Figure 6.15) and the points B and F on it, and to re-stating it by replacing Euclid's expressions 'the rectangle contained by −, − ' and 'the square on −' with 'the product of − and − ' and 'the square of −'. What is important, however, is the use he makes of this proposition. Since in the "known geometric problem" to which al-Khwārizmī's equation has been reduced, the product of EA and BA (or the rectangle contained by EA, BA) is supposed to be known, and this is also the case with the square of (or on) FB, because this last segment is equal to a half of a given segment, from proposition II.6, it now follows that also the square of (or on) FA is known, which entails that FA itself is known. To solve the problem it is then enough to cut off FB, from FA, since the result of this is just BA, which thus comes to be known.

This argument comes without any explicit construction and its only corre-late on the diagram is the display of point F. It seems then that proposition II.6 is used as a sort of rule of inference, independent of any construction or positional model, and allowing a further reduction of the problem[24].

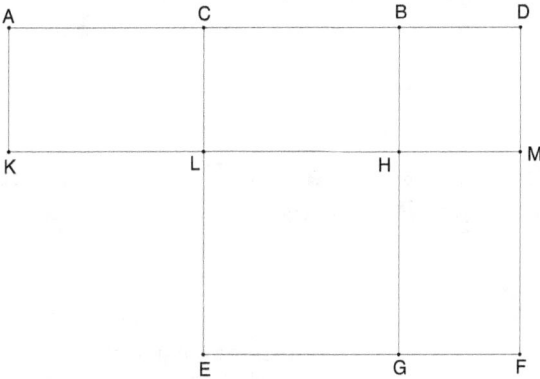

Fig. 6.16: The solution of proposition VI.30 of the Elements suggested by Thābit's solution of al-Khwārizmī's first trinomial equation

For Euclid's proof of this proposition to work, the square EGHL (Figure 6.16) has to be taken to be the square on CB. This is a further clue for understanding Euclid's expressions of the form 'the square of α' and 'the rectangle by α, β', and Thābit's corresponding ones of the form 'the square on α' and 'the product of α and β' as referring to any square having as side a segment equal to α, and to any rectangle having as sides two segments equal to α and β, respectively. The fact that Thābit feels no need to represent the squares of (or on) FB and FA, together with his shift from the former expressions to the latter ones suggests that he understands these expressions just in this way.

In my notation, his understanding of proposition II.6 is thus expressed by rephrasing it under the form of the following equality

$$R(a + b, b) + S\left(\frac{a}{2}\right) = S\left(\frac{a}{2} + b\right),$$

or, by appropriate replacements,

$$R(x, a + x) + S\left(\frac{a}{2}\right) = S\left(\frac{a}{2} + x\right). \tag{6.17}$$

---

[24] The exact sense in which I speak here of rule of inference will be clarified in section 6.5, below, p. 225.

His argument then reduces to a comparison of this last equality with the equality (6.16), so as to get the second of the two equalities (6.13), to which the problem is finally reduced.

The conclusion of this argument clearly parallels the one which al-Khwārizmī reaches through the consideration of the second of his geometrical models. Still, it is justified by the inspection of no geometrical model, and merely depends on an application of proposition II.6, understood as a purely quantitative theorem used as a rule of inference.

6.5 Early-Modern Algebra and Purely Quantitative Theorems and Problems

As observed before, if a is taken to be the same segment as b, condition (6.12), and thus al-Khwārizmī's equation understood as a geometric problem, reduces to condition (6.10), and the second of the two equalities (6.13) is transformed in the second of the two equalities (6.14). Hence, when applied to the problem stated by proposition VI.30, Thābit's argument suggests a quite easy way to solve this problem: according to proposition I.47 (the Pythagorean theorem), it is enough to cut the given segment AB in its middle point C (Figure 6.17), to construct the right angled rectangle ABD, whose sides BD is equal to CB, or a half of AB, and to cut off a segment FD equal to such a side from the hypotenuse of this triangle. The remaining segment AF satisfies condition (6.10) for $a =$ AB: if IA = AF, the square AFHG and the rectangle IJBA taken together are equal to the square KLBA. Hence, if the point E is taken on AB so that AE = AF, AB is cut in extreme and mean ratio, as required.

This provides, in the same time, the (unique) solution of proposition VI.30, and a quite simple positional model for the purely quantitative problem associated to it, whose solutions include, beside AE, also, AF, AG, AI, GH, FH, BJ, and any other segment equal to them. The relevant point here is that this easy way of solving proposition VI.30 only appears if this proposition is converted into such a purely quantitative problem, and this last problem is then reduced to an easier one by relying on proposition II.6, understood in turn as a purely quantitative theorem used as a rule of inference. Furthermore, to prove that the solution that one gets this way is appropriate, one has to rely either on this last proposition so understood, or on some analogous theorem.

In a nutshell, my point is that the passage from Euclid's understanding of proposition VI.30 and his way to solve it, to its conversion into a purely quantitative problem, which is solved in turn through an appeal to a purely quantitative theorem used as a rule of inference, manifests a structural feature

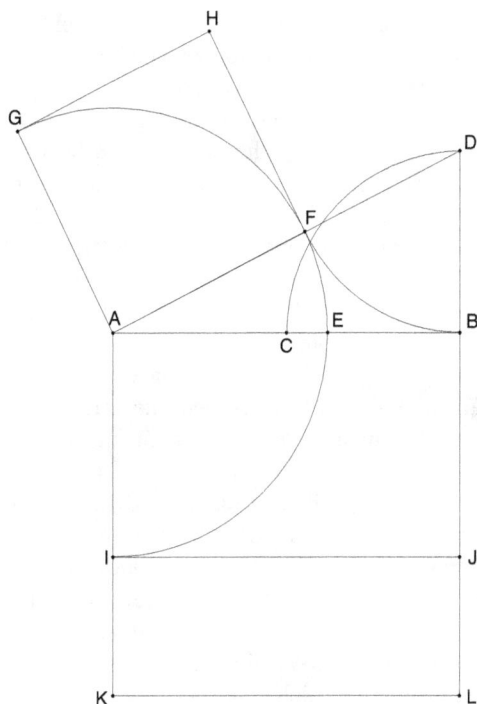

Fig. 6.17: Thābit's solution to cut a given segment AB in its middle point C

of early-modern algebra, which is both essential to it and independent of any literal formalism.

Of course, there is much more in early-modern algebra. Still, this much more is, in my view, dependent on such an understanding of geometrical problems and theorems and on this way of solving them through appropriate reductions which are not suggested by the consideration of appropriate positional models of them, but rather suggest these models. In other terms, this is a necessary condition that made possible the development of early-modern algebra and the adoption of its literal formalism.

So, a natural question arises: what more has been needed for the shaping of early-modern algebra? Or better: what is there more in it than in an argument like Thābit's? Suppose one transcribes this argument using a skillful notation, like the one I have used above. This would be certainly not enough to reproduce the early-modern algebraic setting. But why not? What would be still lacking?

Let me conclude my paper by sketching an answer to these questions.

According to H. Bos, early modern geometry presents a peculiar form of problematic analysis that is made possible by the introduction of the algebraic formalism, especially thank to Viète, and that he calls 'analysis by algebra' ([(Bos, 2001)], pp. 97-98). Its crucial difference from Pappusian analysis is, for Bos, that it requires rephrasing the problems to be solved through a system of equations written in such a formalism. There is no doubt that the development of early-modern algebra goes together with the adoption of a non-Pappusian kind of analysis, which is clearly described by Viète in the *Isagoge* [(Viète, 1591)]. Still, it seems to me that Vietian analysis is a particular form of a more general kind of analysis, crucially different from Pappus's, which is, as such, perfectly independent of Viète's formalism, and depends, instead, on the conceptual transformation I have described above.

Elsewhere [(Panza, 2007)], I have coined the term 'trans-configurational' for this kind of analysis, since it consists in transforming the configuration of data and unknowns that comes with the problem to be solved, into another configuration. The starting configuration is fixed through a system of purely quantitative conditions that known and unknown magnitudes are supposed to meet, then analysis operates on this system of conditions through appropriate rules of inference, so as to transform it in a new equivalent, but essentially different system of conditions, expressing a new configuration of data and unknowns. Typically, this final system of conditions suggests a simple positional model for the given problem (which is not suggested by the original one). The synthesis (which is nothing but the solution of the given problem) consists then in the construction of the sought after elements of this model starting from the given ones.

Thābit's previous arguments is an example of trans-configurational analysis. Consider the particular case corresponding to Euclid's proposition VI.30. Once this proposition is converted into a purely quantitative problem, it requires constructing a mean proportional x between any segment equal to a given segment a and any other segment segment $a - x$ resulting from taking away a segment equal to x from a segment equal to a, that is, a segment x meeting the following condition

$$a : x = x : a - x.$$

Trans-configurational analysis transforms this problem as follows:

- For proposition VI.16, it reduces it to the problem of constructing a segment x such that
$$S(x) = R(a, a - x).$$

- For proposition II.2 (*i. e.*, according to equality (6.9)), this last condition reduces to the other one:
$$S(x) + R(a, x) = S(a)$$

- For proposition II.3 (*i. e.*, according to equality (6.15)), this reduces in turn to:

$$R(x, a + x) = S(a)$$

- Finally, for proposition II.6 (*i. e.*, according to equality (6.17)), this reduces to:

$$S(a) + S\left(\frac{a}{2}\right) = S\left(x + \frac{a}{2}\right),$$

which immediately suggests a simple positional model, and then a solution.

This piece of analysis is structurally akin to the Vietian one, but it does not rely on the literal formalism of early-modern geometry. As I have observed above, the notation used to expound this argument enters only for its semantic function, that is, it is only employed for the purpose of avoiding too long and tiresome expressions and phrases: any step in this argument could be rephrased in a very long and cumbersome, but essentially unmodified way without using this notation, but rather a non symbolic language like Euclid's or Thābit's.

The answer I suggest for the previous questions depends on a further clarification of this point. The symbols that my notation consists of do not only denote objects that one could also denote in a different way, but they overall compose complex expressions that transform into each other not because of some syntactical rules, but rather because of theorems that concern these objects and are proved by reasoning on them in a way that is independent of the use of this notation. Consider, as an example, the last step of the previous argument. It is not licensed by any syntactical rule relative to such a notation, but it rather depends on the admission of equality (6.17), which is proved by reasoning on an appropriate configuration of rectangles and squares. By saying that proposition II.6 enters the previous argument as a rule of inference, I do not mean that this proposition provides any such syntactical rule, but simply that it allows to pass from a certain equality to another.

The literal formalism of early-modern algebra works in an essentially different way. The signs composing it are also used for their syntactical function. This means that they are part of a syntactical system providing appropriate rules for transforming the expressions composed by these symbols into each other. These rules express the properties of a number of operations, and these properties are independent of the objects to which these operations are supposed to apply (that is, they are the same whatever these objects might be: numbers, segments, or any other sort of quantities), to the effect that they can be expressed without having to care about the semantic function of the relevant signs, that is, their power to denote some objects. The use of literal signs for their syntactical function is, I argue, the essential feature of the new formalism adopted in early-modern algebra.

Of course, this means neither that the same signs were previously adopted for their semantic function, nor that they did not also have this last function within such a formalism. As a matter of fact, the former is only very partially true. The latter is plainly false, instead. What is relevant is rather that the two functions of signs, the semantic and the syntactic, are integrated within the formalism of early modern algebra: these signs both denote objects (according to some appropriate interpretation of them), and compose expressions that transform into each other according to some syntactical rules, *i. e.* rules that directly apply to these signs independently of their power to denote any object.

But, if this so, one could object, there is no room to argue, as I just did, that the shift in conception that I accounted for was a necessary condition for the adoption of the literal formalism of early modern algebra. The emergence of such a formalism, one could argue, only depended on the fixation of appropriate syntactical rules applying to appropriate signs; and just insofar as these rule are syntactical in the previous sense, this could only be independent on any way of conceiving geometrical objects and their relation.

This objection is clearly flawed, however, and for a quite trivial reason. After all, the features of the syntactical system involved in the literal formalism of early modern algebra were just motivated (as it happens for any syntactical system to be used in mathematics) by the purpose of making some interpretations of this formalism possible. This system was just conceived in order to provide a tool for dealing with arithmetical and/or geometrical objects. The fact that the rules involved in it are syntactical, that is, directly apply to the relevant signs independently of their power to denote any object, does not entail that they were not conceived for the purpose of providing such a tool. Now, it is easy to understand that it was only insofar as geometrical objects came to be conceived as pure quantities – that is, as possible *relata* of a system of purely quantitative conditions – that a geometrical interpretations of a literal formalism as that of early-modern algebra (intrinsically unsuitable for expressing the positional relations of geometrical objects) could be licensed. This is just the reason why I argue that the adoption of such a formalism was made possible by the shift in conception that I have tried to describe[25].

---

[25] One could perhaps argue that things should be seen the other way around, that is, that it was the rephrasing of some geometrical problems (like that stating by proposition like VI.30) using a literal formalism like that of early modern algebra that forced the understanding of these problems as a purely quantitative ones. But, for such a line of argumentation to be plausible, one should also explain what motivated this rephrasing. A possible response could be that this rephrasing was motivated by the desire of stating (and solving) geometrical problems by using a language analogous to that of arithmetic. The choice between my hypothesis and this alternative one should be ultimately justified by detailed textual considerations that I can only suggest to undertake here. Still, it seems to me that the evidence I offered in my paper, relative both to Euclid's *Elements* (in section 6.3) and to al-Khwārizmī's and Thābit's treatises (in section 6.4) is at least enough to show that the understanding of geometrical problems as a purely quantitative ones is not

At this point, another story should be told: a story accounting for the way in which such a formalism was conceived, so as to be appropriate for integrating the semantic and the syntactical functions of the signs involved in it. This is a quite complex story, that has been told many times in different ways. Of course, I would have my own way to tell it[26]. But this is not something I can do here.

Acknowledgements The present paper is based on a talk which, besides to the Ghent's conference *Philosophical Aspects of Symbolic Reasoning in Early Modern Science and Mathematics*, I also presented in some other occasion, including the Oberwolfach meeting *History and Philosophy of Mathematical Notations and Symbolism*, held in October 2009. I thank all the people attending these lectures, for different questions and comments. I'm particularly grateful, for useful discussions and advises, to Hourya Benis-Sinaceur, Karine Chemla, Massimo Galuzzi, Albrecht Heeffer, Antoni Malet, Sébastien Maronne, Roshdi Rashed, Maarten Van Dyck and Lauren Heeffer.

References

[(Bos, 2001)] Bos, Henk, 2001. *Redefining geometrical exactness. Descartes' Transformation of the Early Modern Concept of Construction.* New York, NY: Springer.

[(Heath, 1926)] Heath, Thomas L., 1926. *The Thirteen Books of Euclid's Elements, translated from the text of Heiberg, with introduction and commentary. (3 vols.)* Cambridge: Cambridge University Press.

[(Luckey, 1941)] Luckey, Paul, 1941. "Ṭābit b. Qurra über den geometrischen Richtigkeitsnachweis der Auflösung der quadratischen Gleichungen". *Sächsischen Akademie der Wissenschaften zu Leipzig. Mathematisch-physische Klasse. Berichte* **93**, 93–114.

[(Netz, 2004)] Netz, Reviel, 2004. *The Transformation of Mathematics in Early Mediterranean World: From Problems to Equations: From Problems to Equations.* Cambridge, New York, etc.: Cambridge University Press.

[(Oaks, 2010)] Oaks, Jeff, 2010. "Al-Khayyam's scientific revision of algebra". Preprint. University of Indianapolis.

[(Oaks, 2011)] Oaks, Jeff, 2011. "Geometry and proof in Abū Kāmil's algebra". In M. Abdeljaouad (ed.) Actes du dixieme colloque Maghrebim sur l'histoire des mathématiques arabes: Tunis, 29-31 Mai, 2010, (to appear in 2011).

[(Panza, 2005)] Panza, Marco, 2005. *Newton et les origines de l'analyse: 1664-1666.* Paris: Blanchard.

[(Panza, 2007)] Panza, Marco, 2007. "What is new and what is old in Viète's *analysis restituita* and *algebra nova*, and where do they come from? Some reflections on the relations between algebra and analysis before Viète's". *Revue d'Histoire des Mathématiques*, 13, 83–153.

---

only independent of the use of a literal formalism like that of early modern algebra, but is also suggested (independently of any concern for such a literal formalism) by the way Euclid expresses himself and reasons, and was explicitly at work many centuries before the early-modern age.

[26] Concerning this matter, I can here only refer the refer to the Introduction of my

[(Panza, 2010)] Panza, Marco, 2007. "Rethinking Geometrical Exactness". *Historia Mathematica*, **38 (1)**, (to appear).

[(Rashed, 1999)] Rashed, Roshdi and B. Vahabzadeh, 1999. *Al-Khayyām mathématicien*. Paris: Blanchard.

[(Rashed, 2007)] Rashed, Roshdi (ed., tr.), 2007. Al-Khwārizmī, *Le Commencement de l'algèbre*. Paris: Blanchard.

[(Rashed, 2009)] Rashed, Roshdi (ed., tr.), 2009. *Thābit ibn Qurra: Science and Philosophy in Ninth-Century Baghdad*. Berlin: Walter de Gruyter GmbH & Co.

[(Rosen, 1831)] Rosen, Frederic, (ed. tr.) 1831. *The algebra of Mohammed ben Musa*. London: Printed for the Oriental Translation Fund. and sold by J. Murray.

[(Viète, 1591)] Viète, François, 1591. *In artem analyticem isagoge. Seorsim excussa ab Opere restituaeæ mathematicaeæ analyseos, seu algebra nova*. Tournon: apud Iametium Mettayer typographum regium.

[Woepcke, 1851)] Woepcke, Franz, 1851. *L'algèbre d'Omar al-Khayyāmī*. Publiée, traduite et accompagnée d'extraits de manuscrits inédits par F. Woepcke. Paris: Duprat.

Chapter 7
The geometry of the unknown: Bombelli's *algebra linearia*

Roy Wagner

Abstract This paper studies the ways algebra and geometry are related in Bombelli's *L'algebra*. I show that despite Bombelli's careful adherence to a from of homogeneity, he constructs several different ways of relating algebra and geometry, building on Greek, Arabic, abbacist and original approaches. I further show how Bombelli's technique of reading diagrams, especially when representing algebraic unknowns, requires a multiple view that makes lines stand for much more than the diagrams present to an untrained eye. This multiplicity reflects an exploratory approach that seeks to integrate the algebraic and geometric strata without reducing one to the other and without suppressing the idiosyncrasies of either stratum.

Key words: Renaissance algebra, geometric algebra, Bombelli, geometric representation

7.1 Introduction

7.1.1 Scope, purpose and methodology

In the year 1572 Rafael Bombelli's *L'algebra* came out in print. This book, which is well known for its contribution to solving cubic and quartic equations

---

Parts of the research for this paper were conducted while visiting Boston University's Center for the Philosophy of Science, the Max Planck Institute for the History of Science and the Edelstein Center for the History and Philosophy of Science, Medicine and Technology in the Hebrew University in Jerusalem.

by means of roots of negative numbers, contains significant work on the relation between algebra and geometry. The manuscript included two additional geometry books, of which only a few fragments were incorporated into the print edition.

This paper will analyse what Bombelli called *"algebra linearia"*:[1] the inter-representation of geometry and algebra in Bombelli's *L'algebra*. I will map out the various ways that geometry and algebra justify, instantiate, translate and accompany each other. The purpose of the analysis is to point out the multiple relations between the two mathematical domains as rigorously set out in Bombelli's text, and the multiple vision that's required for understanding Bombelli's geometric diagrams.

This paper runs as follows. After a historical introduction and a review of Bombelli's notation, the second section of this paper will conduct a survey of Bombelli's geometric representation of algebra and its original regimentation of homogeneity considerations. I will then show how, despite the rigour of Bombelli's geometric representation of algebra, this representation does not dictate a one-to-one correspondence between algebraic and geometric elements, but rather allows different translations and interpretations to coexist.

The third section will then study the various functional relations between algebra and geometry. I will show that, building on Greek, Arabic, abbacist[2] and original practices, Bombelli's geometric representations sometimes serve to justify algebraic manoeuvres, sometimes to instantiate algebra in a different medium of representation, sometimes to connect different algebraic expressions through a common geometric translation, and sometimes simply to provide an independent accompaniment for algebra (this classification may serve as a framework to understand Renaissance juxtapositions of algebra and geometry).

The fourth section will go on to show how Bombelli's geometric representations of algebra confront negative magnitudes, expressions involving roots of negatives, and algebraic unknowns and their powers. This section will demonstrate that Bombelli implicitly hypothesized a co-expressivity of algebra and geometry, and that this hypothesis helped endorse questionable algebraic entities, generated new hybrid geometrico-algebraic entities and practices, and rendered geometric and algebraic signs polysemic and multi-layered. The overall result of this study will be a complex picture of multiple relations between algebra and geometry, which does not reduce or subject the one to the other, but assumes that they are deeply related, and builds on this relation to produce new mathematical practices.

---

[1] This term is best translated as 'algebra in lines', to avoid confusion with contemporary linear algebra.

[2] As the abbacists – the arithmetic teachers for Italian traders youth – had very little to do with the abacus as instrument, I follow other researchers in retaining the 'bb' spelling.

My approach is semiotic and mostly intrinsic. I explore Bombelli's work as an inspiring example of practicing mathematics rigorously, but without a confined, pre-charted ontology. Bombelli's hybridisation of algebra and geometry, expanding the boundaries of each without reducing one to the role of a servant to the other, is an instructive example of the open horizons of mathematical constructions of meaning.

7.1.2 Bombelli and L'algebra

Not much is known about Bombelli's life. According to [Jayawardene (1965), Jayawardene (1963)] he was born around 1526 and died no later than 1573. He was an engineer and architect involved in reclaiming marshland and building bridges. There is no record that he studied in a university, but he was obviously a learned man, so much so that a scholar from the university of Rome invited him to cooperate on a translation of the works of Diophantus.

The writing of the manuscript draft of *L'algebra*, Bombelli's only known publication, took place during a long pause in the Val di Chiana marsh reclaiming, which Bortolotti [Bombelli (1929)] dates to the early 1550s and [Jayawardene (1965)] to the late 1550s.

The manuscript [Bombelli (155?)], which was uncovered by Ettore Bortolotti, is divided into five books. The first three books of the manuscript appeared with revisions in the 1572 print edition [Bombelli (1572)]. Bortolotti published a modern edition of the remaining two books with an introduction and comments [Bombelli (1929)], which was later combined with the 1572 edition [Bombelli (1966)].[3]

The first book of *L'algebra* is a treatise on arithmetic, including root extraction and operating on sums of numbers and roots (mostly binomials, but some longer sums as well). The second book introduces the unknown (*Tanto*), presents an elementary algebra of polynomials up to and including division, and goes on to systematically present techniques for solving quadratic, cubic and quartic equations, following the discoveries of dal Ferro, Tartaglia, Cardano and Ferrari. The third book is a collection of recreational problems in the abbacist tradition together with problems borrowed directly from Diophantus. The problems are solved using the algebraic techniques taught in the second book.

---

[3] In referring to Books III, IV and V, I use problem number and section number (the 1929 version is available online, so it makes more sense to use section numbers than the page numbers of the out-of-print 1966 edition). In references to Books I and II the page numbers of the 1966 edition are used, as there is no numerical sectioning. The translations from the vernacular Italian are my own.

The fourth book, which did not make it to print at the time, concerns what Bombelli calls *algebra linearia*, the reconstruction of algebra in geometric terms. It opens with some elementary Euclidean constructions, and then builds on them to geometrically reproduce the main techniques of Book II and some problems of Book III. Book V treats some more traditional geometric problems in both geometric and algebraic manner, goes on to teach some basic practical triangulation techniques, and concludes with a treatise on regular and semi-regular polyhedra. Book IV is not entirely complete. Many of the spaces left for diagrams remain empty. Book V is even less complete, and its sections do not appear in the manuscript table of contents.

There are some substantial differences between the manuscript and the print edition. Several sections that appear as marginalia in the manuscript were incorporated as text into the print version of books I and II. Some of the geometric reconstructions of the unpublished Book IV were incorporated into the first two books. The print edition also has a much more developed discussion of roots of negative numbers, and introduces some new terminology and notation that will be addressed below. Book III went through some major changes. Problems stated in terms of commerce in the manuscript were removed, and many Diophantine problems were incorporated (for a full survey of these changes see [Jayawardene (1973)]). The introduction to the print edition states an intention to produce a book that appears to be based on the manuscript Book IV, but this intention was never actualized (Bombelli died within a year of the print publication).

According to Bortolotti's introduction, *L'algebra* seems to have been well received in early modern mathematical circles. Bortolotti quotes Leibniz as stating that Bombelli was an "excellent master of the analytical art", and brings evidence of Huygens's high esteem for Bombelli as well [Bombelli (1929), 7–8]. Jean Dieudonné, however, seems less impressed with Bombelli's achievements and renown [Dieudonné (1972)]. Note, however, that the geometry books (IV and V) were not available in print, and are unlikely to have enjoyed considerable circulation.

The vast majority of technical achievements included in Bombelli's printed work had already been expounded by Cardano. The exceptions include some clever tinkering with root extraction and the fine tuning of techniques for solving cubics and quartics. Bombelli's achievements in reconciling algebra and geometry, which are the subject of this paper, were not published in print at the time. But Bombelli's one undeniable major achievement is the first documented use of roots of negative numbers in order to derive a real solution of a polynomial equation with integer coefficients. He is not the first to work with roots of negative numbers, but he is the first to manipulate them extensively beyond a basic statement of their rules.

However, judging Bombelli's book through the prism of technical novelty does not do it justice. Indeed, Bombelli explicitly states in his introduction that he is presenting existing knowledge. He explains that "in order to remove finally all obstacles for the speculative theoreticians and the practitioners of this science" (algebra) ... "I was taken by a desire to bring it to perfect order".[4] In fact, the main text that Bombelli sought to clarify was Girolamo Cardano's *Ars Magna*.

Despite this strictly pedagogical aim, Bombelli does occupy a special place in the history of algebra. While Bombelli's explicitly mentioned sources (Leonardo Fibonacci, Oronce Finé, Heinrich Schreiber, Michael Stifel, Luca Pacioli, Niccoló Tartaglia and Girolamo Cardano) fail to include two centuries of vernacular Italian algebra developed in the context of abbacus schools, Bombelli is, in a sense, the last proponent of the abbacist tradition. He is the last important and innovative author to organize his work around rules for solving first a comprehensive list of kinds of equations, and then around a much looser collection of recreational problems borrowed from the abbacist reservoir. Bombelli, like his sixteenth century predecessors, adds much to the knowledge of past abbacists, but in terms of practices, terminology and problems he is a direct descendent of their tradition.[5]

7.1.3 Notation

The name of the unknown in Italian abbacus algebra is usually *cosa* (thing), and occasionally *quantità* (quantity). Bombelli writes in his manuscript that he prefers the latter, but uses the former, because that's the received practice. In the print version, following Diophantine inspiration, Bombelli changes the name of the unknown to *Tanto* (so much, such; I retain these Italian terms in this paper in order to maintain a distance from modern practice). The second power is called *Censo* in the manuscript (Bombelli prefers *quadrata*, but again follows received practice) and *potenza* in the print version. *Cubo* is the third power, and *Censo di Censo* or *potenza di potenza* is the fourth. Higher powers are treated and named as well, but are not relevant for this paper.

Bombelli's manuscript notation for powers of the unknown is a semicircle with the ordinal number of the power over the coefficient. The print edition reproduces this notation in diagrams of calculations, but, due to the limitations of print, places the semicircle next to the coefficient in the running text.

So a contemporary $5x^2$ would be rendered as $5 \overset{2}{\smile}$ in print and as $\overset{2}{\underset{\smile}{5}}$ in the

---

[4] "per levare finalmente ogni impedimento alli speculativi e vaghi di questa scientia e togliere ogni scusa a' vili et inetti, mi son posto nell'animo di volere a perfetto ordine ridurla" [Bombelli (1966), 8].

[5] The authoritative survey of abbacus algebra is still that of [Franci & Rigatelli (1985)]. A comprehensive catalogue of abbacist algebra was compiled by [Van Egmond (1980)].

manuscript. The manuscript Book II accompanies plain numbers with a $\overset{0}{\smile}$ above them, but this is almost entirely discarded in the print edition and in the other manuscript books.

Bombelli uses the shorthand *m.* for *meno* (minus) and *p.* for *più* (plus). The modern edition replaces those signs with the contemporary − and +, but otherwise respects the notations of the 1572 edition. I follow this practice here as well.

7.2 Elements of *Algebra Linearia*

In this section I will discuss Bombelli's inter-representation of algebra and geometry. The first subsection will raise the issue of relations between algebra and geometry in the abbacist tradition. The second subsection will consider the problem of homogeneity[6] in Cardano and Bombelli, and describe Bombelli's method of preserving homogeneity without committing himself to a fixed translation of algebraic powers into geometric dimensions. The third subsection will present Bombelli's algebraic-geometric translation system, but will also show that Bombelli's geometric representations of algebra were not restricted to this system. Along the way we will encounter a couple of preliminary examples of geometric signs that function on several levels at once. The point of this section is that Bombelli's practice is highly rigorous, but still allows for different algebraic-geometric interpretations to coexist.

7.2.1 *Letting representation run wild*

The second book of *L'algebra*, which deals with the solution of polynomial equations up to the fourth degree, concludes with Bombelli's "reserving it for later, at my leisure and convenience, to give to the world all these Problems in geometric demonstrations".[7] The third book, a collection of algebraic problems solved by the techniques of Book II, concludes with a more elaborate statement: "I had in mind to verify with geometrical demonstrations the working out of all these Arithmetical problems, knowing that these two sciences (that is Arithmetic and Geometry) have between them such accord that

---

[6] Homogeneity here means never adding or equating geometric entities of different dimensions. If a practice is homogeneous, the constructed geometrical object will be invariant with respect to the choice of units of measurement.

[7] "riserbandomi poi con più mio agio e commodità di dare al mondo tutti questi Problemi in dimostrationi geometriche" [Bombelli (1966), 314].

the former is the verification of the latter and the latter is the demonstration of the former. Never could the mathematician be perfect, who is not versed in both, although in these our times many are those who let themselves believe otherwise; how they deceive themselves they will clearly recognize when they will have seen my former and latter work; but because it is not yet brought to such perfection as the excellence of this discipline requires, I decided I want to consider it better first, before I were to present it for the scrutiny of men".[8] Unfortunately, Bombelli died soon after, and never got to perfect this part of his work, which remained incomplete in manuscript.

Of course, the notion of relating algebra or arithmetic and geometry was hardly new. The tradition of representing numbers by lines goes back to Euclid's diagrams in his arithmetic books, if not earlier. But while Euclid made an effort to set apart the theories of ratios between general homogeneous magnitudes, of ratios between numbers, and of relations between lengths, areas and volumes (books V, VII and XI of the *Elements* respectively), a renaissance author such as [Cardano (1968), 28] could refer simultaneously to V.19, VII.17 and XI.31 to explain why, if three cubes equal 24, then one cube equals 8.

But this conflation emerged much earlier. Abbacus algebra imported Arabic geometric diagrams that had already been interpreted as referring to numbers.[9] Among the abbacists, the most emblematic 'algebraic geometer' was the artist Piero della Francesca. Piero's geometry dealt with diagrams with some line lengths given numerically. It used *cosa* terms to model unknown lengths of lines, translate information concerning these lengths into polynomial equations, and derive their solutions using abbacist rules. His work was so influential, that Luca Pacioli practically imported it as is into his work.[10]

The algebraic entities — the *cosa*, *censo* and *cubo* (the unknown, its square and its cube) — were sometimes (but by no means generally!) interpreted as geometric in introductory presentations and some applications. As 14th century abbacist Maestro Dardi wrote, "The *cosa* is a linear length and is root of the *censo*, and one says *cosa*" (literally, thing) "because this name

---

[8] "io fussi di animo di provare con dimostrationi Geometriche l'operatione di tutti questi problemi Arimetici, sapendo che queste due scientie (cioè l'Arimetica e Geometria) hanno intra di loro tanta convenientia che l'una è la prova dell'altra e l'altra è la dimostration dell'una, nè già puote il Matematico esser perfetto il quale in ambedue non sia versato, benchè a questi nostri tempi molti siano i quali si danno a credere altrimente; del che quanto si ingannino all'hor chiaramente lo conosceranno quando che l'una e l'altra mia opera havranno veduta; ma perchè non è per ancora ridutta a quella perfettione che la eccellentia di questa disciplina ricerca, mi son risoluto di volerla prima meglio considerare, avanti che la mandi nel conspetto de gli huomini" [Bombelli (1966), 476].

[9] For the earlier roots of the arithmetisation of Greek geometry see Reviel Netz' highly remarkable *The transformation of mathematics in the early Mediterranean world* [Netz (2004)].

[10] See Gino Arrighi's introduction to [della Francesca (1970)].

can be attributed to all things of the world in general. The *censo* would be a surface breadth and is the square of the cosa" ... "The *cubo* would be a corporeal magnitude, the body of which includes in itself the length of the *cosa* and the surface of the *censo*, and is called *cubo* according to the arithmetic of Boethius" ... "meaning such aggregation of numbers".[11] The algebraic terms here come from geometry and arithmetic, but have to do generally with "all things of the world". Indeed, the one interpretation of algebraic unknowns common to all abbacists is the economic one: they are used to represent monetary units and quantities of merchandize.

Since the *cosa* can be anything, practitioners were not always bound to respect the above allocation of algebraic terms to geometric dimensions. In fact, the notion of dimension was probably not terribly rigid in a culture that used such term as "quadro chubico" for volume [Paolo dell'Abbaco (1964), 128], and sometimes measured volumes in terms of square *braccia* (*braccia* is a unit of length). By the time we reach Cardano we see one and two dimensional diagrams for algebraic cubic terms, and a square whose one side represents the second power of an unknown and the other its fourth power [Cardano (1968), 21,52,238].

The correspondence between geometric and algebraic dimensions is not fixed in the work of Bombelli as well. He is apt to say such things as "the rectangle .i.l.g. will be a cube and the rectangle .i.l.f. will be 6 $\smile$",[12] denying any consistent relation between the algebraic and geometric hierarchies of powers and dimensions. He follows this conduct in his Book IV as well, where algebraic cubes can be drawn as squares with one side an algebraic square and the other an algebraic *cosa* (e.g. §28). While Book II does not attempt to geometrically instantiate quartic and biquadratic equations, Book IV represents fourth powers as squares whose sides are algebraic squares (§§43,46).[13]

7.2.2 *Regimenting representation*

But a geometric representation of algebra, which has any regard for classical practice (as renaissance mathematics obviously had) must confront the issue of homogeneity. Cardano, for one, included a highbrow warning against the

---

[11] "La cosa è una lunghessa lineale ed è radicie del censo, e diciensi cosa perchè questo nome cosa si può atribuire a tutte le cose del mondo gieneralmente. Lo censo sie una anpiessa superficiale ed è quadrato della cosa, e diciesi censo da cerno cernis che sta per eleggiere, inperciò che el censo eleggie lo meçço proportionale in tra la cosa e'l cubo. Lo cubo sie una grossessa chorporale lo cuj chorpo inchiude in sè la lunghessa della cosa e lla superficie del censo, ed è ditta cubo sicondo l'arismetrica di Boetio da questo nome cubus cubi che tanto vuol dire quanto agreghatione di numerj" [Dardi (2001), 37–38].

[12] "il paralellogramo .i.l.g. sarà un cubo ed il paralellogramo .i.l.f. sarà 6 $\smile$" [Bombelli (1966), 229].

[13] The diagrams are missing from the manuscript treatment of these sections, but the text clearly shows that the diagrams inserted by Bortolotti are correct in this respect. Recent

misunderstandings that failure to respect homogeneity may bring: "it is clear that they are mistaken", explains Cardano, "who say that if BH, for instance, is the value of" the unknown[14] "and GF is 3, the rectangle" formed by BH and GF "will be 3BH or triple BH. For it is impossible that a surface should be composed of lines" [Cardano (1968), 34]. But Cardano's way of actually dealing with homogeneity in his geometric proofs of algebraic rules was simply to turn a blind eye to its requirements. After making this very warning, Cardano ignores the difference between multiplying a line by a number and by another line, and very casually adds numbers, lines, areas and volumes as if they were all homogeneous [Cardano (1968), e.g. 76,65,124].

Bombelli's practice, however, is much more carefully regimented. When Bombelli illustrates geometrically the solution of the problems 'Cose equal number' and 'square(s) equal number' (§§22–24), he carefully equates products of lines with rectangles, and homogeneously applies rectangles to lines[15] or reduces them to squares.[16] Nevertheless, when constructing a line representing one of Euclid's special binomials (sums of number and square root), Bombelli has no problem writing "Let the line .a. be 16 square number, and the line .b. 12 non square number" ... "and let .c. be 144, the square of .b., and .d. 192, product of .a. and .b." (§56).[17] Does this mean that Bombelli, too, violated the requirements of homogeneity?

The answer is negative. Bombelli explicitly developed the means that would allow him to represent the product of lines as a line, while maintaining homogeneity. For instance, in §98 Bombelli poses the following question, a geometric version of a problem from his manuscript Book III: "Let the line .a.b. be given, which has to be divided into three parts in continued proportion in such a way that having found a line that would be equal to the product of the first and the second, the line .o. being the common measure, and the said line multiplied by the third, it would make a line equal to .g.l.".[18] The product of the first and second lines is a rectangle, and as such can be applied to a line,

---

work by Marie Hélène Labarthe indicate that Pedro Nuñez applied similar practices.

[14] The English translation has x for Cardano's res, which I'd rather avoid.

[15] To apply a rectangle to a line is to construct another rectangle of the same area with one side equal to the given line.

[16] This accords with the notion of "strict homogeneity" attributed by [Freguglia (1999)] to Bombelli, but as we shall see below, this attribution has to be qualified and applied very cautiously.

[17] "Sia la linea .a. 16, numero quadrato, et la linea .b. 12, numero non quadrato, et minore de la .a., et sia la .c. 144, quadrato della .b., et la .d. 192, moltiplicatione dell'.a. in .b.".

[18] "Sia data la linea .a.b. la quale si habbia a dividere in tre parti continoe, et proportionali in tal modo, che trovato una linea che sia pari alla potentia della prima nella seconda, essendo la linea .o. la commune misura, et la detta linea moltiplicata via la terza faccia una linea pari alla .g.l.". Note that the common measure could also be understood as unit measure. Indeed, Bombelli writes a few lines below that "the .o. is 1 always by rule" ("et la .o. 1, sempre per regola").

namely to the common measure .o.. The other side of this applied rectangle represents the product of the two original lines. The common measure need not be carried explicitly with the product line, but it is there implicitly as the other side of a rectangle, for otherwise homogeneity is violated. This practice makes the product line a hybrid between an independent line and a rectangle named by its non-common side. The final requirement posed by the problem may be simultaneously viewed as an equation confronting two lines (the product of the three parts and .g.l.), or as an equation between two boxes, where the first has one side equal to .o., and the second has two sides equal to .o. (and one equal to .g.l.). A line no longer simply stands for a line.

In other problems the common measure might not be mentioned under that name, but is still there. When in §111 Bombelli asks to find the line representing the square of a given line, he uses one of the given lines of the problem as a common measure. Indeed, when he verifies the solution by assigning lengths to lines, he assigns this line the length 1, and says that this is the value one should "always take".[19] But this "always" is immediately qualified. Geometrically, we may choose our unit as we wish, but arithmetically, our chosen common measure might already be given a non unit numerical length. And so Bombelli explains that if the line used as geometrical unit is not of length 1, then a common measure should be derived by rescaling[20] this given line according to its given length so as to produce a line of length 1. If we now perform the construction with this newly derived unit as the common measure, geometric practice will accord with the arithmetic score of line lengths.

We see here that Bombelli takes care to maintain homogeneity on the geometric level, and at the same time remains faithful to arithmetic considerations, to which homogeneous geometry in itself is blind. This practice distributes the length of a line between different co-extensive registers. Geometrically, the common measure should "always" be interpreted as 1. But arithmetically it may be assigned any value. The common measure ties together two mathematical orders, and is allowed to function on both levels, accepting and validating their difference. A single line — that of the common measure — may represent two values in two contexts at once.[21]

---

[19] "sempre pongasi la .a.b. essere 1".

[20] I use the term *rescaling* to refer to reducing or extending a line or an entire geometric structure by a given ratio.

[21] It is interesting to note that Bombelli's virtuosity concerning homogeneity confused even as deep and insightful an editor as Ettore Bortolotti (his rampant whiggishness notwithstanding). In the notes to §110 Bortolotti says that Bombelli's rectangle .e.b.f. should be a box, because it equals what Bortolotti reconstructs as X^3. But in fact Bortolotti's X^3 is for Bombelli not a box whose sides are equal to an unknown X, but a square whose one side is the *cosa* (Bortolotti's X), and the other side the square of the *cosa* reduced to a line, with .b.f. serving as unit measure (in Bortolotti's terms: $X^2/\overline{bf}$). In Bortolotti's terms, then, the correct algebraic model for the term equated with .e.b.f. would be $X^3/\overline{bf}$, and homogeneity is indeed maintained if .e.b.f. is a rectangle. This shows how subtle Bombelli's practice

7.2.3 The vicissitudes of regimentation

Bombelli's geometric representation of algebra is provided with a more or less systematic setting. After reviewing some elementary Euclidean constructions (e.g. bisecting a line, reducing a rectilinear surface to a square, applying a rectangle to a line), Bombelli sets out to instantiate arithmetic and algebraic operations in geometric terms. Addition and subtraction are represented as concatenating and cutting off lines (§§15,16); the product of lines is represented as the rectangle that they contain (§17); dividing lines is done by introducing a common measure and using similar triangles (*Elements* VI.12) to draw a line whose ratio to the unit is as the ratio of the divided to the divisor (§18). As Bortolotti observed, this practice of division would later be reinvented by Descartes. Earlier in the text, in the context of applying a rectangle to a line, Bombelli uses a construction based on *Elements* III.35 (chords in a circle cut each other proportionally). This construction, too, is used further on to implement division with a common measure.

Next Bombelli considers root extraction (§19). This is performed either by Euclid's semicircle construction of the geometrical mean between a given line and a common measure (*Elements* VI.13, without explicit reference) or the variation of the same construction, where the unit is taken on the given line, rather than appended to it (a variation later used by Descartes in his *Regulae*). Bombelli then relates this operation to reducing a rectangle or a sum of squares to a square. Cubic roots (§20) are then extracted through a trial-and-error method based on constructing similar right angle triangles that Bombelli, following Barbaro's commentary on Vitruvius, attributes to Plato (or his disciples) [Bombelli (1966), 47,228], and that is known from Pappus's *Mathematicae Collectiones*.[22] Bombelli promises two methods for the extraction of cubic roots, but provides only one. A second method appears in Book I of the print edition, and is based on superposing right-angled rulers [Bombelli (1966), 48]. Bombelli relates the extraction of a cubic root to finding two mean proportionals between a given line and a common measure as well as to reducing a box to a cube.

These building blocks allow Bombelli to translate arithmetic operations into geometric ones. However, we should note, as did Giusti [Giusti (1992), 311–312], that the use of these building blocks is not terribly consistent. We must acknowledge that the strategy of reducing rectangles to lines and employing explicit common measures is in fact not central to Bombelli's work.[23]

---

of regimenting homogeneity was, and how evolved was Bombelli's skill of tacitly retaining the two-dimensionality represented by a single line.

[22] See [Giusti (1992), 305–306] for more details.

[23] While I am puzzled by Giusti's claims that modeling division via proportional triangles "is somewhat hidden" and that division by intersecting chords "is abandoned"

Bombelli usually maintains homogeneity in the classical way, which does not require such manoeuvres. We should also note that Bombelli's practice here is in fact in line with the choices of Descartes, who would later develop the same technology to deal with homogeneity, but, again like Bombelli, would rarely use it to work outside the classical practice of homogeneity [Bos (2001), 299]. Van Ceulen, the other pre-Cartesian author who used a unit measure to reduce a product of lines to a line, had his contemporary editor insert a note insisting that the product of two lines is an area [Bos (2001), 156]. We see that bypassing homogeneity by common measures was hard to digest, even when it was explicitly introduced in a rigorous manner. In fact, the haphazard approach represented by Cardano enjoyed more popularity, as those who didn't mind homogeneity followed its path, while those who did mind homogeneity were uncomfortable even with the Cartesian approach.[24]

But Bombelli's deviations from his expository building blocks for instantiating algebra in geometric terms are not restricted to issues of homogeneity management. For example, Bombelli frequently uses the Pythagorean theorem, which is not mentioned in the expository part. In the context of geometrically instantiating algebraic solutions to problems imported from his manuscript Book III, Bombelli's approach is sometimes more creative than a strict combination of his expository building blocks would suggest. For instance in §94 (figure 7.1), when algebra instructs to divide the square on .b.c. by twice .a.c., rather than actually double .a.c. and divide using intersecting chords or proportional triangles as in the introduction, Bombelli draws the line .c.e. whose square is half the square on .b.c., and then takes advantage of the fact that .c.e. is a mean proportional between .a.c. and the result of the division, drawing this result (.g.c.) using one of the diagrams for producing a mean proportional.

The point of this example is not to nit-pick on Bombelli's commitment to the procedures presented in his exposition. I include these details to provide an example of how Bombelli follows the respective idiosyncratic practicalities of algebra and geometry, and ends up with different procedures even when geometry is there supposedly to reproduce an algebraic practice.[25]

[Giusti (1992), 312], it is true that these diagrams rarely depend on an explicit assignment of the role of a unit to any particular segment.

[24] Newton, for instance, wrote: "Multiplication, division and such sorts of computation are newly received into geometry, and that unwarily and contrary to first design". Quoted in [Roche (1998), 79].

[25] At the same time, in some cases of reconstructing algebraic solutions geometrically the diagrams follow algebraic orders very strictly. In §99, for example, the algebraic rule requires doubling one of the given quantities, and then dividing it by 2 (when Bombelli constructed the quadratic equation that solved the algebraic version of this problem, twice the given quantity appeared as a coefficient in the equation, and the solution rule then required this coefficient to be halved — hence the subsequent multiplication and division by

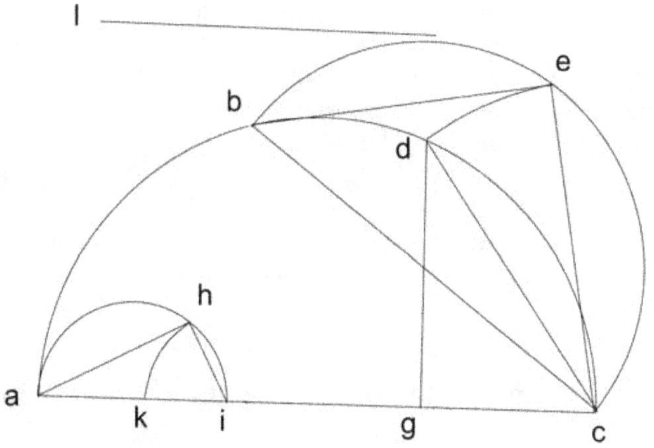

Fig. 7.1: [Bombelli (1929), §94]: geometric idiosyncracies with respect to algebra

These reservations notwithstanding, Bombelli's concern with regimenting homogeneity and representing algebra by geometry demonstrates two major points of my overall argument. First, a rigorous representation of algebra by geometry does not necessarily entail establishing univocal links between the two. Indeed, we saw above how algebraic powers could correspond to various geometric dimensions, how arithmetic operations received various geometric interpretations (e.g. root extraction interpreted as reducing a rectangle to a square and as finding a mean proportional with a common measure), and how the algebraic and geometric strata of expression maintained their own irreducible particularities (as in the description of §94 above, where a geometric construction, meant to represent an algebraic procedure, depended on the opportunities provided by the geometric diagram, rather than on the introductory gemetrico-algebraic 'dictionary'). Second, the interaction of geometry and algebra forces each stratum to express more than it had before the en-

2). The geometric reconstruction follows the algebraic rule so faithfully, that it insists on reproducing the subsequent and mutually-cancelling doubling and halving. A further interesting attempt to express arithmetic information geometrically occurs when extracting the root of binomials. In §63, when subtracting two roots whose ratio is rational (in Bombelli's language, the numbers under the root are to each other as "a square number to a square number" — "come da numero quadrato a numero quadrato"), the roots are modeled as a a square removed from a square. But in the next section, where the ratio between the subtracted roots is not rational, they are modeled as a rectangle removed from a square.

counter. We do not have a simple double expression, where each geometric sign expresses a geometric entity and an algebraic one. What we have is a relative sliding of multiple algebraic and geometric interpretations under the same geometric sign. Indeed, as we saw in the previous subsection, a single geometric line could represent a line or a rectangle, a geometric unit measure or an arbitrary arithmetic length.[26]

We will return to this latter point in the fourth section of this paper. But first let's expand the former point, and explore the functional (rather than semantic) relations that Bombelli constructs between geometry and algebra.

7.3 The functional relations between geometry to algebra

The first geometric representation of an algebraic operation in Bombelli's Book II is preceded by the following statement: "And while this science is arithmetic (as it was called by Diophantus the Greek author and the Indians), nevertheless it does not follow that one can't *provare* it all by geometric figures (as does Euclid in the second, sixth, tenth)" books of the *Elements*.[27] According to the convincing arguments of Sabetai Unguru and David Rowe

---

[26] It is interesting to recall here Lacan's reaction in his *Instance of the letter* to de Saussure's representation of the signifier and the signified strata as two parallel horizontal wave figures correlated by vertical lines. Rather than horizontal strata of 'upper water' (signifier) and 'lower water' (signified) anchored to each other by the vertical dashes, Lacan suggests reading the diagram as one horizontal stream flowing under another, while vertical drops of rain flow between them [Lacan (2006)].

[27] "E benchè questa scientia sia Arimetica (come la chiamano Diofante Autore Greco e li Indiani) però non resta che il tutto non si possi provare per figure Geometriche (come fa Euclide nel secondo, sesto, decimo)" [Bombelli (1966), 184–185]. Bombelli rarely quotes explicitly from Euclid — almost all explicit quotations are found in the first 18 sections of Book IV, where Bombelli introduces his basic geometric constructions. From the reference to Euclid's VI.12 in Bombelli's §18 we can infer that Bombelli used either a Greek version or an edition of Zamberti's Latin translation from the Greek (Campanus's Latin translation from an Arabic source has the Greek VI.10 as his VI.12). Nevertheless, Bombelli's diagram in §18 is not a reproduction of the Greek diagram, but an ad-hoc diagram adapted to his specific needs. In fact, Bombelli's list of algebraic sources (quoted above) suggests a thorough bibliographic research, and if this research extended to geometry too, it is likely that Bombelli consulted several versions of the *Elements*, and was not committed to any one particular edition. Medieval translators and commentators had already conflated arithmetic and algebra (e.g. Barlaam's commentary on Book II), but Bombelli's reduction of Euclid's binomials to sums of roots or of a number and a root is closest to what we observe in Tartaglia's Italian translation, which is based on an integration of Campanus' and Zamberti's translations, but which takes a further step toward an arimethization of Euclidean geometry [Malet (2006)]. Bombelli's step, in turn, towards such arithmetization is bolder still, as he considers Euclid's entire Book X as covered by the arithmetic of his own Book I [Bombelli (1966), 9].

[Unguru & Rowe (1981–1982)], Euclid saw things quite differently. But for the Renaissance revisionists the *provare*-bility of arithmetic by geometry was taken for granted. What must be carefully analysed here, however, is the precise use that Bombelli makes of the verb *provare*.

Provare could potentially mean 'prove', 'demonstrate', 'test', 'verify' and a myriad of shades in between. In this section we will follow the different ways that Bombelli relates geometry and algebra. In some cases geometry provides an independent *justification* of algebraic manoeuvres. In others geometry serves as *instantiation* or reconstruction of algebraic operations in lines rather than characters. In yet other cases a geometric diagram serves as a common pivotal *translation* of distinct algebraic expressions, which ties them together as equivalent. Finally, we must not neglect the cases where geometric diagrams serve as *accompaniment* for algebraic problems, either to manifest two approaches to a single problem without an attempt to relate them, or to supply a visual accompaniment without any specific functional purpose. I will review these themes in detail so as to demonstrate the multiplicity inherent in Bombelli's relating of algebra and geometry. I hope that this multiplicity may serve as a guiding framework for understanding geometry-algebra relations in wider contexts as well.

7.3.1 Geometric justification of algebra

One aspect of the relation between geometry and algebra in Bombelli's text goes back to the Arabic sources. This is the justification of algebraic rules for solving quadratic equations by geometric diagrams. In §49, for example, Bombelli justifies a rule for squaring the sum of roots by drawing a square, whose sides are the sum of those roots (see figure 7.2). The square is divided into the square of the first root (.g.e.), the square of the second (.c.e.) and the two rectangles, each representing the product of the roots (.i.e. and .a.e.). The mode of the explanation is so geometric that the congruence between the two rectangles is established not by arithmetic commutativity, but by Theorem II.4 of the *Elements*. Nevertheless, this geometric mood is complemented by assigning numerical values to the lines, without which the diagram could not specifically represent a sum of roots.

Similarly, many of the rules of Books I and II concerning operating on binomials and solving equations receive (at least an attempted) geometric justification in Book IV. These diagrams are not new, and go back to Cardano, the abbacists, Latin sources and Arabic sources. Quite a few of these diagrams were included in the print edition of Books I and II. Geometry appears to be a required footing for algebra to stand on.

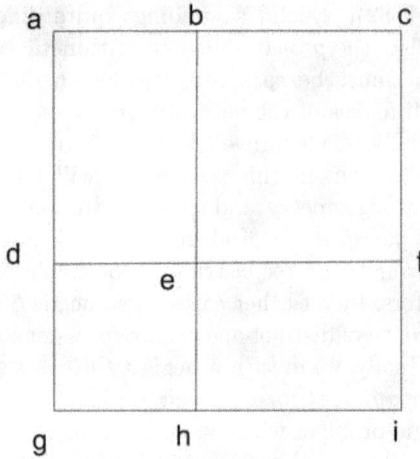

Fig. 7.2: [Bombelli (1929), §49]: geometric justification of an arithmetic rule

Bombelli extends this geometric footing to specific algebraic problems imported from the manuscript Book III as well. §75, for example, translates the division of 20 *scudi* in two parts, such that $\frac{2}{3}$ of the first part equals $\frac{3}{4}$ of the second [Bombelli (155?), 17r], into the following geometric problem. Given a line, divide it into two parts, such that if we take a portion of the first part according to the ratio between the given lines .a. and .b., and a portion of the second according to the ratio between the given lines .c. and .d., then the resulting portions turn out equal. The geometric solution that Bombelli provides does not imitate the procedure of the algebraic solution from the manuscript Book III, and its justification depends not on arithmetic, but on elementary proportion theory (*Elements* VI.12, which is not referenced explicitly here, but is referred to earlier in Book IV). The geometric analysis is followed by supplementing the diagram with numerical data identical to those in the problem from the manuscript Book III. The end result is of course the same as well, but the intermediary numbers involved in the Book IV solution are different from those in Book III, as the geometric solution is not a replication of the algebraic moves. Geometry thus provides an independent verification for algebra.

7.3.2 Geometric instantiations of algebra

If we go on to §78, however, we get another side of the story. The problem here is to divide a line .a. in two parts, such that the rectangle made of one part and the given line .b. equals the rectangle made of the other part and the given line .c..

First we are presented with a properly geometric solution. On the left hand part of figure 7.3 the line .d.f. equals .a., .d.g. equals .c. and .g.h. equals .b.. The lines .g.e. and .f.h. are parallel. Euclidean proportion theory guarantees that .e. divides .d.f. in the required way. If we were to plug in the numerical values from the original algebraic question into the diagram, and derive the same result that we had obtained algebraically, we would obtain, as above, a geometric verification of the algebraic result.

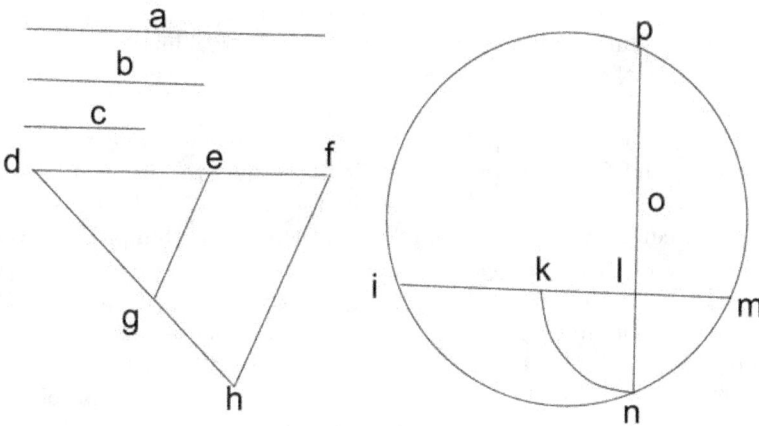

Fig. 7.3: [Bombelli (1929), §78]: geometric justification and geometric instantiation

But Bombelli does not do that. Instead, Bombelli writes: "and since the rule of algebra shows another solution for it, I didn't want to hold back from presenting it".[28] Bombelli's next solution for the same problem depends on the right hand part of Figure 7.3. Here .i.l. equals .a., .l.m. equals .c., and .l.p. equals the sum of .b. (equal to .l.o.) and .c. (equal to .o.p.). Euclidean circle

---

[28] "Et perchè la regola de l'algebra ne mostra un'altra dimostratione non ho voluto restare di metterla".

theory (*Elements* III.35), interpreted by Bombelli as a means of implementing division, yields that .l.n. equals .a. times .c. divided by the sum of .b. and .c.. Setting .k.l. equal to .l.n., claims Bombelli, obtains the desired partition of the given line.

What is the justification for this claim? Here Bombelli makes no geometric attempt to show that the rectangle contained by .i.k. and .l.o. equals that contained by .k.l. and .o.p.. What Bombelli actually did here is take the algebraic solution procedure from the corresponding problem in the manuscript Book III [Bombelli (155?), 119r], and implement it geometrically. Reformulating Bombelli's solution anachronistically and replacing his numbers by the letters a, b and c (but remaining faithful to his step by step reasoning), the algebraic solution says that the parts are x and $a - x$, and since we require that $bx = c(a - x)$, our equation reduces to $(b + c)x = ac$, which yields $x = \frac{ac}{b+c}$. The diagram simply applies Bombelli's geometric mechanism for dividing ac by $b + c$ to produce the result .l.n..

Instead of justifying the construction geometrically, Bombelli verifies that the constructed line represents the correct result by providing numerical values for the lines. In the example above .a. is given the value 11, .b. is 6 and .c. is 4. Therefore .l.p. is 10, .l.n. is $\frac{11 \cdot 4}{10}$ (note that this is an arithmetic claim, and not a claim immediately expressible in Euclidean terms!), and the resulting parts are $4\frac{2}{5}$ and $6\frac{3}{5}$. This result is explicitly verified to satisfy the terms of the problem, namely, that multiplying the former part by .b. and the latter by .c. yields equal results. But, I emphasise, there is no attempt here to argue geometrically why the rectangle contained by .k.l. and .b. equals the rectangle contained by .k.i. and .c.. If there is any justification for this procedure, it is to be found in the manuscript Book III algebraic solution (compounded by the endorsement of intersecting chords as modelling division).

In fact, for the most part, Bombelli's geometric solutions of problems imported from Book III reproduce his algebraic constructions, rather than justify them by an independent geometric procedure. In some cases the algebraic rule is quoted explicitly, while in others only a reference to the relevant problem of the manuscript Book III is included. If there are explanations added, these are usually clarifications as to how the geometric manipulations mimic the algebraic procedure (e.g. end of §§94,95), and not geometric arguments for the soundness of the solution.

This double gesture — *justification* and *instantiation* — is not restricted to problems imported from Book III, but occurs in the context of rules for solving polynomial equations as well. In [Bombelli (1966), 195] and in the manuscript §25, the solution of the quadratic is justified by the standard diagram borrowed from Arabic sources (figure 7.4 left). But next to this well known diagram appears a new diagram, which, as far as I can tell, is new to the abbacist and the Latin contexts. This diagram geometrically constructs the solution of the quadratic equation, rather than justify it. Indeed, if we write the equation anachronistically as $x^2 + bx = c$, then the area .e.f. represents the number c, and the side of the area .o. represents $\frac{b}{2}$. By the Pythagorean theorem the line .i.e. is the root of the sum of .e.f. and .o. $\left(\sqrt{\left(\frac{b}{2}\right)^2 + c}\right)$. Taking .i.p. equal to .i.e., and subtracting the side of .o. (which, recall, is $\frac{b}{2}$), we obtain .p.g., which we can anachronistically write as the well known solution formula $-\frac{b}{2} + \sqrt{\left(\frac{b}{2}\right)^2 + c}$. I emphasize again: the left hand diagram explains geometrically why the standard rule actually solves the equation; the right hand diagram instantiates the rule by a geometric construction that follows its arithmetic steps, but makes no attempt to justify the claim that it is in fact the sought line.

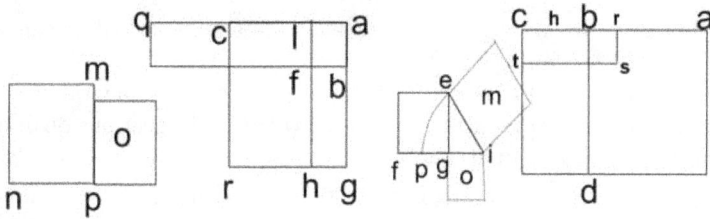

Fig. 7.4: [Bombelli (1966), 195–196]: from geometric justification to geometric instantiation

7.3.3 Common geometric translation of distinct algebraic entities

But to reduce Bombelli's geometric representations to the distinct roles of justification and instantiation of algebra would be inattentive to the details. In the sections dealing with the solution of cubics and quartics (§§32–48) the geometric representations typically do not quite do either.

In the case of cubics, for example (a treatment adapted from Cardano), the algebraic cube is represented by a geometric cube, and the various other algebraic elements (numbers, *Cose*, squares) are represented by slabs added to the cube or removed from it. When areas or volumes are said to equal or cancel each other — unlike in the case of summing roots quoted above, and in contrast with Cardano's preferred practice — Bombelli's argument does not depend on Euclidean theorems, but on an arithmetic calculation of the sizes of the areas or volumes involved. Then, the transformations required to solve the cubic (change of variable or completion to cube) are presented as different algebraic models for capturing the same geometric diagram. What we have here is not a geometric construction *instantiating* the bottom line of the algebraic solution, nor a *justification* of algebra by a strictly geometric argument, but a geometric diagram that binds different algebraic expressions by serving as their common *translation* — a geometric pivot that enables superposing different algebraic expressions on each other. For example, in §31, the diagram serves to show that the solution of the equation '1 $\overset{3}{\smile}$ equals 6 $\overset{1}{\smile}$ +9' equals the sum of the solution of '$\frac{1\overset{6}{\smile}+8}{1\overset{3}{\smile}}$ equals 9' and of 2 divided by that solution; in §35 the diagram serves to show that the solution of the equation '1 $\overset{3}{\smile}$ +6 $\overset{2}{\smile}$ equals 81' is related to the solution of the equation '1 $\overset{3}{\smile}$ equals 12 $\overset{1}{\smile}$ +65' by a shift of 2 (the text drags a silly scribal or calculation error and has 64 for 65).[29]

An even more explicit expression of the role of geometric diagrams as pivots binding different algebraic expressions is provided in the sections dealing with transforming quadratic equations. For example, in [Bombelli (1966), 204] the equation '$\overset{2}{\smile}$ +6 $\overset{1}{\smile}$ equals 16' is related to the equation '$\overset{2}{\smile}$ equals 6 $\overset{1}{\smile}$ +16' not by algebraically substituting $\frac{6}{1\overset{1}{\smile}}$ for $\overset{1}{\smile}$, but by showing that both equations translate the same diagram of a rectangle of area 16 whose one side is longer than the other by 6 — the difference lying in the side modeled as $\overset{1}{\smile}$. But this translation of algebraic language into geometry is useful precisely because it is *not* one-to-one. Different algebraic texts (the original and reduced equations) are translated by the same geometric figure — only the parts of the figure are modeled differently by each algebraic interpretation. The single reference or translation for the two different algebraic texts serves as a pivot

---

[29] [Freguglia (1999)] also studies Bombelli's geometric representations of solutions of algebraic equations. He states that for second and third degree equations Bombelli provides geometric step-by-step justifications of the algebraic processes of solution as well as geometric constructions of the solutions that do not justify the algebraic rule. For fourth degree equations, Freguglia claims, Bombelli only provides a partial geometric justification of the algebraic solution process. But as far as I can see, this analysis is imprecise. For second degree equations Bombelli provides a geometric *justification* of the solution and then a distinct geometric *instantiation* (see figure 7.4); for third degree equations he provides a geometric *translation* of the original equation to a reduced one and a distinct geometric *accompaniment* that constructs a solution independently of the algebraic procedure (see below); and for fourth degree equations Bombelli provides only a geometric *translation* of the original equation to a reduced one. Freguglia links Bombelli's different representation strategies to the problem of homogeneity and of geometric representation of high powers of the unknown. But my analysis of Bombelli's treatment of homogeneity undermines this explanation. In fact, Bombelli had the means to provide all kinds of geometric representations to all relevant equations. His choices had more to do with complexity than with inherent limitation of representation techniques.

that identifies the distinct algebraic equations beyond their formal differences, and allows for *seeing* them as solving the same problem.

But there's another way in which geometry can tie together different algebraic objects: it can serve as a common representation that bridges their conceptual divisions. In §§38–42, for example, a single diagram accompanies five different reductions of equations of the form 'cube, squares and *Cose* equal number' to simpler cubic equations. Each reduced form is of a different kind, in parallel to the treatment in Book II. The fact that a single diagram manages to represent all cases weakens the organising principle structuring Book II, which considers different forms of cubic equations as distinct cases that require separate treatment.

A similar impact is had by §44 and §45, which share the same diagram, and where the latter, claims Bombelli, "is no different from the former except in that the square number .u. is added to the square .e.f., and in the other subtracted".[30] That a common diagram can represent different problems that differ by a sign renders these problems less obviously distinct. The point here is not simply that a single 'general' diagram can represent several 'specific' algebraic cases; the point here is that a single diagram can tie together algebraic elements that are considered essentially different (in the former example different kinds of cubic equations, in the latter addition and subtraction). Such common geometric representations may have helped Bombelli (on top of the exhaustion expressed explicitly in the conclusion of his Book II) to eventually let go of treating separately all different reductions of quartic equations and of distinguishing all subtractive and additive changes of variable required for these reductions.

Another instance of geometry bridging algebraic differences occurs in the context of the principle setting irrational and rational numbers apart as differing in 'nature' [Bombelli (1966), 13]. Indeed, when Bombelli explains how to geometrically extract a square root, he notes that it "doesn't suffer the difficulties that it suffers in numbers; because one will always find the root of any given line" (§19).[31] Given that for Bombelli any quantity can be represented by a line, the division between numbers that have discrete roots and those that do not becomes much less substantial.[32]

7.3.4 Geometric accompaniment of algebra

So far we saw that geometry sometime served as a *justification* for algebra, sometimes as *instantiation* of an algebraic solution procedure, and sometimes

---

[30] "non è differente da la prima se non in questo, che il quadrato numero .u. si aggiungeva con il quadrato .e.f., et in questo si cava.".

[31] "il quale creatore non patisce le dificultà, che pate nel numero; perchè sempre si troverà il creatore d'ogni preposta linea, essendo noto la comune misura".

[32] See [Wagner (2010)] for an overview of Bombelli's division of numbers according to their

as a *translation* or pivot tying together different algebraic terms. But these functions, which are not always easy to set apart, do not exhaust Bombelli's ways of relating algebra and geometry.

Problems §123–124, for example, both require to find the point .c. equidistant from .e. and .a. (left hand side of figure 7.5). The solution of §123 models the line .b.c. as an unknown, and uses the Pythagorean theorem to derive a quadratic equation for it. The solution in §124 uses the right hand diagram, and constructs the solution geometrically, having nothing to do with the algebra (the trick is to construct .k.c., an orthogonal bisector of .e.g., and use the Pythagorean theorem). This independent geometric construction is not accompanied by numerical values for verifying that it, too, would lead to the same solution as the algebraic argument. It is simply there as a counterpoint, accompanying the algebra, without any attempt to project it on the algebra or tie them together (another example for a geometric accompaniment independent from the algebraic solution is the construction of solutions to cubic equations by means of lines, discussed in section 4.2 below).

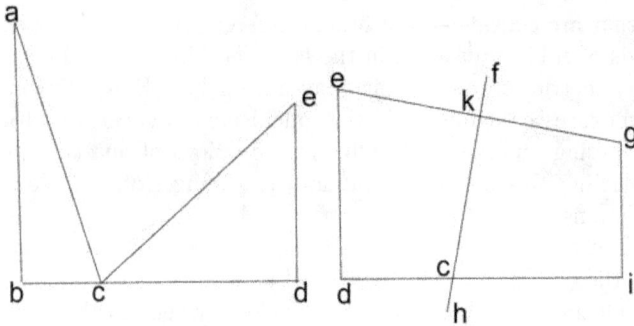

Fig. 7.5: [Bombelli (1929), §§123–124]: independently accompanying algebraic and geometric solutions of a geometric problem

In other cases the relation between geometry and algebra is weaker still. I'm referring, for example, to the demonstration of rules for adding and subtracting roots. While in figure 7.2 we saw one of these rules represented by a partitioned square, the diagrams for the other rules (figure 7.6) are nothing but disconnected lettered lines. Book III also has one of those odd diagrams, where three lines of equal lengths marked .a., .b. and .c. accompany a Diophantine problem (problem 140).

---

natures and of the undermining of this very division in his algebraic practice.

b _____

c _____

d _____

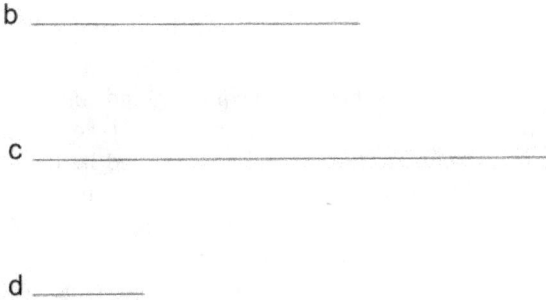

Fig. 7.6: [Bombelli (1929), §51]: non-functional geometric diagram accompanying an algebraic rule

These examples perhaps appear more odd than they should. The Euclidean tradition insisted on drawing diagrams for all problems and theorems, even arithmetic problems and theorems, and even those where the diagrams were nothing but a bunch of independent lettered lines. Ian Mueller suggested that in such cases "the diagram plays no real role" ... "except possibly as a mnemonic device for fixing the meaning of the letters" [Mueller (1981), 67]. Reviel Netz asserts that arithmetic diagrams "reflect a cultural assumption, that mathematics *ought* to be accompanied by diagrams" [Netz (1999), 42]. To appreciate the role of diagrams in Bombelli's text, we must not neglect this tradition of diagrams, which are there to illustrate, not to argue.[33]

One more accident allows us to better appreciate the role of diagrams in Bombelli's work. In §72 the triangle .d.e.g. is given the numerical side lengths 1 for .d.e., 8 for .d.g. and 4 for .e.g.. These data contradict the triangle inequality, as $1 + 4 < 8$, and therefore cannot describe a genuine Euclidean triangle. But since the geometric instantiation was not practiced as a faithful illustration of the data (many other diagrams are disproportional compared to the numerical data), Bombelli could go on with the example unhindered.

---

[33] For an elaboration of this point see my Deleuzian analysis of classical Greek diagrams [Wagner (2009b)].

7.3.5 So what is the relation between algebra and geometry?

If the multiplicity of relations between geometry and algebra documented above appears confusing, it is probably because we take too much critical distance. We must remember: algebra had already had its diagrams when it came to Italy. The Latin tradition, as well as some abbacus treatises, retained the geometric diagrams that accompanied the Arabic solution of quadratic equations and led up to a geometric understanding of the solution of the cubic equation. On the other hand, many abbacists set algebra and geometry apart, and developed the former independently. When this maturing algebra came across geometry again, it brought about an arithmetic understanding of some of Euclid's geometric books and algebraic solutions of geometric problems.

It is therefore not surprising that Bombelli, who inherited all these different traditions, came to piece geometry and algebra together in various different ways, and, according to his own quotation above, felt that algebra and geometry belonged together. This togetherness of algebra and geometry did not depend on a unique, one-directional relation of *justification, instantiation, translation* or *accompaniment*. For people like Bombelli, it seems, algebra was geometry's younger — but not for all that entirely dependent — sister, and these siblings were believed to play best when they were allowed to play together. Bombelli was intent on exploring and diversifying the playgrounds available for the common games of geometry and algebra. For an author as mathematically proficient as Bombelli, this didn't come at the expense of rigour.[34]

7.4 The geometry of what's not quite there

Bombelli's *algebra linearia* becomes much more challenging when we consider not only its various relations with algebra's established entities, but also its

---

[34] But that's a very local interpretation. Seen from a wider epistemological-historic perspective, the conjunction of geometry and algebra may have had to do with more wide ranging trends. For one, the regulated projection of one domain of signs on another is, drawing on Kristeva, a semiotic technique for producing an effect of truth (*vraisemblance*) since the birth of the novel in the 15th century (see *La productivité dite texte* in [Kristeva (1969)]; for how this theory works in a mathematical context see my [Wagner (2009a)]). Moreover, the emphasis on geometrically observing algebra, rather than just symbolically writing it, can be seen as related to the role of vision in the epistemology of early modern science. I don't pursue either direction here, because Bombelli does not provide enough textual evidence to substantiate either interpretation. But Bombelli did not write in scientific isolation, and the general trends of the period should not be ignored.

original manner of representing more challenging algebraic objects whose existence or status was not yet properly settled. Here we can see the genuine synergy inherent in the core of *algebra linearia*: its treatment of negative entities, their roots, algebraic unknowns and their powers. A proper understanding of Bombelli's approach and achievements depends on this hitherto neglected aspect of his work. A careful study of this aspect will show us how Bombelli's bringing together of geometry and algebra generated new mathematical practices, hybrid geometrico-algebraic entities, grounds for endorsing questionable mathematical entities, and multiple ways of reading a given sign.

The first subsection will deal with the geometric representation and endorsement of negative magnitudes. The second subsection will deal with the endorsement of expressions involving roots of negative numbers and with the underlying implicit hypothesis of co-expressivity between algebra and geometry. The last two subsections will study the hybrid gemetrico-algebraic practices built around representations of unknown magnitudes, and the multiple vision of geometric signs that these practices depended on.

7.4.1 The geometry of missing things

I argue in [Wagner (2010)] that the Renaissance *meno* cannot be properly reduced to subtraction. But in the context of geometric representations of algebra the subtraction interpretation seems more defensible, although not entirely exhaustive. Cardano, for one, uses the term 'add negatively' for turning from addition to something between a subtraction and an addition of negative geometric elements.[35] Bombelli's practice is similarly ambiguous. When he says that some areas are *meno*, they can be, in that context (e.g. §§26,27,46,47), interpreted as subtracted geometric magnitudes rather than as added negative geometric magnitudes.

One of the most interesting points of ambiguity occurs around the diagram reproduced in figure 7.7. Bombelli explains that "from the square .g.h.o. one removes the square .r.o.s." (where the latter is bigger than the former). "And because we're missing the gnomon .h.r.p.g. and we have the surface .f., it is necessary that the said gnomon be equal to that surface".[36] It is clear that Bombelli allows for the subtraction of a larger area from a smaller one, and that the result is a "missing" area. But this missing area is positive, in as much

---

[35] "addito per m̃.", where m̃ is short for *minus* [Cardano (1968), e.g. 54,97]. Note that the latter expression was revised in the later editions of the book so as to use unambiguous subtraction talk.

[36] "del quadro .g.h.o. si levi il quadro .r.o.s." ... "E perchè ci manca il gnomone .h.r.p.g. et habbiamo la superficie .f., di necessità bisogna che il detto gnomone sia pari ad essa superficie" [Bombelli (1966), 202–203].

as it can be compared to another positive area. Here geometric representation turns what's missing into something present. By way of a similar geometric treatment, the absence or presence of the gnomon becomes a relative position, rather than an ontological characteristic of a geometric element. Again we witness the role of diagrams as pivots, translating different arithmetic entities into a common referent.[37]

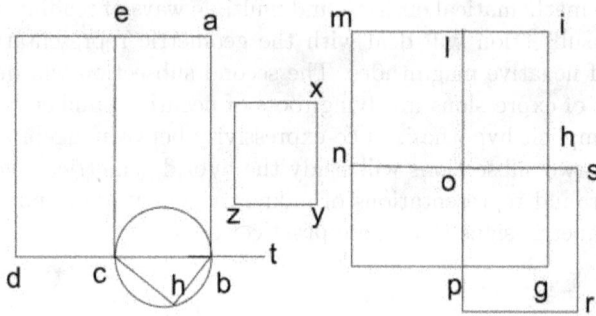

Fig. 7.7: [Bombelli (1966), 203]: subtracting a larger area from a smaller one

7.4.2 The geometry of sophistic things

The first appearance of roots of negative numbers in Book II occurs in the context of quadratics. A solution involving a root of a negative number is presented on a par with another "sophistic method",[38] that of simply replacing the impossible subtraction in the solution rule by an addition.[39] A few pages

---

[37] This representation should be contrasted with Cardano's diagram illustrating a situation where 40 were to be subtracted from 25, if one applied the usual solution rule to the quadratic equation 'square plus 40 equals 10 things'. There, all that the diagram shows is a square of area 25 built on half a line of length 10 [Cardano (1968), Ch. 37].

[38] "modo sofistico" [Bombelli (1966), 201].

[39] This approach might be explained by the work of Marco Aurel, who taught and published in Valencia. In his *Despertador de ingenios, Libro Primero de Arithmetica Algebratica* (1552), he offers as solution to the irreducible case of $x^2 + c = bx$ the negative counterpart of Bombelli's "sophistic" suggestion, namely $-\frac{b}{2} - \sqrt{\frac{b^2}{4} + c}$ (this is, of course, anachronistic notation). If one postulates that the product of two isolated negative numbers is negative (a postulate that Bombelli took up in his manuscript; see my [Wagner (2010)]), one indeed

later, in the context of the bi-quadratic without real solutions, solutions involving roots of negatives are not mentioned at all, not even under the mark sophistry. It is simply claimed that "such case could not be solved because it concerns the impossible".[40]

But when we turn to cubic equations, Bombelli's approach is very different. The context here is equations of the kind 'Cube equals *Tanti* and number'. Applying Dal Ferro's and Tartaglia's solution rule for some instances of such equations — Cardano's so called irreducible case — yields an expression involving roots of negative numbers. Bombelli's most renowned achievement is his endorsement and analysis of such solutions. In the manuscript, Bombelli justifies his approach on pragmatic grounds: such solutions arise from the same rule that worked in the reducible cases, and Bombelli can often enough transform them into correct real solutions [Bombelli (155?), 72v]. Bombelli's prime example (in anachronistic notation) is the equation $x^3 = 15x + 4$, whose solution, according to the received solution rule and Bombelli's method for extracting cubic roots of binomials, is $\left(2 + \sqrt{-121}\right)^{1/3} + \left(2 - \sqrt{-121}\right)^{1/3} = \left(2 + \sqrt{-1}\right) + \left(2 - \sqrt{-1}\right) = 4$. But the derivation of correct real solutions cannot alone account for Bombelli's endorsement of solutions involving roots of negative numbers, because a few pages further on such a solution is endorsed, even though Bombelli can't rewrite it as a real solution [Bombelli (155?), 76v].[41]

By the time the print edition was ready, Bombelli could provide another reason for endorsing solutions for cubic equations that involve roots of negative numbers. He explains: "and although to many this thing will seem eccentric — for I too was of this opinion some time ago, having the impression that it should be more sophistic than true — nevertheless I sought until I found the demonstration, which will be written down below, that indeed this can be shown in lines, that moreover in these operations it works without any difficulty, and often enough one finds the value of the *Tanto* as number".[42]

---

obtains a correct solution (this observation is derived from a talk by Fàtima Romero Vall-honesta delivered at the PASR conference in Ghent on August 27, 2009). Bombelli's reference to this way of solving quadratic equations, as well as his manuscript reference to the negative result of the product of isolated negative numbers, might suggest that the "certain spaniard" ("certo spagnuolo") mentioned among Bombelli's sources in his introduction was not the Portuguese Pedro Nuñez, as asserted by Bortolotti, but in fact Marco Aurel.

[40] "Ma se non si potrà cavare il numero del quadrato della metà delle potenze, tal capitolo non si potrà agguagliare per trattarsi dell'impossibile" [Bombelli (1966), 207].

[41] In contemporary terms the solution is $\left(3 + \sqrt{-720}\right)^{1/3} + \left(3 - \sqrt{-720}\right)^{1/3} - 3$. Bombelli's techniques do not allow him to simplify the cubic root in this expression.

[42] "Et benchè a molti parerà questa cosa stravagante, perchè di questa opinione fui ancho già un tempo, parendomi più tosto fosse sofistica che vera, nondimeno tanto cercai che trovai la dimostratione, la quale sarà qui sotto notata, sì che questa ancora si può mostrare in linea, che pur nelle operationi serve senza difficultade alcuna, et assai volte si trova la valuta del Tanto per numero" [Bombelli (1966), 225].

But what exactly is this demonstration "in lines"? Bombelli's argument is as follows. First Bombelli shows that if one has a sum of cubic roots of the anachronistic form $\left(a + \sqrt{-b}\right)^{1/3} + \left(a - \sqrt{-b}\right)^{1/3}$, then a cubic equation can be constructed, which, given the received solution rule, would produce that sum as its solution [Bombelli (1966), 226]. Then Bombelli shows geometrically that such cubic equations must have solutions [Bombelli (1966), 227–228]. Since he can show that the questionable algebraic expression solves an equation, and since, moreover, he can draw a solution for that same equation geometrically, Bombelli concludes that the geometrically found "length of the *Tanto* will also be the length of" the sum of "the two cubic roots above".[43] Bombelli concludes that the algebraic solution of the equation must coincide with the geometric construction, and that the latter therefore validates the former.

This argument, however, is not without its difficulties. The first difficulty is one that is explicitly addressed by Bombelli. Indeed, Bombelli's geometric solution is a planar construction, which, given a segment and an area representing the coefficients of the equation, a unit measure, and two right angled rulers, yields a segment representing the solution. Bombelli rejects possible objections to the use of right angled rulers by noting that a planar solution to a solid problem must use advanced tools[44] and by relying on the authority of no less than Plato and Archytas.

A second difficulty, which Bombelli fails to address, concerns the conditions of viability of the geometric construction. The solid diagram for solving cubic equations [Bombelli (1966), 226], derived from the one introduced by Cardano, fails to solve the irreducible case, as Bombelli and Cardano explicitly note. The novelty in Bombelli's planar diagram is precisely that it circumvents this difficulty. However, Bombelli makes no effort to show that his diagram is indeed constructible for all possible coefficients of the relevant kind of cubic equation. This is all the more unsettling, as the next kind of cubic equation that Bombelli treats does not always have a positive solution, but while Bombelli is well aware of this fact, and provides a precise arithmetic solvability condition [Bombelli (1966), 231], he makes no attempt to point out the geometric obstruction restricting his construction. Bombelli does not raise the question of whether his former geometric construction is or is not restricted by obstructions, and leaves his claim of general solvability without critical examination.

There's a further difficulty that Bombelli fails to address, which concerns the correspondence between the arithmetic and geometric representations of the solution. Bombelli is well aware of negative solutions of cubic equations. He uses such solutions to derive positive solutions of other equations, and

---

[43] "e trovata che si haverà la longhezza del Tanto sarà ancora la longhezza delle due R.c. legate proposte" [Bombelli (1966), 226].

[44] See [Bos (2001), Ch.3–4] for the context of such an argument.

sometimes even considers them independently.[45] But Bombelli never explicitly raises the possibility that his questionable sum of cubic roots might capture a negative solution, rather than the positive solution that he constructed geometrically.

Bombelli is obviously eager to make his solution for the cubic acceptable. To that end he is willing to use questionable construction methods, avoid dealing with the viability conditions of his diagrams, and suppress negative solutions. But most important here is that on top of all that Bombelli takes another crucial step on this already shaky ledge. Bombelli uses his argument above to endorse an algebraic entity, which his geometric construction does not actually draw. Nowhere does the diagram pick up roots of negative numbers, either directly or inside the cubic root of a binomial. In the language we introduced above, the geometric construction *accompanies* the algebraic solution, rather than *justify*, *instantiate* or even *translate* it. The diagram does indeed construct a line satisfying the terms of the equation, but its construction has nothing to do with the algebraic rule of solution and its roots of negative numbers. The speculation that the algebraic solution is identical to the line constructed in the diagram is snuck in through the back door without an explicit account.

The relation between geometry and algebra here can be qualified in a finer manner, if we observe Bombelli's remark that a certain cubic equation allowed him to trisect an angle, and that this fact led him to keep attempting (in vain) to transform that equation into one that he could solve without roots of negative numbers. His conclusion was that "it is impossible to find such general rule" for solving cubics without roots of negatives.[46] The point here is the tension that Bombelli expresses between being able to draw a solution and not being able to write it down in traditional arithmetic terms. This tension is something that Bombelli finds so hard to sustain, that he concludes by allowing an expression involving roots of negative numbers as an algebraic representation of the geometric solution. Recall, in contrast, that in the case of quadratic equations, where solutions involving roots of negative numbers could not be drawn geometrically or reduced to verifiable real solutions, Bombelli rejected them as sophistry. Bombelli's underlying conviction thus reveals itself:

---

[45] See [Wagner (2010)] for details.

[46] "Sì che (quanto al mio giuditio) tengo impossibile ritrovarsi tal regola generale" [Bombelli (1966), 245]. The discussion probably refers to §135 of Book V, where the trisection of an angle for the construction of a regular nine-gon is reduced to a cubic equation. This is not the same cubic equation as cited in Book II, and there's no attempt in Book V to solve this cubic equation with or without roots of negative numbers, but we must recall that Book V is the least complete among the books of *L'algebra*. Note also that this algebraic-geometric reflection, unlike the one concerning the general plane geometric solution of 'Cube equals *Tanti* and number', was already present in Bombelli's manuscript [Bombelli (155?), 88r], and may have therefore factored into his original endorsement of solutions involving roots of negative numbers.

if you can draw it, you should be able to express it algebraically, even if the expression looks like gibberish. To put it into a catch phrase: 'What you see is what you say'.

If we are to understand Bombelli's articulation of the algebra-geometry nexus, we should acknowledge the intimacy granted here to algebra and geometry. It's not just about justification, instantiation, translation or accompaniment. It's not just about two strata of expression interacting with each other in order to bring the best out of both. It's about a deeply underlying assumption of co-expressivity: what's expressible in the one domain, should be expressible in the other. Geometric visibility does not only guarantee reality, it should also guarantee algebraic expressibility. Without such underlying assumption, Bombelli's argument above would not have forced him to integrate roots of negative numbers into his mathematics.

The logic that short-circuits visibility and expressibility is, of course, an issue that deserves an independent tracking across the history of science. But here we'll restrict ourselves only to validating that it works in both directions. The other direction of the principle that 'what you see is what you say' is the principle that what you say should be visible as well. According to this principle algebraic unknowns should be expressible geometrically. And this is where we're turning next.

7.4.3 The geometry of the unknown: the rule of three

Geometric modelling of unknown magnitudes is not new. The very diagrams that since Al-Khwarizmi accompanied the solution of the quadratic equation did just that. The unknown line was represented by an arbitrary line, and so were lines representing known magnitudes, without necessarily respecting the proportions between the eventual value of the unknown and the given known values. But Bombelli is interested in doing better than that.

Let's go back to figure 7.4. The left hand side is the typical diagram traceable to Arabic sources, where .a.b. is the unknown line and .b.g. is a known line. The right hand side of the diagram, as explained above, is not another justification of the algebraic solution, but a geometric instantiation of the solution rule. The unknown square .a.b.d. (standing for the square of the unknown) and rectangle .d.b.c. (standing for the known coefficient .b.c. times the unknown) are to be equated to the known .f.g.e. (a number). The construction derives .p.g. as the value of the unknown. Since this is obviously disproportional with respect to .a.b., the previous representation of the unknown, Bombelli redraws the right hand side of the right hand diagram as the rectangle .r.s.t.

This strategy is not typical of the tradition, nor is it typical of Bombelli. It stands out more as an expression of a problem than as a way of dealing with it. When including known and unknown lines in the same diagram, and then using the known lines to derive a representation of the unknown, the unknown can no longer be drawn arbitrarily. Bombelli is concerned with this problem and attempts to derive a more satisfactory form of representation.

Bombelli's alternative representation is as ingenious as it is simple, and the fact that it was not systematically replicated (as far as I know) was, I feel, a genuine loss to techniques of mathematical representation. One of the most interesting examples occurs in figure 7.8. This geometric problem is presented by Bombelli as an analogue of Problem 41 from the manuscript Book III, which reads: "Two people have money, and find a purse in which there was as much money as the first person had, and this first person says to the other: if I had the purse, and $\frac{1}{3}$ of yours, and 2 more, I would have twice as much as your remainder. The second says: if I had the purse, by giving you 4 of mine, I would have two times and a half as much as you. One asks how much money was in the purse and how much each of them had".[47] In the geometric version of Book IV §86, the money of each person is replaced by an unknown line, $\frac{1}{3}$ is replaced by the ratio between the lines .a. and .b., and 2 and 4 are replaced by the lines .c. and .e. respectively.

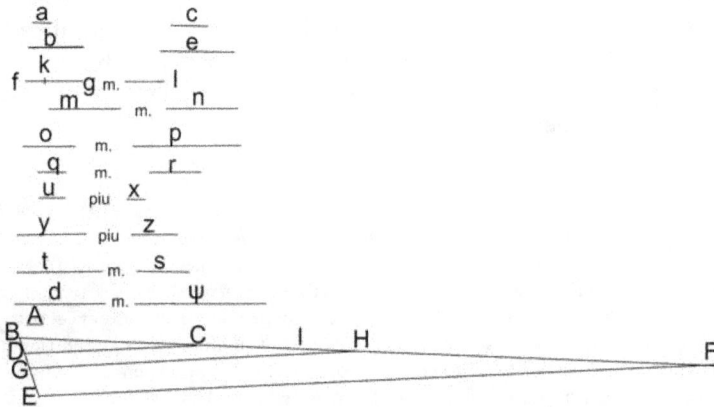

Fig. 7.8: [Bombelli (1929), §86]: adding known and unknown lines

---

[47] "Due hanno denari, et trovano una borsa ne la quale era tanti denari quanti haveva il primo, et dice esso primo à l'altro: se io havessi la borsa con $\frac{1}{3}$ de tuoi, et 2 puì," [sic] "io haveria duo tanti del tuo rimanente. Dice il secondo: se io havessi la borsa, con il dare a te 4 de miei, io haverei due volte et mezzo quanto te. Si domanda quanti denari era ne la borsa et haveva ciascuno da se" [Bombelli (155?), 122r].

The solution begins by selecting a line .f.g. "as one wishes, which would play the role that the *cosa* plays for numbers",[48] and which stands for the second person's original money (note that Bombelli mentions the algebraic version only after the geometric statement and solution are concluded; I conflate the two problems to bring out the analogy and make things easier to follow). But Bombelli warns that "the unknown lines can never be added or subtracted from the known".[49] And so in figure 7.8, when .f.g. is rescaled according to the given ratio .a. : .b. to form .f.k., and then the known .c. is subtracted from the remainder .k.g. (this represents taking $\frac{1}{3}$ of the second person's money and 2 more), .f.k. (the result of the rescaling) is drawn on .f.g., but the subtracted .l. (which equals the known .c.) is drawn separately. If one continues the geometric modelling of the protagonists' manipulations and wades through their geometric representations,[50] one eventually obtains that the unknown .d. minus .y. is to equal the known .z. plus .ψ.

Only at this point does Bombelli bring known and unknown lines into contact via the bottom part of the diagram. On the 'unknown axis' (this is *not* Bombelli's term) we draw .B.D. equal to the difference between the unknown .d. and .y., and .B.E. equal to our original unknown .f.g.. On the 'known axis' we draw .B.C. equal to the sum of .z. and .ψ.. Now, as .B.C. is supposed to equal .B.D., drawing the parallel lines .D.C. and .E.F. yields .B.F., which, according to Euclidean proportion theory, is the sought value of the unknown .f.g..

Bombelli does not justify his procedure here, because he has already done so earlier, in §73, where this rescaling technique was explicitly related to the rule "*of three*".[51] In the context of this example, the rule of three applies as follows. After the geometric construction is concluded, Bombelli assigns to the known lines the original numerical values from Book III, and to the unknown

---

[48] "Pigliasi una linea a beneplacito la quale farà l'effetto che fa la cosa nel numero".

[49] "Et notasi, che mai le linee incerte, non si possano aggiungere nè cavare con le certe".

[50] This goes as follows. The difference between .k.g. and .l. corresponds to the second person's remainder in the first person's narrative. To retrieve what the first person originally had according to his narrative, double the latter to produce the difference between the unknown .m. an the known .n., subtract from the latter the unknown .f.k. and the known .c. to get the difference between the unknown .o. and known .p., and then divide in half to get the difference between the unknown .q. and known .r.. Now that we have a representation of the first person's original money, we follow the second person's narrative. We add to the latter the known .e. to get the sum of the unknown .u. and the known .x., which corresponds to the first person's money at the end of the second person's narrative. Rescaling by the ratio of 5 : 2 we get the sum of the unknown .y. and the known .z., which corresponds to the second person's money at the end of his own narrative. According to that narrative, this sum must also correspond to the result of adding .f.g. to .q. minus .r. (represented as the unknown .t. minus the known .s.) and then removing the known .e.. This final magnitude is represented by the unknown .d. minus the known .ψ.. The bottom line is that the unknown .y. and known .z. equal the unknown .d. minus the known .ψ..

[51] "la regola della proportione chiamata *del tre*".

line .B.E.=.f.g. he assigns the tentative numerical value 3. Then, following the steps of the geometric construction, .B.D. ends up being $\frac{3}{4}$ and .B.C. ends up as $9\frac{1}{2}$. Now, since $\frac{3}{4}$ (.B.D.) is actually supposed to stand for $9\frac{1}{2}$ (.B.C.), then, according to the rule of three, .B.E. is actually supposed to stand for $\frac{3\cdot 9\frac{1}{2}}{\frac{3}{4}} = 38$, which is the length of .B.F..

Just as the arbitrary numerical choice 3 was a place holder for the correct solution 38, so is the line .B.E.=.f.g. a place holder for to correct solution .B.F.. The number 3 didn't simply stand for itself, and neither did the line .f.g.. This form of representation enables a line to stand for its possible rescaling with respect to other given lines. As long as we keep the known magnitudes and unknown magnitudes along different 'axes' or 'dimensions', there's no risk of error. The point is that just as the rule of three (as I argue in [Wagner (2010)]) evolved into an abbacist practice of seeing numbers as other numbers, here we learn to see lines as other lines.

The double vision required to understand the diagram extends to the side-by-side setting of known and unknown lines in the upper part of the diagram. The two scales are confronted, but not confused. And this separation occurs in the numerical instantiation that follows the geometric construction as well. Given the numerical values he assigns, Bombelli obtains that .t. minus .s. is "$4\frac{1}{2}$ *meno* 3", but this is never interpreted as $1\frac{1}{2}$, because $4\frac{1}{2}$ is our place holder for an unknown magnitude, while 3 is a known magnitude. Indeed, the equivalent term in the algebraic solution of Book III is $1\frac{1}{2}\underset{\smile}{} - 3$ (recall that .f.g., which stands for the unknown, was assigned the value 3 in the arithmetical verification of the geometric version of this problem, and so $1\frac{1}{2}\underset{\smile}{} = 4\frac{1}{2}$). Moreover, in §85, solved by a similar technique, a certain difference is reconstructed as "2 *meno* 3", but this difference is not marked as negative, as the 2 is a place holder for an unknown magnitude. Indeed, in the diagram, the lines representing this difference have a positive difference, as is the case if we substitute the final result for these lines.

The bottom line of this geometric representation technique is that one is trained in seeing lines as tentative place holders and as representing more than their visible lengths, not only with respect to a rescaling of the entire diagram (as is already the case in Euclidean diagrams), but also relative to other lines in the same diagram. What you get is more than what you see. Lines belong to various systems of relations with respect to other lines, and these relations are invisible for an eye not trained in this hybrid practice of algebra in lines.

The confrontation of known and unknown lines described above starts as reproducing cossist algebra, but then ends up as a geometric representation of the rule of three. Similarly, §§119–120 juxtapose in the context of the same problem cossist algebra, geometric rescaling and a quadratic version of the rule of three (where rescaling a given magnitude by a certain factor rescales another magnitude by that factor squared; here this applies to the relation between the side of a triangle and its area). While in terms of arithmetic/algebraic practice the rule of three and linear cossist algebra are distinguishable, Bombelli's geometric representation makes it difficult to set them apart. This technique of representation serves as a common reference that helps bind the rule of three with algebra, and may have contributed to the eventual absorption of the older practice into the newer one.

7.4.4 The geometry of the unknown: cosa

But another step is required if one is to render cossist algebra geometrically visible. According to one interpretation, geometric representations of powers of the *cosa* were restricted to the line-square-cube scale. Beyond that, algebra was not geometrcially instantiated.[52] However, as we saw above, not only is Bombelli willing to represent higher algebraic powers by lower geometric dimensions, he also has the technique to rigorously reduce higher geometric dimensions to lines. This is combined explicitly in §21, where an elegant spiral-like diagram starts from a unit segment and reconstructs each power of the *cosa* as the geometric mean between the previous power and the next.

The geometric representation of higher powers in Books IV and V of *L'algebra* is treated in several ways. The first treatment occurs in the context of geometric translations of algebraic solutions of polynomial equations, following the Arabic sources and Cardano's diagrams discussed above (see Subsection 7.3.3). In this treatment higher powers of the *cosa* correspond to higher geometric dimensions. The second treatment concerns 'arithmetic geometry', where geometric problems are presented in terms of numerical line lengths. There the unknown line is modeled as *cosa*, and the geometric situation is expressed by an algebraic problem, which is then solved using algebraic rules. This was an abbacist practice whose state of the art exposition belongs to Piero della Francesca (Bombelli's innovation here is restricted to the often included geometric instantiation of the algebraic solution, see Subsection 7.3.2).

But Bombelli's third geometric treatment of *cosa* powers, however rare, is the most interesting. The four problems §§102,104,122 and 131 go beyond the two forms of geometric treatment mentioned in the previous paragraph. These geometric problems are analysed without recourse to specific numerical values, they model higher powers of *cosa* as lines, and they manage to provide not

---

[52] See, for example, [della Francesca (1970), 91], where Piero makes this comparison explicit for the first three powers of the *cosa*, reconstructs the fourth power as "two squares" ("doi quadrati"), but then neglects a further geometrisation of the higher powers that he names and defines.

only a geometric instantiation of the algebraic solution, but also a geometric
representation of the equations themselves and of their algebraic reduction to
a canonical form. This modeling technique requires a delicate and specialized
way of seeing.

Let's have a close look at §102.[53] The problem asks for three lines, which
are represented in figure 7.9 as follows. The unknown .d. is the first line sought
by the problem. .e. is twice .d., and together with .f., which equals the given
.a., constitutes the second line sought by the problem. The third line sought
by the problem is such that Rect(third line, .b.)[54] (where .b. is given) equals
the rectangle formed by the previous two lines, namely Rect(.d., .e. plus .f.).
Finally, the ratio between .d. and the sum of all three sought lines should
equal the ratio between the given .c. and .G.C..

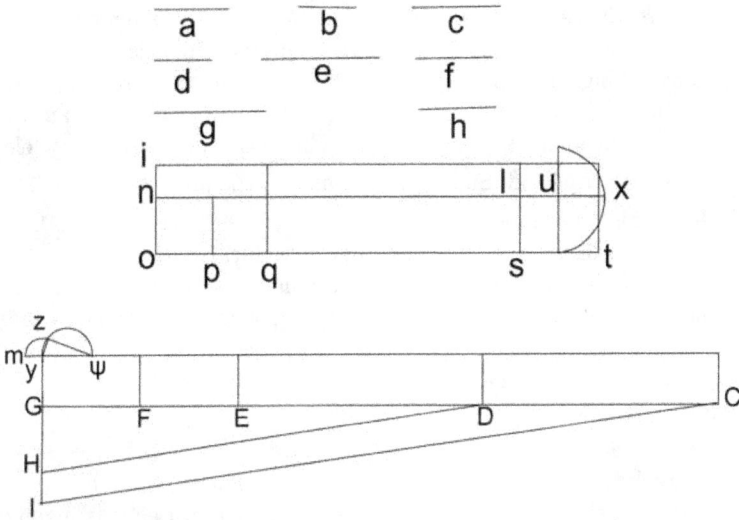

Fig. 7.9: [Bombelli (1929), §102]: multiplying known and unknown lines

To make things easier to handle, Bombelli represents the unknown .d. as
equal in length to the known .b. This choice is possible because .d. and .b.

---

[53] §104 starts with the same approach, but then skips directly to a geometric instantiation
of the algebraic solution. §§122 and 131 have blank spaces for diagrams in the manuscript,
and the missing diagrams are difficult to reconstruct from the text.
[54] By 'Rect(line1, line2)' I refer to the rectangle built from the two lines. This is not
Bombelli's notation, and I use it here to make things easier to follow. The same goes for
the 'Sum(line1, line2, ...)' and 'line1:line2' notations.

belong to different scales — unknown and known lines. That they appear to be
equal doesn't commit us to assume that they actually are. As long as they're
kept apart, any choice is as good as any other. Now that .b. equals .d. "in
length",[55] the third line required above (such that Rect(third line, .b.) equals
Rect(.e. plus .f., .d.)) is equal "in length" to .e. plus .f., and is represented
by .g. plus .h.. But Rect(.g. plus .h., .b.) must equal Rect(.e. plus .f., .d.)
not only in area but also in scale. Since .e. and .d. are of the scale of the
unknown (corresponding to the algebraic *cosa*), and .f. and .b. are known
(corresponding to numbers), .g. must be scaled like the second power of the
unknown (algebraic *censo*), and .h. must be of the scale of the unknown (*cosa*).
This way both rectangles are sums of *censi* and *cose*. In the corresponding
algebraic problem 92 of the manuscript Book III [Bombelli (155?), 134v], the
line .d. is modeled as $\overset{1}{\smile}$, .e. plus .f. as $2\overset{1}{\smile}+4$ and .g. plus .h. as $2\overset{2}{\smile}+4\overset{1}{\smile}$ (.b.
has no explicit algebraic equivalent, and serves as the geometric unit measure).

We can already see the multiple vision, a geometrico-algebraic hybrid,
required to understand the relations between lines in this diagram. Now
let's see how things are put together. Recall that the problem requires
that the ratio .d.:Sum(.d., .e. plus .f., .g. plus .h.) equal the known ratio .c.::G.C.. In problem 92 of Book III The ratio .c.::G.C. is modeled as
20 : 300, and the ratio .d.:Sum(.d., .e. plus .f., .g. plus .h.) as $\overset{1}{\smile}$: $2\overset{2}{\smile}$
$+7\overset{1}{\smile}+4$. This translates to an equality between Rect(.d., .C.G.) and
Rect(.c., Sum(.d., .e. plus .f., .g. plus .h.)). The former rectangle is the bottom rectangle of the diagram, where .G.y. equals the unknown .d.. The latter rectangle is the middle one in the diagram, where .o.i. equals the given
.c., .o.q. equals the *censi* .g., .q.s. equals the *cose* .d., .e. and .h., and .s.t.
equals the known .f.. In the algebraic version of Book III the equation of
the rectangles is mirrored by the cross multiplication yielding the equation
$40\overset{2}{\smile}+140\overset{1}{\smile}+80=300\overset{1}{\smile}$.

Now the task is to compare the two rectangles, balancing at the same time
their areas and scales. Bombelli's trick here is elementary as it is ingeniously
elegant. Both rectangles are rescaled according to the known ratio between
.b. and .c.. But the middle rectangle is rescaled along the known vertical
dimension, reducing its height from .o.i. to .o.n., whereas the lower rectangle
is rescaled along the known horizontal dimension, reducing its length from
.G.C. to .G.D.. Now both rectangles have the same height in terms of length
(but not in terms of scale). The middle rectangle is divided into *censi*, *cose*
and known areas, whereas the bottom rectangle is strictly *cose*. Note that this
normalisation move does not have an algebraic equivalent in the solution of
the algebraic version in Book III. As with the choice of .d. equal in length to
.b., it is a geometric artefact of the geometrico-algebraic hybrid.

---

[55] "in lunghezza"

Since the middle rectangle includes the *cose* rectangle .q.s.l., which is of the same scale as the lower rectangle, this can be subtracted from both rectangles, leaving the rectangles .o.q.n. and .s.t.x. in the middle and .E.G.y. below. This manoeuvre reflects the subtraction of *cose* terms from both sides of the algebraic equation to obtain an equation of the form *censi* and numbers equal *cose*. The resulting equation in problem 92 of Book III is $40 \overset{2}{\smile} + 80 = 160 \overset{1}{\smile}$.

The last normalization step is to reduce the *censi* to a single square. Algebraically this is the division of the quadratic equation by its leading coefficient, 40. Geometrically this is the rescaling of all rectangles so that .o.q.n. becomes a square, that is, rescaling by the ratio .b.:.o.q., which, due to the representation of .b. as equal to .d. and the equality between .o.q. and twice .d., is exactly $\frac{1}{2}$. Note that the extra geometric rescaling and the choice of .b. for the role of geometric unit measure led to a difference between the geometric and algebraic rescaling: in the algebraic model we rescale once by a factor of $\frac{1}{40}$, whereas in the geometric model we first rescaled along one dimension by the ratio of .b.:.c. ($\frac{1}{20}$) and then along the other dimension by the ratio of .b.:.d. ($\frac{1}{2}$). The resulting equation is $1 \overset{2}{\smile} + 2 = 4 \overset{1}{\smile}$

We are finally in a situation where the single *censo* .n.p.o. plus the known .s.l.u. equal the *cose* .F.G.y.. This is a normalized equation, and can be treated geometrically as in the left hand side of figure 7.4 above. But Bombelli skips directly to the geometric instantiation of the solution rule as in the right hand side of figure 7.4. He draws the known .ψ.y., half the coefficient of the *cose* (2 in the algebraic equation), finds u.x., the root of the known rectangle .s.l.u. ($\sqrt{2}$ in the algebraic equation), and then copies it as .ψ.z. to form the right angle triangle .y.ψ.z.. Now .y.z. is the root of the difference between the squares of .ψ.z. and .ψ.y. ($\sqrt{2^2 - 2}$), which is copied as .y.m.. Finally this root is added to the known .ψ.y.. The end result is .ψ.m. ($2 + \sqrt{2^2 - 2}$), the sought value of the original unknown .d..

It is, however, crucial to note that despite the analogies with the algebraic model, the geometric treatment includes no assignment of arithmetic values to line lengths. The solution is algebraico-geometric, but strictly non-arithmetic.

The point of pursuing this last example in such detail is of course to bring out the multiple vision of lines according to their lengths and scales that's required to pursue Bombelli's diagrams correctly on the arithmetic, algebraic and geometric levels. I made a point of highlighting the shifts that algebraic practice undergoes when Bombelli translates it into a geometrico-algebraic hybrid. It is this multiple, co-expressive, but not entirely congruent vision of several mathematical strata, which renders Bombelli such an insightful mathematician.

7.5 Conclusion

Together with a recognition of the hybridisation of algebraic and geometric points of view in the above diagrams, it is important to appreciate the multiplicity and deferral inherent in each of the two perspectives of this double vision. In subsection 7.4.3 we saw that a segment/number may not stand for its length/value, but as a place holder for another; in subsection 7.4.4 we saw that a line may carry a scale or dimension that turns its apparent relation to other lines problematic, but not for all that meaningless (recall that scale-independent equality of lengths and areas did play an important role in the last solution above); in subsection 7.2.2 we saw that a line can be seen a an implicit rectangle whose other side is a unit measure, and that the geometric unit role can be played by any line, but this is no longer the case where arithmetic values are imposed on lines. All this is compounded by the *sine qua non* of algebra: that the unknown stand by definition for a deferred value.

Interesting as they are, Bombelli's inventive geometric representations are a minor and isolated strategy. I do not mean to say that they have never been repeated by later mathematicians; I mean that they have been used locally, sketchily, heuristically and without a rigorous articulation. The canonical co-representation of algebra and geometry is the one emanating from Descartes: instead of Bombelli's single unknown line that can take the place of various different values and be interpreted as belonging to various different scales, Descartes' successors draw all possible lines together, so to speak side by side. To put this last claim more clearly, note that in Bombelli's representation each side of the equation $40 \overset{2}{\smile} + 140 \overset{1}{\smile} + 80 = 300 \overset{1}{\smile}$ is represented by two rectangles, whose scales are yet to be determined, and therefore stand for a range of possible values. In the post-Cartesian representation, however, each of the vertical lines that cover the space between the curve $y = 40x^2 + 140x + 80$ and the x-axis captures one possible value of the term $40x^2 + 140x + 80$, each of the vertical lines that cover the space between the curve $y = 300x$ and the x-axis captures one possible value of the term $300x$, and the intersection of the curves captures the one value of x for which the former is equal to the latter. Instead of representing arbitrarily the desired length and deferring the determination of its scale, we post-Cartesians put the infinity of possible representations side by side, and extract the one that's required. With the post-Cartesian representation, multiple vision is disambiguated. Instead of seeing one thing as many, we review all possibilities at a glance.

The transition from Bombelli's representation technique to the post-Cartesian one, which dominated early modern mathematics, can perhaps be situated in a more general context: the production of modern techniques for exhaustive representation and observation rather than a representation through a token

instance.[56] But the view of many underdetermined things under one sign is
not for all that abandoned. When one extends one's scope, one can recognise
that the pair of post-Cartesian curves above may contract the representation
of many different empirical phenomena; that it may serve as a token for a general technique rather than a specific example; that it may be read as hiding
several compressed dimensions (the x can always be reinterpreted, for example, as a projection of two independent variables onto their sum). The story
of mathematical representation is a story played with hybrids, contractions
and disambiguations. The range of hybridizations and contractions is never
confined or charted in advance, and the disambiguation is never entirely exhaustive.

The transition from Bombelli's *algebra linearia* to post-Cartesian analytic
geometry is also a process of normalising the relations between geometry and
algebra. Whereas Bombelli simultaneously built on several traditions of representations (Greek, Arabic, Latin and abbacist), and observed relations of
justification, instantiation, translation and accompaniment.(which were mixed
far more casually than this crude division suggests), the following centuries
tended toward a more foundational approach. Rigorous practices were developed that tried to impose a single, well regimented set of relations between
algebra and geometry. Not that foundational attempts ever suppressed the
practice of plural relations between algebras and geometries, but they often
suppressed an explicit account of these multiple relations, as well as the idiosyncratic residues that the isomorphisms constructed to reduce one to the
other kept leaving behind.

Why is this important for historians of mathematics? When 19th and most
20th century historians considered Greek mathematics, all they could see was
geometric algebra — geometry was conceived of as a technique for representing
(perhaps even concealing) algebraic knowledge. This view was so thoroughly
entrenched that historians who reacted against it had to dislodge not only the
claim that classical Greek geometry was a coded algebra, but also the weaker
claim that Greek geometry contained traces of algebraic thinking. Some of the
arguments used to establish such claims show that if one allows some traces of
algebra into our interpretation of Greek geometry, one will end up algebraising
one's interpretation to an extent that effectively recreates geometric algebra.[57]
But historians should be careful of extending such arguments beyond their intended scope. Outside the classical Greek framework (and occasionally at its
boundaries) geometry can be algebraic to various different extents without

---

[56] I'm referring here to recent work by Lorraine Daston presented in her talk at the History
of Science Society meeting in Pittsburgh in 2008, but the connection I'm making is yet to
be historically validated.

[57] I'm referring, of course, to the ground breaking work of Sabetai Unguru and his col-
leagues, e.g. [Unguru & Rowe (1981–1982)], which went as far as asserting the *death* of geometric algebra in Hebrew, Greek and Latin at the end of the paper.

necessarily recreating geometric algebra. One does not simply have two possibilities: pure geometry and geometric algebra. As Bombelli shows, there's very much between these options and beyond.

Bombelli's case is a fine demonstration of the fact that one can rigorously handle various different ways of relating algebra and geometry without giving up on the specificities of either. Geometry and algebra are conceived as co-expressive; what the one can show the other should be able to say and vice-versa. But this co-expressivity does not turn into a hierarchy, a reduction or a complete isomorphism. Each algebraic sign and each geometric line stands for more than a single value chosen from among a confined space of interchangeable choices. The techniques of translation are many, and leave idiosyncracies behind. They require specialised multiple vision and lead to hybrid and expanded algebraic and geometric practices. But, most importantly, these multiplicities are not a problem — they are, precisely, (what enables finding) solutions.

References

[Paolo dell'Abbaco (1964)] Arrighi, Gino (ed.), 1964. Paolo dell'Abbaco, *Trattato d'aritmetica*, Pisa: Domus Galilæna.

[della Francesca (1970)] Arrighi, Gino (ed.), 1970. Piero della Francesca, *Trattato d'abaco* Dal codice ashburnhamiano 280 (359* - 291*) della Biblioteca Medicea Laurenziana di Firenze. A cura e con introduzione di Gino Arrighi. (Testimonianzi di storia della scienza, 6), Pisa: Domus Galilæna.

[Bombelli (155?)] Bombello, Rafaello. 155?. *L'algebra*. Manuscript B.1569 of the Biblioteca dell'Archiginnasio, Bologna.

[Bombelli (1572)] Bombelli, Rafael. 1572. *L'algebra*. Bologna: Giovanni Rossi.

[Bombelli (1929)] Bortolotti, Ettore (ed.), 1929. Bombelli, Rafael, *L'algebra, Libri IV e V.* Bologna: Nicola Zanichelli. Available at http://a2.lib.uchicago.edu/pip.php?/pres/2005/pres2005-188.pdf

[Bombelli (1966)] Bortolotti, Ettore (ed.), 1966. Bombelli, Rafael, *L'algebra, Prima edizione integrale*. Milano: Feltrinelli.

[Bos (2001)] Bos, Henk, 2001. *Redefining geometrical exactness. Descartes' Transformation of the Early Modern Concept of Construction*. New York, NY: Springer.

[Dieudonné (1972)] Dieudonné, Jean. 1972. Review of J.E. Hofmann "R. Bombelli — Erstentdecker des Imaginären", Praxis Math., **14(9)**, 225–230. *Math Review* MR0484973 (58 #4832a)

[Dardi (2001)] Franci, Raffaella (ed.), 2001. Maestro Dardi, *Aliabraa argibra* dal Manoscritto I.VII.17 della Biblioteca Comunale di Siena, Siena: Dipartimento di Matematica Roberto Magari dell'università di Siena

[Franci & Rigatelli (1985)] Franci, Rafaella & Laura Toti Rigatelli. 1985. "Towards a history of algebra from Leonardo of Pisa to Luca Pacioli", *Janus* **72(1–3)**, 17–82.

[Freguglia (1999)] Freguglia, Paolo. 1999. "Sul principio di omogeneità dimensionale tra Cinquecento e Seicento", *Bollettino dell'Unione Matematica Italiana B*, **(8)2-B**, 143–160.

[Giusti (1992)] Giusti, Enrico. 1992. "Algebra and geometry in Bombelli and Viète", *Bollettino di storia delle scienze matematiche*, **12**, 303–328.

[Høyrup (2007)] Høyrup, Jens, 2007. *Jacopo Da Firenze's Tractatus Algorismi and Early Italian Abbacus Culture*. Basel: Birkhäuser.

[Jayawardene (1963)] Jayawardene, S.A. 1963. "Unpublished Documents relating to Rafael Bombelli in the archives of Bologna", *Isis*, **54(3)**, 391–395.

[Jayawardene (1965)] Jayawardene, S.A. 1965. "Rafael Bombelli, engineer-architect: some unpublished documents of the Apostolic Camera", *Isis*, **56(3)**, 298–306.

[Jayawardene (1973)] Jayawardene, S.A. 1973. "The influence of practical arithmetics on the algebra of Rafael Bombelli", *Isis*, **64(4)**, 510–523.

[Kristeva (1969)] Kristeva, Julia, 1969. *Semeiotike*. Paris: du Seuil.

[Lacan (2006)] Lacan, Jacques, 2006. *Ecrits, The First Complete Edition in English*. New York, NY: W.W. Norton.

[Malet (2006)] Malet, Antoni. 2006. "Renaissance notions of number and magnitude", *Historia Mathematica*, **33**, 63–81.

[Mueller (1981)] Mueller, Ian, 1981. *Philosophy of Mathematics and Deductive Structure in Euclid's Elements*. Cambridge, MS: Dover Publiactions.

[Netz (1999)] Netz, Reviel, 1999. *The Shaping of Deduction in Greek Mathematics*. Cambridge: Cambridge University Press.

[Netz (2004)] Netz, Reviel, 2004. *The Transformation of Mathematics in the Early Mediterranean World*. Cambridge: Cambridge University Press.

[Roche (1998)] Roche, John J, 1998. *The Mathematics of Measurement. A Critical History*. London: Athlone Press

[Unguru & Rowe (1981–1982)] Unguru, Sabetai & David Rowe. 1981–1982. "Does the quadratic equation have Greek roots", *Libertas Mathematica*, **1**, 1–49, **2**, 1–62.

[Van Egmond (1980)] Van Egmond, Warren, 1980. *Practical Mathematics in the Italian Renaissance: A Catalog of Italian Abbacus Manuscripts and Printed Books to 1600*. (Istituto e Museo di Storia della Scienza, Firenze. Monografia N. 4). Firenze: Istituto e Museo di Storia della Scienza.

[Wagner (2009a)] Wagner, Roy. 2009. $S(z_p, z_p)$: *Post-Structural Readings of Gödel's Proof*. Milano: Polimetrica

[Wagner (2009b)] Wagner, Roy. 2009. "For some histories of Greek mathematics", *Science in Context*, **22**, 535–565.

[Wagner (2010)] Wagner, Roy. 2010. "The natures of numbers in and around Bombelli's *L'algebra*", *Archive for the History of Exact Sciences*, (to appear).

[Cardano (1968)] Whitmer, T.R. (trans.), 1968. Cardano, Girolamo, *The Rules of Algebra (Ars Magna)*, Mineola, NY: Dover.

Part III
Mathesis universalis and *charateristica universalis*

Chapter 8
The "merely mechanical" vs. the "scab of symbols": seventeenth century disputes over the criteria of mathematical rigor

Douglas M. Jesseph

Abstract This paper deals with seventeenth-century understandings of rigorous demonstration. Although there was a widely-shared concept of rigor that has its origins in classical Greek sources, philosophers in the early modern period were divided over how to characterize the ultimate foundation for mathematics. One group (whom I term the "geometric foundationalists") held that seeming physical concepts such as space, body and motion, were properly foundational. The other group (whom I call "algebraic foundationalists") claimed that the true foundations of mathematics must be abstract notions of quantity which were the subject of algebra, or even a more general *mathesis universalis* that encompassed all reasoning about number and measure. The geometric foundationalists faced the objection that they had introduced "merely mechanical" or insufficiently abstract principles into the foundations of mathematics. In contrast, the algebraic foundationalists needed to rebut the accusation that they based mathematics on a "scab of symbols", or empty notation divorced from anything real or substantial. I argue that this episode offers some useful insights into general questions about foundations, and can help us understand what is at stake in disputes over foundational issues, as well as how such disputes rise to prominence and then fade away.

8.1 Introduction

One of the things that philosophers find most intriguing about mathematics is the rigor of its demonstrations. A properly developed mathematical theory is one which starts with clearly conceived and well-understood first principles, then unfolds via deductively valid inferences to reach conclusions that, however *outré* or surprising they may seem to untutored common sense, are every bit as secure as the first principles on which they depend. This picture of mathematics has convinced many philosophers that it is the pinnacle of human intellectual achievement, something that can serve as a bulwark against the challenges of skepticism, and whose deductive structure might serve as a model for philosophical theorizing.

We can see this dynamic at work in the famous anecdote concerning Thomas Hobbes's discovery of mathematics as related by his friend John Aubrey:

> He was forty years old before he looked on geometry; which happened accidentally. Being in a gentleman's library Euclid's Elements lay open, and 'twas the forty-seventh proposition in the first book. He read the proposition. "By God," said he, "this is impossible!" So he reads the demonstration of it, which referred him back to such a proof; which referred him back to another, which he also read. *Et sic deinceps*, that at last he was demonstratively convinced of that truth. This made him in love with geometry. (Aubrey 1898, p. 322)

As my concern here is with the interaction between philosophy and mathematics in the seventeenth-century, it is fitting to begin with an anecdote involving Hobbes, even if his mathematical misadventures featured numerous outrageously poor attempts to solve such problems as the squaring of the circle or the duplication of the cube.[1]

The issue I wish to discuss concerns seventeenth-century understandings of what it is that makes mathematical (and perhaps more specifically geometric) demonstrations rigorous. Although there was a widely-shared general characterization of rigor that has its origins in classical Greek approaches to the philosophy of mathematics, philosophers and mathematicians in the early modern period were divided over whether the ultimate foundation for mathematics should be sought in seemingly physical concepts such as space, body and motion, or in more abstract notions of quantity which were the subject of algebra, or perhaps a more general *mathesis universalis* that encompassed all reasoning about number and measure. Those who sought to ground the whole of mathematics in concepts such as space and motion had to confront the objection that they had introduced "merely mechanical" or insufficiently abstract principles into the foundations of mathematics. On the other hand, proponents of a *mathesis universalis* needed to rebut the accusation that, in doing so, they based the whole science on a "scab of symbols," or a jumble of empty notation divorced from anything real or substantial.

I begin with a brief overview of the classical Greek point of view on the subject, as exemplified in the works of Euclid and his neo-Platonist commentator Proclus Diadochus. I then consider the "geometric foundationalism" propounded by Isaac Newton and Isaac Barrow (among others), which seeks to ground all of mathematic in fundamental geometric notions. I will then inves-

---

[1] For an account of Hobbes's failed campaign for mathematical glory, see Jesseph (1999)

tigate the "algebraic foundationalism" put forward by John Wallis, Descartes, and others. In the end, I think this episode offers some useful insights into general questions about foundations, and it can help us understand what is at stake in disputes over foundational issues, as well as how such disputes rise to prominence and then fade away.

8.2 Euclid, Proclus, and the Classical Point of View

Classical Greek authors took pure mathematics to consist of two essentially different subjects: arithmetic on the one hand and geometry on the other.[2] The applied mathematical sciences, which include such things as astronomy, music, optics, mechanics, were taken to differ from the pure sciences because the pure part is – to quote Proclus – "concerned with intelligibles only," while the applied part of mathematics "work[s] with perceptibles and [is] in contact with them" (Proclus 1970, p. 31). Classically conceived, then, the pure sciences of geometry and arithmetic are distinguished from one another by having different objects. The object of arithmetic is discrete quantity ("number" or ἀριθμός in the Greek), while the object of geometry is continuous quantity ("magnitude" or μέγεθος in Greek). Thus understood, the two branches of pure mathematics can be developed in separate ways and there is no question of attempting to "reduce" one to the other or to show that one is in some sense the genuine foundation of all mathematics. This pluralistic approach to the philosophy of mathematics does not presume or require that the different mathematical sciences be reduced to some underlying foundational science.

This does not, however, imply that classical authors always conceived of the two branches of mathematics as of equal importance. Proclus took the view that, although pure geometry was entirely independent of arithmetic, geometry nevertheless "occupies a place second to arithmetic," because (in keeping with his neo-Platonist epistemology) everything knowable must ultimately be expressed in terms of commensurable ratios (Proclus 1970, p. 39). The principal point here is that classical authors saw no relation of dependence between the truths of arithmetic and the truths of geometry.

This becomes clearer in the presentation of the theory of ratios in Euclid's *Elements*. Book V develops the theory of proportions in the context of continuous magnitudes, and Book VII offers an account of ratio and proportion as applied to discrete quantities or "numbers" (ἀριθμόι). It was, of course, well understood in the ancient world that the theory of proportions in Book

---

[2] On the Greek classification of the geometrical sciences, see Heath (1931), Klein (1968, Part I) and Knorr (1981)

VII could be taken as just a special case of the theory developed in Book V. Aristotle remarked that

> it might seem that the proportion alternates for things as numbers and as lines and as solids and as times – as once it used to be proved separately, though it is possible for it to be proved of all cases by a single demonstration. But because all these things – numbers, lengths, times, solids – do not constitute a single named item and differ in sort from one another, it used to be taken separately (*Posterior Analytics* 74a 18-22)

Thus, rather than assuming that there is one underlying notion of proportion that applies equally to continuous magnitudes and discrete multitudes, the tradition in Greek geometry was to develop the two theories separately. This simply reflects, and indeed reinforces, the notion that geometry and arithmetic are autonomous sciences.

Given that they are separate sciences, arithmetic and geometry will have to have separate first principles. The first principles of a demonstrative science were traditionally distinguished into three classes: axioms, definitions, and postulates. Axioms or "common notions" apply to any science whatever and include such general principles as "things which are equal to the same thing are equal to one another" or "if equals be subtracted from equals, the remainders are equal" (*Elements* 1, Ax. 1, 3). These apply universally to any subject matter, but they are insufficiently specific to develop a particular science. Definitions are principles specific to a given science and set out the essential properties of the objects of that science. So, in the case of arithmetic we have the definition of a unit as "that by virtue of which each of the things that exist is called one," and a number as "a collection of units" (*Elements* 7, Def. 1, 2). In the case of geometry, we require definitions such as those of point, line, surface, and angle, which are the proper object of geometric investigation. Finally, postulates are demands that a certain construction be admitted or effected, such as the geometric postulate "to describe a circle with any centre and distance" (*Elements* 1, Pos. 3).

Aristotle famously declared that the first principles upon which demonstrative sciences depend must themselves be indemostrable and known by a kind of immediate, non-inferential understanding (*Posterior Analytics* 72b19-21).[3] Of course, not just any first principles are permissible in mathematics, for the obvious reason that the security of the derived theorems can be no greater than that of the principles from which the theorems are derived. The result is Aristotle's famous requirement that the first principles of demonstration must be "true, and primitive and immediate and more familiar than and prior to and explanatory of the conclusion" (*Posterior Analytics*, 70b 21-23).

---

[3] On Aristotle and demonstrative science see Smith (1995, pp. 47-53) and Heath (1931, pp. 194 - 201).

We can therefore summarize the classical notion of mathematical rigor as follows: there are two mathematical sciences, geometry and arithmetic, distinguished from one another by the fact that they investigate the properties of very different kinds of objects: continuous, infinitely divisible magnitudes on the one hand, and discrete collections of units on the other. Both sciences must be developed demonstratively from first principles, which include very general axioms applicable to any science whatever, along with definitions and postulates that are specific to either geometry or arithmetic. Any claim that can be added as a theorem to either science must be shown to follow from these first principles by a truth-preserving deductive process.

Two features of this criterion of rigor are worth noting. The first of them is what I term the "purity constraint" which requires that the objects of pure mathematics must be abstracted from any messy features of the material world. As Proclus put the matter in his discussion of the Euclidean definitions of such terms as 'point', 'line', 'boundary', and 'surface':

> in the forms separable from matter the ideas of the boundaries exist in themselves and not in the things bounded, and it is because they remain precisely what they are that they become agents for bringing to existence the entities dependent upon them... Matter muddies ... the precision [of the forms]; the idea of the plane gives the plane depth, that of the line blurs its one-dimensional nature and becomes generally divisible, and the idea of the point ends by becoming bodily in character and extensible together with the thing it bounds. For all ideas when they flow into matter ... are filled with their substrates: they forsake their natural simplicity for alien combinations and extensions. (Proclus 1970, 87)

The notion of "purity" at work here is likely to be difficult to articulate in any detail, but the fundamental idea is this: the objects of pure mathematics cannot be confused with the messy objects of ordinary sense-experience. In part, this demand arises from the notion that nothing in the material world answers exactly to the dimensionless points or breadthless lines of geometry or the number seventeen in arithmetic. Another source of the purity constraint is presumably the intuition that the truths of pure mathematics must hold regardless of the structure or contents of the actual world, and so we should demand that the objects of pure mathematics be strictly segregated from the spatio-temporal realm.

A second feature of the classical notion of rigor, however, is the thought that rigorous demonstrations must proceed from principles that express causes. This "causality constraint" as I term it, is well expressed in the demand that a genuine demonstration must identify causes. When Aristotle required that the first principles of demonstration must be "true, and primitive and immediate and more familiar than and prior to and explanatory of the conclusion" (*Posterior Analytics* 70b 21-23), the requirement that the principles explain the conclusion was typically understood to mean that they in some sense cause the conclusion. Aristotle and the Aristotelian tradition characterized a demonstration satisfying the causality constraint as ἀπόδειξις τõυ διότι, as opposed to ἀπόδειξις τõυ ὅτι, a distinction which is typically rendered in English as the distinction between "a demonstration of the fact" and "a demonstration of the reasoned fact." The point of the distinction is that an ἀπόδειξις τõυ ὅτι shows merely that something is the case, while an ἀπόδειξις τõυ ὅτι shows why the fact must obtain by constructing a syllogism whose premises exhibit the causes of the conclusion.

It should be clear that this characterization of rigorous demonstration involves something of a tension between the purity constraint on the one hand and the causality constraint on the other. To be "pure" a mathematical theorem must lack any physical content, but to be truly "causal" the theorem must be demonstrated from principles that articulate causes. Since the best understood model of causation involves objects in the physical world interacting with one another, it seems natural to suppose that satisfying the causality constraint must tell against the purity constraint and *vice versa*. Aristotelianism could navigate this tension by employing a wider array of causal principles than those we acknowledge today. Because the Aristotelian philosophy takes individual substances as composites of form and matter, its methodology can distinguish between formal, material, efficient, and final causes. Thus, a causal explanation in the Aristotelian tradition can include a reference to a substance's form (the formal cause), its matter (the material cause), the process that produced it (the efficient cause), and the end or purpose for which it was produced (the final cause) (Aristotle, *Metaphysics* 1013a).

Given this broad conception of causation, it is no great leap to see mathematical demonstrations as concerned with formal causes: the form or essence of the circle (as expressed in its definition) is causally responsible for the theorem that every triangle inscribed in a semicircle is a right triangle. Now, it is no secret that Aristotelian philosophy came under serious pressure in the early modern period, and one consequence of the rise of the mechanistic "new philosophy" was the gradual demise of this concept of formal causality. The rise of an alternative to Aristotelianism led, I claim, to the conflict I now wish to consider: one that pitted those who were accused of employing "merely mechanical" principles against those whose abstract demonstrations were denounced as nothing but a "scab of symbols."

8.3 The Mechanical Style of Thought in Seventeenth Century Mathematics

In an essay entitled "Der mechanistische Denkstil in der Mathematik des 17. Jahrhunderts" (Breger, 1991) Herbert Breger has drawn attention to the role of the mechanistic "new philosophy" in setting the agenda for mathematicians of the seventeenth century. Breger cites a variety of instances in which the mathematical research of the seventeenth century was profoundly influenced by the rise of the mechanistic paradigm, and I think that one salient instance of this is the conflict that mechanism engenders between the purity constraint and the causality constraint. Mechanism, roughly speaking, is the thesis that the phenomena of the natural world are to be explained in terms of the motions and impacts of material bodies. The rise of mechanism as an explanatory model therefore involves an abandonment of the Aristotelian notions of formal or final causality in favor of a philosophy that emphasizes matter and motion. Nature, as understood by the mechanists, has no intrinsic teleology, and the appeal to substantial forms is viewed with something between suspicion and derision. The question of interest, then, is what becomes of the traditional demand that genuine demonstrations must articulate causes? Or, to put the matter another way, what does a demonstration τοῦ διότι look like if we abandon the Aristotelian doctrine of four causes and content ourselves with material and efficient causality?

The answer, not surprisingly, is that the demise of formal causality demands that the basic definitions of mathematical concepts (and specifically those in geometry) be understood as brought about by the motions. This kinematic account of magnitudes takes a line or curve, for example, to be the path traced by a point in motion through space. A circle, then, is conceived as something traced by the revolution of a line about one of its endpoints, and surfaces are understood to be produced by the motion of lines. The kinematic approach to geometric magnitudes is well-summarized by Newton in the introduction to his treatise *On the Quadature of Curves*:

> I don't here consider Mathematical Quantities as composed of Parts *extreamly small*, but as *generated by a continual motion*. Lines are described, and by describing are generated, not by any apposition of Parts, but by a continual motion of Points. Surfaces are generated by the motion of Lines, Solids by the Motion of Surfaces, Angles by the Rotation of their Legs, Time by a continual flux, and so in the rest. These *Geneses* are founded upon Nature, and are every Day seen in the motion of Bodies. (Newton 1964 - 67, 1: 141)

The emphasis here on the "geneses" and "generation" of magnitudes is important, as it provides the basis for the claim that demonstrations which employ the kinematic conception of magnitudes satisfy the causality constraint.

Newton was hardly the originator of this doctrine. Isaac Barrow, Newton's predecessor in the Lucasian chair of mathematics at Cambridge, held essentially the same view, and his *Geometrical Lectures* are dominated by it. As Barrow explained, "Among the ways of generating magnitudes, the primary and chief is that performed by local motion, which all [others] must in some sort suppose, because without motion nothing can be generated or produced" (Barrow 1860 1: 159). I term this approach to mathematics and its demonstrations "geometrical foundationalism" because it takes geometry

as a more fundamental science than arithmetic, and it seeks to ground all of
mathematics in geometric concepts. As Barrow stated the view

> number (at least that which the mathematician contemplates) does not differ in the
> least from that quantity which is called continuous, but is formed wholly to express
> and declare it. And neither are arithmetic and geometry conversant about diverse
> matters, but equally demonstrate properties common to one and the same subject,
> and from this it will follow that many and great advantages derive to the republic of
> mathematics. (Barrow 1860, 1: 47)

As a matter of fact, the definition of geometric objects by motion has a classical pedigree. A salient instance is Euclid's definition of a cone, which reads
"When a right triangle with one side of those about the right angle remains
fixed is carried round and restored again to the same position from which it
began to be moved, the figure so comprehended is a cone" (*Elements* 11, Def.
18). Likewise, there were many "special" curves such as the spiral, conchoid,
and quadratrix which were defined in terms of the compound motions of lines
or points.[4]

Nevertheless, the appeal to motion was regarded with some suspicion in the
classical period, as it was thought to introduce an unnecessarily "mechanical"
aspect into geometry that is inconsistent with the desired purity of the subject. Plutarch gives a memorable account of the status of such "mechanical"
definitions:

> For the art of mechanics, now so celebrated and admired, was first originated by
> Eudoxus and Archytas, who embellished geometry with its subtleties, and gave to
> problems incapable of proof by word and diagram, a support derived from mechanical
> illustrations that were patent to the senses. For instance, in solving the problem
> of finding two mean proportional lines, a necessary requisite for many geometrical
> figures, both mathematicians had recourse to mechanical arrangements, adapting to
> their purposes certain intermediate portions of curved lines and sections. But Plato
> was incensed at this, and inveighed against them as corrupters and destroyers of the
> pure excellence of geometry, which thus turned her back upon the incorporeal things
> of abstract thought and descended to the things of sense, making use, moreover, of
> objects which required much mean and manual labour. For this reason mechanics
> was made entirely distinct from geometry, and being for a long time ignored by
> philosophers, came to be regarded as one of the military arts. (Plutarch 1949–59, 5:
> 472–3).

The concern here is that a "merely mechanical" approach to mathematics
violates the purity constraint by introducing an unacceptable dependence of
mathematics on the merely physical. Obviously, not all seventeenth-century
mathematicians were worried by this sort of apparent dependence, but there
was no shortage of thinkers who objected to the unwarranted intrusion of

---

[4] On such curves and the problems they were introduced to solve, see Heath (1931, Chapter
VII) and Knorr (1986).

"plainly physical" concepts such as motion into the foundations of mathematics.

A case in point can be drawn from the controversy between Hobbes and John Wallis. Hobbes was second to none in his insistence that the fundamental principles of mathematics must express causes, and his enthusiastic embrace of materialist mechanism led him to conclude that the whole of geometry should be re-written to put mechanical principles at its foundation. As he put the issue in discussing his own program for demonstrating geometric results by considering compound motions of points, and lines:

> "But," you will ask, "what need is there for demonstrations of purely geometric theorems to appeal to motion?" I respond: first, all demonstrations are flawed, unless they are scientific, and unless they proceed from causes, they are not scientific. Second, demonstrations are flawed unless their conclusions are demonstrated by construction, that is, by the description of figures, that is, by the drawing of lines. For every drawing of a line is motion: and so every demonstration is flawed, whose first principles do not contain the definitions of motions by which figures are described. (Hobbes 1839-45, 4: 421)

Wallis objected that such an approach introduces physical principles into the foundations of mathematics, thereby contaminating it. He asked, for instance, "what need is there for the concepts of body or motion [in these foundations], since the concept of a line or curve can be understood without them?" (Wallis 1655, p. 6). As Wallis explained, such physical notions are "plainly accidental, nor do they pertain to the essences [of the things defined], so it is strange to find motion in the definitions of geometry" (Wallis 1655, p. 7).

Wallis was not alone in disparaging the attempt to introduce alien physical concepts into mathematics. Descartes distinguished between "truly geometrical" curves that could be expressed as polynomial equations in two unknowns and what he termed the "merely mechanical" curves such as the cycloid. These mechanical curves did not admit of a "precise and exact" measure (according to Descartes) and could only be defined by a composition of curvilinear and rectilinear motions which could not be the object of a properly geometric theory.[5] The concern here is that something merely mechanical must inevitably be imprecise or inexact because mechanical considerations fail to respect the purity constraint.

Newton emphatically rejected this attitude by arguing that mechanics, properly understood, is as precise and pure a science as any. In Newton's view, the failure to appreciate the abstract and perfectly precise nature of mechanics has given rise to the erroneous opinion that geometry must not introduce anything mechanical into its foundations:

> Both the genesis of the subject-matter of geometry, therefore, and the fabrication of its postulates pertain to mechanics. Any plane figures executed by God, nature

---

[5] On Descartes and mathematics see Bos (2001, especially chapters 24-26) and Sasaki

or any technician you will are measured by geometry on the hypothesis that they are exactly constructed.... Geometry makes the unique demand that [its objects] be described exactly. It has now, however, come to be usual to regard as geometrical everything which is exact, and as mechanical all that proves not to be of the kind, as though nothing could possibly be mechanical and at the same time exact. But this common belief is a stupid one, and has its origin in nothing else than that geometry postulates an exact mechanical practice in the description of a straight line and circle, and moreover is exact in all its operations, while mechanics as it is commonly exercised is imperfect and without exact laws. It is from the ignorance and imperfection of mechanicians that the common opinion defines mechanics. On this reasoning a thing would be the more mechanical the more imperfect it was. Posit a mechanical thing to be perfect and you will correct the error. (Newton 1967-86, 7: 289)

The program of "geometric foundationalism" thus finds itself ultimately committed to the principle that mechanical concepts properly belong to the "pure" science of mathematics and are the basis for its capacity to provide true demonstrations. We may now turn to a consideration of a very different program for the foundations of mathematics, namely "algebraic foundationalism."

8.4 The Idea of a *Mathesis Universalis*

I have been arguing that the "mechanical style" of thought in seventeenth-century mathematics results from an emphasis on the causality constraint at the expense of the purity constraint; another trend in seventeenth-century mathematics works from the other direction and stresses the purity of the first principles while downplaying any notion of mathematical causality. The trend I have in mind here can be usefully associated with the term *mathesis universalis*, or "universal mathematics." Rather than arising from concerns about the adequacy of the Aristotelian model of causation and demonstration, the notion of *mathesis universalis* was a consequence of the development of algebra in the sixteenth century.[6] As more complex problems were posed and solved, the power of algebra became more evident, and some were encouraged to see in it a tool that could be applied to the solution of all manner of mathematical problems. Indeed, some went so far as to claim that algebra was the genuine foundation of all mathematics. It was, however, far from obvious whether algebra had an appropriate object, and it was unclear where it might be placed in the traditional classification of the sciences. The theory of equations can be applied to all quantities – discrete or continuous – and the success of algebraic methods undermined the traditional division of mathematics into

---

(2003).

[6] See Rabouin (2009) on the concept of *mathesis universalis* and its origins.

geometry and arithmetic. And yet, since the fundamental operations of addition, subtraction, multiplication, and division seem to derive from arithmetic, some thinkers conceived of algebra as a generalized "arithmetic with letters" in which symbols or "species" representing arbitrary quantities were employed in developing the "arithmetic of species."

Joseph Raphson's *Mathematical Dictionary* contains an instructive account of the understanding of algebra and its development into a general *mathesis universalis*:

> *Algebra*, from *Al* in Arabic, which signifies Excellent, and *Gerber* the Name of the supposed Inventor of it, is a Science of Quantity in general, whence it has also got the name among some of *Mathesis Universalis*, & is chiefly conversant in finding Equations, by comparing of unknown and known Quantities together whence also by some it is called the Art of Equation, and is distinguished into

> [Algebra]*Numeral*, which is the more ancient and serves for the Resolution of Arithmetical Problems: For these see *Diophantus*

> [Algebra]*Specious*, or the new *Algebra*, which is also called *Logistica Speciosa*, and is conversant about Quantity denoted by General or Universal Symbols, which are commonly the letters of the Alphabet; and serves indifferently for the Solution of all Mathematical Problems, whether Arithmetical or Geometrical. (Raphson 1702, 2-3)

Raphson was somewhat late to the party, writing in the very early eighteenth century, and his fanciful etymology for the term 'algebra' suggests that his understanding of the history of the subject could use some improvement. Nevertheless, he summarizes the seventeenth-century understanding of the development of algebra as a kind of mathematical "superscience" more abstract and universal than arithmetic or geometry. Indeed, in Henry Billingsley's 1570 translation of Euclid we can find a reference to "that more secret and subtill part of Arithmetike commonly called Algebra" (Billingsley 1570, 229) which contains a very abstract treatment of quantity in general.

Among the authors who embraced the notion of a *mathesis universalis*, Descartes and John Wallis figure quite prominently, although one cannot overlook the contributions of François Viète and Leibniz. A strong commitment to what I have termed the "purity constraint" is fundamental to the notion of *mathesis universalis*. Wallis, in his work entitled *Mathesis Universalis* stresses that the true principles of mathematics concern only quantity in general, and argues that "when time, place, ... and even motion or weight" are said to be quantities, the term 'quantity' must be taken in a "broad sense" that is not really suited to pure mathematics (Wallis 1693-99, 1: 17). In Wallis's understanding, "we call those parts of mathematics 'pure' which treat of quantity considered absolutely, so far as it is abstracted from matter" (Wallis 1693-99, 1: 18). The result is that

> we say there are two pure mathematical disciplines, namely arithmetic and geometry the one of which is concerned with discrete quantity, or number, and the other of continuous quantity, or magnitude. But of these one is indeed more and the other

less pure: for the subject of arithmetic is purer and more universal that the subject
of geometry; and it also has more universal speculations which are applicable to the
matter of geometry equally and to all others. (Wallis 1693-99, 1: 18).

It might seem that Wallis takes arithmetic to be the true *mathesis universalis*, but in fact he sees arithmetic as a kind of special case of a more
fundamental science of quantity. Geometry, in Wallis's view, is less pure and
abstract than arithmetic, but arithmetic itself falls under the scope of algebra
or *mathesis universalis*.

The difficulty for proponents of this grand fundamental algebraic science
is that the emphasis upon the abstract and pure nature of their supposed
mathesis universalis raises the suspicion that it amounts to nothing more
than an empty formalism devoid of any specifically mathematical content.
Barrow is a case in point. He objected that algebra "is really no science
at all" (Barrow 1860, 1: 59). The reason for this harsh judgment is that
algebra "has no object distinct to itself, ... but only delivers a kind of artifice
for designating magnitudes and numbers by certain notes or symbols, and
collecting and comparing their sums and differences, founded on Geometry
or arithmetic; and so it constitutes no part of mathematics distinct from
geometry or arithmetic, but is entirely contained in them" (Barrow 1860, 1:
46). On Barrow's view, algebra amounts to little more than a collection of rules
for manipulating symbols, but it lacks any genuine mathematical content.

Such considerations lead quite naturally to the objection that employing
algebraic reasoning in a geometric context offers no real advantage over traditional methods. If algebraic techniques have governing principles and do not
simply proceed by guesswork, then these principles must be vindicated by
appeal to geometric considerations. Put another way, this sort of objection
reasons that if an algebraic manipulation in an equation is legitimate, it must
correspond to a proper geometric construction; but in that case, the most
that algebra can offer is a collection of symbols that might abbreviate the
writing of a proof, but cannot really show the genuine reasons why the proof
is correct.

Barrow was adamant on this point and devoted many pages to arguing
that geometry was the real foundation of all mathematics, which could be
developed without recourse to arithmetical or algebraic techniques.[7] Newton,
too, held that geometric (or, indeed mechanical) concepts have a kind of
explanatory primacy. Although he was well aware of the power of algebraic
techniques, Newton came to view Cartesian analytic geometry with some
suspicion, and praised the power of ancient methods which, he declared,

Men of recent times, eager to add to the discoveries of the ancient, have united the
arithmetic of variables with geometry. Benefitting from that, progress has been broad
and far-reaching if your eye is on the profuseness of output, but the advance is less

---

[7] On Barrow's mathematics, see Mahoney (1990).

of a blessing if you look at the complexity of the conclusions. For these computations
... often express in an intolerably roundabout way quantities which in geometry are
designated by the drawing of a single line. (Newton 1967-86, 4: 421)

Hobbes framed essentially the same objection as follows:

> What else do the great masters of the current symbolics, Oughtred and Descartes,
> teach, but that for a sought quantity we should take some letter of the alphabet,
> and then by *right* reasoning we should proceed to the consequence? But if this be an
> art, it would need to have been shown what this *right* reasoning is. Because they do
> not do this, the algebraists are known to begin sometimes with one supposition and
> sometimes with another, and to follow sometimes one path, and sometimes another...
> Moreover, what proposition discovered by algebra does not depend upon [theorems
> in Euclid]? Certainly, algebra needs geometry, but geometry has no need of algebra
> (Hobbes 1839-45, 4: 9-10)

This sort of objection lies behind Hobbes's dismissal of Wallis's *Treatise of
Conic Sections* as a work "covered over with the scab of Symboles," and
offering "no knowledge neither of Quantity, nor of measure, nor of Proportion,
nor of Time, nor of Motion, nor of any thing, but only of certain Characters,
as if a Hen had been scraping there" (Hobbes 1839-45, 1: 316, 330).

Proponents of "algebraic foundationalism" could, of course, point to the
success of their new theories in solving outstanding problems as evidence
that they had identified not only the true foundation of mathematics but
also the means of solution for a vast range of problems. Leibniz, for instance,
famously dreamed of a universal characteristic that could reduce any problem
whatever to a calculation, with the result that disputes could be resolved and
the sciences advanced.

8.5 Conclusion: What Became of the Dispute?

If the story I have been telling is anywhere near the truth, there was a genuine
dispute among seventeenth-century philosophers and philosophically-minded
mathematicians about the criteria of rigorous demonstration. One group,
which I have called the "geometric foundationalists," stressed the notion that
demonstrations must proceed from true causes and demanded that a truly rigorous demonstration be grounded in geometric or even physical notions such
as space, time, motion, and body. The opposing group – the "algebraic foundationalists" – deemed the proposed geometric foundation as insufficiently
pure because it was contaminated by extraneous "merely mechanical" concepts. The algebraic foundationalists sought to base all of mathematics on a
very abstract (and we might note, completely "pure") science of quantity in
general that had no admixture of any specific physical content. The objection

to this program was that its very abstractness or "purity" rendered it incapable of being the foundation of anything more than a set of rules for the manipulation of symbols.

One striking fact about this dispute is that it is no longer a topic of interest in the philosophy of mathematics. Indeed, for as much as it was a live issue in the seventeenth century, the quarrel between geometrical foundationalists and algebraic foundationalists seems to have disappeared without a trace. This raises the obvious question of why and how the dispute was eventually resolved, and what its demise can tell us about how issues in the philosophy of mathematics come to prominence and then fade away. The dispute did not die as a result of some kind of "killer argument" that one side could launch against the other. That is to say, the question was not decided on the basis of some purely technical or strictly mathematical grounds. Instead, the conflict seems to have been reformulated in a rather different struggle over how best to explicate the foundational concepts in the calculus.

In other words, I propose that the opposition between the geometric foundationalists and their algebraic counterparts was subsumed in a later opposition over what course to take in providing a solid foundation for the mathematical study of continuous variation that we now know as the calculus. By the end of the seventeenth century there were two very different approaches to the subject – on the one hand the Newtonian "method of fluxions" and on the other the Leibnizian *calculus differentialis*. The Newtonians held that the best way to understand the mathematics of continuous variation was to appeal to the experience of bodies in motion. The method of fluxions was conceived as a tool for modeling all instances of continuous change in a theory whose basic concepts (in Newton's words) "are founded upon Nature, and are every Day seen in the motion of Bodies" (Newton 1964-67, 1: 141). The Leibnizians, in constrast, conceived their *calculus differentialis* as ultimately justified by appeal to very abstract and general principles of magnitude that could extend reasoning about finite increments and ratios to the case of infinitesimal increments and their ratios. Such a science went beyond anything available in mere geometry and had no dependence on anything involving sense or imagination.

The conflict between these two ways of considering the calculus was sharpened by the nasty priority dispute between Newton and Leibniz.[8] The Newtonians denounced the scandalously lax standards of the decadent Continentals, whose leader, they claimed, had stolen the perfectly rigorous Newtonian calculus and transformed it into a bundle of self-contradictions concealed behind empty symbolic notation. The Leibnizians retorted that their British opponents were incapable of abstract thought, and were "mere empirics" who required that mathematical theories be closely tied to physical notions. By 1710, then, the earlier conflict between geometric and algebraic foundational-

---

[8] See Bertoloni-Meli (1993) on the Newton-Leibniz dispute.

ism had been re-cast as a battle between the British method of fluxions and the Continental calculus of infinitesimal sums and differences.

This suggests that one salient means whereby foundational disputes can disappear. They need not necessarily be resolved through argument and counterargument, but can be "relocated" when the relevant mathematical context changes. As worries about the proper formulation of the calculus and the rigor of its procedures came to the fore, they displaced earlier concerns about how best to understand the model of rigorous demonstration.

References

1. Aubrey, John, 1898. *Brief Lives, chiefly of contemporaries*, Oxford: Oxford University Press.
2. Barrow, Isaac, 1860. *The Mathematical Works*, ed. William Whewell, 2 vols., Cambridge: Cambridge University Press.
3. Bertoloni-Meli, Domenico, 1993. *Equivalence an Priority: Newton versus Leibniz*, Oxford: Oxford University Press.
4. Billingsley, Henry, 1570. *The Elements of Geometrie of the most ancient Philosopher Euclide of Megara*, London: John Daye.
5. Bos, Henk J. M., 2001. *Redefining Geometrical Exactness: Descartes' Transformation of the Early Modern Concept of Construction*, Berlin and New York: Springer Verlag.
6. Breger, Herbert, 1991. "Der mechanistische Denkstil in der Mathematik des 17. Jahrhunderts". In H. Hecht (ed.) *Gottfried Wilhelm Leibniz im philosophischen Diskurs über Geometrie und Erfahrung*, Berlin: Akademie-Verlag, pp. 15-46.
7. Heath, Sir Thomas L., 1931. *A Manual of Greek Mathematics*, Oxford: Oxford University Press.
8. Hobbes, Thomas, 1839-45. *Thomae Hobbes Malmesburiensis Opera Philosophica*, ed. William Molesworth, 5 vols, London: Longman.
9. Jesseph, Douglas M., 1999. *Squaring the Circle: The War between Hobbes and Wallis*, Chicago: University of Chicago Press.
10. Klein, Jacob, 1968. *Greek Mathematical Thought and the Origin of Algebra*, Cambridge, Massachusetts: M. I. T. Press.
11. Knorr, Wilbur Richard, 1981. "On the early history of axiomatics. The interaction of mathematics and philosophy in Greek antiquity". In J. Hintikka, D. Gruender, and E. Agazzi (eds.) *Theory change, ancient axiomatics, and Galileo's methodology*, Dordrecht and Boston: Kluwer, pp. 146-186.
12. Knorr, Wilbur Richard, 1986. *The Ancient Tradition of Greek Geometric Problems*, Basel, Boston, and Berlin: Birkhäuser Verlag.
13. Mahoney, Michael S., 1990. "Barrow's Mathematics: Between Ancients and Moderns", in Mordechai Feingold (ed.), *Before Newton: The Life and Times of Isaac Barrow*, Cambridge: Cambridge University Press, pp. 179-249.
14. Newton, Sir Isaac, 1964-67. *The Mathematical Works of Isaac Newton* ed. Derek T. Whiteside, 2 vols., New York and London: Johnson Reprint.
15. Newton, Sir Isaac, 1967-1986. *The Mathematical Papers of Isaac Newton*, ed. and trans. Derek T. Whiteside, 8 vols., Cambridge: Cambridge University Press.
16. Plutarch, 1949-1959. *Plutarch's Lives*, ed. and trans. Bernadette Perrin, 5 vols., Cambridge, Massachusetts: Harvard University Press.

17. Proclus Diadochus, 1970. *A Commentary on the First Book of Euclid's "Elements"*. Trans. and ed. Glen R. Morrow. Princeton: Princeton University Press.
18. Rabouin, David, 2009. *"Mathesis Universalis": L'idée de mathématique universelle d'Aristote à Descartes*, Paris: Presses Universitaires de France.
19. Raphson, David, 1702. *A Mathematical Dictionary*, London: Midwinter and Leigh.
20. Sasaki, Chikara, 2003. *Descartes's Mathematical Thought*, Dordrecht and Boston: Kluwer.
21. Smith, Robin, 1995. "Logic". In Johnathan Barnes (ed.) *The Cambridge Companion to Aristotle*, pp. 27-65. Cambridge: Cambridge University Press.
22. Wallis, John, 1655. *Elenchus Geometriae Hobbianæ. Sive, Geometricorum, quæ in ipsius Elemtis philosophiæ, Thoma Hobbes Malmesburiensi, proferuntur, refutatio*, Oxford: excudebat H. Hall, impensis Johannis Crook.
23. Wallis, John, 1693-99. *Opera Mathematica* 3 vols, Oxford: E Theatro Sheldoniano.

Chapter 9
Leibniz between *ars characteristica* and *ars inveniendi*: Unknown news about Cajori's 'master-builder of mathematical notations'

Eberhard Knobloch

Abstract Leibniz's paramount interest in these three disciplines caused him to spend his entire life trying to perfect and organize the *ars characteristica* that is the art of inventing suitable characters, signs; the *ars combinatoria* that is the art of combination; the *ars inveniendi* that is the art of inventing new theorems, new results, new methods. These arts are strongly correlated with each other. There are two famous examples for the usefulness and success of Leibniz's invention of suitable signs in order to foster the mathematical development: 1. The differential and integral calculus, 2. determinant theory. The article focusses on less well-known examples of Leibnizian inventions of mathematical symbolism related to differential equations, products of power sums, number-theoretical partitions, and elimination theory. To that end, Leibniz especially reintroduced numbers instead of letters, consciously deviating from Viète's practice.

Key words: character, differential equations, power sums, partitions, elimination theory.

9.1 Introduction

Leibniz spent his entire life trying to perfect and organize the *ars inveniendi*, the *ars characteristica*, and the *ars combinatoria*. These three arts were, in his view, inseparably connected. The methodically pursued expansion of knowledge, the *ars inveniendi*, was decisively based on the suitable choice of characters, of signs, such as those invented by the *ars characteristica*. Meanwhile, the *ars combinatoria* provided the rules according to which the characters created were to be manipulated to create new knowledge: in his eyes it was a general science that teaches a merely syntactical manipulation of signs (Krämer, 1992, p. 229). His ability of creating suitable notations made Florian Cajori call him the 'master-builder of mathematical notation' (Cajori, 1925).

This paper is meant to confirm Cajori's statement by illustrating less well known examples of Leibniz's mathematical studies. First we will clarify his concept of the *ars characteristica*. Secondly we will discuss some of the ways considered by Leibniz for writing the coefficients of differential equations, his ideas regarding the products of power sums, and finally his so-called explication theory.

9.2 *Ars characteristica* – The characteristic art

Leibniz clearly defined what he meant by the notions of character, *ars characteristica, expressio*. These definitions are worth considering. Let us have a closer look at them (Leibniz, 1688; Poser, 1979, 321):

> Characterem voco, notam visibilem cogitationes repræsentantem.

> Ars characteristica est ars ita formandi atque ordinandi characteres, ut referant cogitationes, seu ut eam inter se habeant relationem, quam cogitationes inter se habent.

> I call 'character' a visible sign representing thoughts.

> The 'characteristic art' is the art of creating and arranging characters in such a way that they reflect thoughts or that they have that relation among one another which the thoughts have among one another.

Thus characters bring about visualizations of intellectual entities. Mathematically speaking a character is a homomorphism ch that maps the set of thoughts T into the set of their representations R:

$$ch : T \to R \text{ so that for } t_1, t_2 \in T$$
$$ch(t_1 \times t_2) = ch(t_1) \times ch(t_2)$$

Leibniz continued:

> Expressio est aggregatum characterum rem quæ exprimitur repræsentantium.

> Lex expressionum hæc est: ut ex quarum rerum ideis componitur rei exprimendæ idea, ex illarum rerum characteribus componatur rei expressio.

> An expression is an aggregate of characters representing the thing that is expressed.

> The law of expressions reads as follows: the idea of a thing that has to be expressed is composed of ideas of (certain) things. The expression of the thing must be composed of the characters of those things.

In other words there are things, ideas (or thoughts) of things and representations or characters. Things are expressed, ideas are represented. Things are expressed by means of the representations of the ideas we have of these things. If an idea (of a certain thing) is composed (of ideas of certain things), the expression of this thing will be composed (of the representations of the ideas of those things).

How to create the characteristic art? Leibniz gave the answer in his "Specimen analyseos novæ", "Example of a new analysis," dating from June 1678 (Knobloch, 1980, 5):

> Artis ergo characteristicæ hæc summa regula est, ut characteres omnia exprimant, quæ in re designata latent, quod numeris, ob eorum copiam et calculandi facilitatem optime fiet. Item et in Geometria magni usus erit, ad situs exprimendos.

> Thus this is the highest rule of the characteristic art that the characters express everything which is hidden in the designated thing. This is best done by numbers because of their variety and suitability for calculation. In geometry, too, it will be very useful in order to express positions.

In the same period when Leibniz wrote these clarifications he elaborated also his "Discours de métaphysique" wherein he explained the relations between ideas and notions saying (Leibniz, 1686, 1572):

> Ainsi ces expressions qui sont dans nostre ame, soit qu'on les concoive ou non, peuvent etre appellées idées, mais celles qu'on concoit ou forme, se peuvent dire notions, conceptus.

> Therefore these expressions which are in our soul whether one conceives them or not can be called ideas. But those which one conceives or forms might be called notions, concepts.

Notions are – according to Leibniz – well understood ideas.

At this point we might resume the main results of the hitherto discussed issue: the characteristic art is an essential part of Leibniz's epistemology: it must serve the *ars inveniendi*, the art of discovering. It should make evident hidden structures, properties, relations etc. Let us consider a Leibnizian, number-theoretical example in order to illustrate the relation between hidden intellectual structures and their representations. Leibniz looked for the distribution law of prime numbers. Among others he worked out the following figure in order to discover it (Leibniz, 1676a, 597):

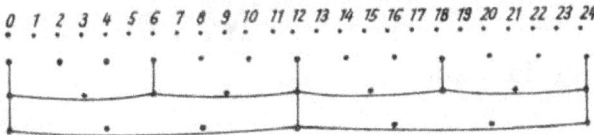

Fig. 9.1: Figure illustrating the distribution of prime numbers (LSB VII, 1, 597)

Leibniz considers the sequence of natural numbers including zero. Line by line the multiples of one, two, three, four etc. are marked by a dot. Perpendicular lines connect dots being one over the other in successive lines. Horizontal lines connect the perpendiculars. He is overwhelmed by the result saying (Leibniz, 1676a, 598):

> Patet hic illustri exemplo artem circa intelligibilia inveniendi præclara theoremata in eo consistere, ut quoniam ipsa pingi aut audiri non possunt pingamus aut audiamus earum repræsentationes, etiamsi non similes, et in iis sensibiles quasdam pulchritudines observemus, quæ in nobis facient intelligi theorema seu proprietatem ipsius rei intelligibilis.

> Here it becomes evident by an illustrious example that the art of discovering famous theorems with regard to intellectual things consists in the following: Because they themselves cannot be painted or heard, we paint or hear their representations even if they are not similar and observe some perceptible beauties in them that will produce the effect in us that a theorem or a property of the intellectual thing itself is understood.

Presumably still in April 1676 Leibniz wrote another study "De characteribus et compendiis" where he referred to this earlier paper (Leibniz, 1676b, 434):

> His recte observatis tam illos confutabimus qui credidere veritates esse sine relatione ad characteres, quam illos, qui credidere non in rebus, sed in characteribus esse veritatem: cum veritas sit in rebus quatenus ad characteres referuntur.

> If this is rightly observed we will refute those who believed that there are truths without relation to characters, and those as well who believed that truth is not in the things but in the characters while truth is in the things in so far as they are referred to characters.

Numbers are especially suitable to make evident hidden structures, properties, relations etc. Leibniz says. Why numbers? For three reasons:

1. They enable us to control calculations.
2. They can express arrangements, orders, relationships between quantities and characters.
3. They facilitate the discovery of laws of progress, continuation, harmonies, in other words they support the mathematical progress.

The following three sections are meant to illustrate this statement.

9.3 Differential equations

Leibniz invented dozens of denotations in order to characterize the relationship between coefficients and the corresponding terms of differential equations: The terms consist of powers of variables and various differentials. These studies remained unknown up to now because nearly all of them are still unpublished and only available in manuscript form. Time and again he changed his solutions and tried to improve them (Knobloch, 1982). We would like to discuss some examples that seem to be especially interesting.

In the manuscript LH 35 XIII,1 sheets 145f. Leibniz uses the following notation:

$$0 = 80dx \underset{dy}{+91xdy} \underset{ydx}{+81xdx} \underset{ydy}{+811xydx} \underset{dy}{+92x^2dy} \underset{y^2dx}{+82x^2dx} \underset{y^2dy}{+921x^2ydy} \underset{xy^2dx}{+...}$$

Leibniz's manner of writing is equivalent to:
$$0 = 80(dx + dy) + 91(xdy + ydx) + 81(xdx + ydy) + 811(xydx + xydy) +$$
$$92(x^2dy + y^2dx) + 82(x^2dx + y^2dy) + 921(x^2ydy + xy^2dx) + ...$$
He explains:

Ubi 9 præfigo, cum differentialis (*dx* vel *dy*) sequitur potentiam inferiorem, alias 8. Nempe 9 præfigo pro x^2y, quia x^2 est superior potentia, et *y* inferior, habetur autem non *dx*, sed *dy*. Ita ex solo aspectu coefficientis iudicari potest cui formæ fuerit præfixus.

There I prefix 9 when the differential (*dx* or *dy*) follows the lower power (i. e. of the variables), otherwise 8. I prefix namely 9 to x^2y because x^2 is the higher power and *y* the lower. Yet, we don't have *dx* but *dy*. Thus on the strength of a mere look at the coefficient we can judge the form it was prefixed to.

Obviously the numerical coefficients consist of two parts:

1. The first part consists of the prefixed number that characterizes the relation between the power of variables and the occurring differentials.
2. The second part consists of the exponents of the powers of variables occurring. They occur strictly symmetrically.

Leibniz says "otherwise 8": This is the negation of "the differential follows the lower power", in other words: If it is not true that the differential follows the lower power Leibniz prefixes 8. There are two possible cases in this respect: either the exponents are equal or the differential follows the higher power. For that reason Leibniz specified the notation in two different ways.

- First, improvement regarding the checking of the calculation: still in the same manuscript Leibniz introduces 'antepræfixæ' saying:

  Sed ut calculus sit tutior, adhibeamus antepræfixas, exprimentes novenarium

  residuum valoris et pro 80   91   81   811   92   82   921   821
        stabit 880 191 481 2811 792 082 6921 1821

  But in order that the calculation be more secure we would like to use 'numbers that precede the prefixed numbers'. They express the excess of nine of the value and

   instead of 80   91   82   811   92   82   921   821
  we will have 880 191 481 2811 792 082 6921 1821

The excesses of nine belong to the coefficients of another differential equation which is compared with that given above.

- Second, improvement regarding the negation of "the differential follows the lower power": in the manuscript LH 35 XIII, 2b sheet 162 prefixes three different numbers 2, 3, 4 in order to distinguish between the three different possible cases:

$$0 = +400\underset{dy}{dx} +210\underset{ydy}{xdx} +310\underset{xdy}{ydx} +411\underset{dy}{xydx} +220\underset{y^2dy}{x^2dx} +...$$

This equation is equivalent to:

$$0 = 400(dy + dx) + 210(xdx + ydy) + 310(ydx + xdy)+$$
$$411(xydx + xydy) + 220(x^2dx + y^2dy) + ...$$

Leibniz explains:

> Ubi notandum in coefficientibus seu numeris fictitiis præfixas esse 2, 3, 4...prout differentialis sequitur literam in combinatione prævalentem vel debiliorem, vel utramque cum scil. neutra prævalet.

> There it has to be remarked that, in the coefficients, fictitious numbers 2, 3, 4 are prefixed depending on whether the differential follows the stronger or the weaker variable in the combination or whether it follows both of them when none of them is stronger.

There are plenty of other notations. I would like to restrict myself to discuss the following three:

- In 1691 Leibniz writes: $001dx+002dy+101xdx+011ydx+102xdy+012ydy+201x^2dx + 021yydx + 111xydx + 202x^2dy... = 0$ (LH 35 VI,2 sheets 1 + 7). This time the coefficients of symmetric terms are different. The first two figures represent the exponents of the powers of the variables x or y. The third, last, figure indicates by 1 or 2 respectively whether the first differential dx or the second differential dy occurs.
- In the manuscript LH 35 XIII,1 sheet 308 Leibniz considers the two equations:

$$100v + 200z + 101dv + 201dz + 120a = 0$$
$$300v + 400z + 301dv + 401dz + 340a = 0$$

The first figure indicates the occurring variable v or z. To that end Leibniz uses different figures 1, 2 or 3, 4 in different equations for the same variables v, z. The third figure indicates by 0 or 1 whether a differential occurs or not,

the first figure indicates which of the two variables is meant. The meaning of the second figure remains unclear: the constant term a is characterized by 1 and 2 or 3 and 4, respectively that occupy the first two places.
- In the manuscript LH 35 XIII,2b sheets 75f. Leibniz needs a system of two equations with three variables x, y, v:

$$110y + 111dy + 120v + 121dv + 100dx = 0$$
$$210y + 211dy + 220v + 221dv + 200dx = 0$$

The first figures 1 or 2 indicate the equation, the second figures 0, 1, 2 indicate the variables, the third figures 0, 1 indicate whether the corresponding differential occurs or not. Yet, this interpretation only applies to dy and dv; dx is characterized by two 0 occupying the third place.

The last two notations do not seem to be satisfactory. No wonder that Leibniz did not stop experimenting in looking for suitable notations.

9.4 Products of power sums

Leibniz was deeply interested in the theory of symmetric functions, especially in power sums.

Since about 1677 to 1679 he was able to represent them by means of the elementary symmetric functions. After writing down these representations of the first nine power sums he remarked (Knobloch, 1976, 195):

Nullam unquam Tabulam numerorum vidi ex qua plura Mysteria pulcherrima duxerim.

I never saw a table of numbers from which I drew more mysteries of the highest beauty.

During the last decade of his life he corresponded with Theobald Overbeck, school-master in Wolfenbüttel, about this subject. In 1714 they mainly dealt with two questions: with the product of power sums and with the reduction of multiform symmetric functions to power sums. Only thanks to suitable notations the close connection of these questions with number-theoretical partitions and their hidden combinatorial structure became evident (Knobloch, 1976, 236-253). Obviously Overbeck elaborated a copy of Leibniz's results.

Let a, b, c etc. be pairwise different variables. Leibniz indicated a power sum $a^m + b^m + c^m + \ldots$ or in general a symmetric function by the first term a^m, putting dots under the term. For technical reasons I replace the dots by $\sum$: $\sum a^m := a^m + b^m + c^m + \ldots$

Step by step he calculated the product of two, three, four, etc. power sums. Then he used an abridged and eventually another abridged notation in order to represent the results. Let us consider the product of three power sums:

$$\left(\sum a^m\right)\left(\sum a^n\right)\left(\sum a^p\right) = \sum a^{m+n+p} + \sum a^{m+n}b^p$$
$$+ \sum a^{m+p}b^n$$
$$+ \sum a^{n+p}b^m + \sum a^m b^n c^p$$

The first abridged representation leaves aside the bases a, b, c etc. because they can be added again in their alphabetical order. Only the combinations of exponents are kept:

$$\left(\sum a^m\right)\left(\sum a^n\right)\left(\sum a^p\right) = mnp + mn \mid p$$
$$+ mp \mid n$$
$$+ np \mid m + m \mid n \mid p$$

The second abridged representation counts the exponents and describes the combinations as number-theoretical partitions:

$$= 3 + ③\, 2 \mid 1 + 1 \mid 1 \mid 1$$

The encircled 3 counts the frequency of the occurrence of the same partition. This special result can be generalized. In order to calculate the coefficients within parentheses the notation makes evident that Leibniz had to solve the following problem:

Let be X a set of k objects. There are N possibilities to distribute them into $d = r_1 + r_2 + ... + r_k$ boxes so that r_1 boxes contain exactly one object, r_2 boxes contain exactly two objects,..., r_k boxes contain exactly k objects. Let $k = 1r_1 + 2r_2 + ... + kr_k$

$$N = \frac{k!}{(1!)^{r_1} \ldots (k!)^{r_k} r_1! \ldots r_k!}$$

The first k powers of the denominator correspond to the fact that we have to suppress all $\nu!$ permutations of exponents that are represented by ν ($\nu = 1, 2, ..., k$). For $a^{m+n} = a^{n+m}$ so that $nm \mid p \mid q$ or $2.1.1$ is the same as $mn \mid p \mid q$.

The second k factors of the denominator take into account the fact that we have to suppress all $r_\nu!$ permutations which are produced by the r_ν sections of the exponent of the same length: $mn \mid q \mid p$ cannot be distinguished from $mn \mid p \mid q$ because of the symmetry of the function.

The second main problem was the reduction of multiform symmetric functions $\sum a^m b^n c^p$ etc. to power sums. In order to solve this problem Leibniz had only to transform the equations of the products of power sums.

Leibniz knew that

$$\left(\sum a^m\right)\left(\sum a^n\right) = \sum a^{m+n} + \sum a^m b^n$$

Hence

$$\sum a^m b^n = \left(\sum a^m\right)\left(\sum a^n\right) - \sum a^{m+n} \tag{9.1}$$

This equation is multiplied by $\sum a^p$, the emerging symmetric functions consisting of two variables are eliminated by means of 9.1 etc. Step by step Leibniz reduced the representations of multiform symmetric functions to power sums again using the two forms of abridgements. For example:

$$\sum a^m b^n c^p d^q = m \mid n \mid p \mid q - mn \mid p \mid q + 2mnp \mid q + mn \mid pq - 6mnpq$$
$$= 1.1.1.1. - 2.1.1. + ② \, 3.1. + 2.2. - ⑥ \, 4.$$

It becomes evident that in the general case of k variables all possible partitions of k have to be enumerated. If $k = 4$, there are five partitions of k namely 1.1.1.1, 2.1.1, 3.1, 2.2, 4. How to find the coefficients and their signs? Let $k = 1r_1 + 2r_2 + ... + kr_k$ be an arbitrary partition of k. The coefficient of k reads:

$$c = (1!)^{r_2}(2!)^{r_3}(3!)^{r_4} \ldots ((k-1)!)^{r_k}$$

The sign rule reads: If there is an even number of even numbers, we have to take $+$, if there is an odd number of even numbers, we have to take $-$.

Let us consider the example discussed above. The last (fifth) partition of 4 is 4. Therefore $k = 4 = 4.1 = kr_k$ or the coefficient is $(3!)^1 = 6$. There is an odd number (that is, one) of even numbers. Therefore we have to take the sign $-$ or the coefficient -6.

Leibniz's notation revealed the relation between the algebraic problem of reducing certain symmetric functions to power sums and the number-theoretical problem of finding the partitions of natural numbers. The formulae he had derived are called after Edward Waring because Waring published them in 1762 and a second time in 1770 (Waring, 1762, pp. 6-8; Waring, 1770, pp. 7-10).

9.5 Leibniz's explication theory dating from 1693/94

A third example for the utility of numbers in order to discover hidden mathematical relationships is Leibniz's elimination theory. A special case of this theory is his explication theory that he elaborated in the years 1693 and 1694 (Knobloch, 1974). About fifty to seventy years later Euler occupied himself with the same problem without using fictitious numbers as coefficients (Euler, 1748 vol. 2, pp. 259-271; Euler, 1750; Euler, 1766).

The aim is the elimination of a common unknown of two polynomials:

$$10x^3 + 11x^2 + 12x + 13 = 0$$
$$20x^3 + 21x^2 + 22x + 23 = 0$$

Leibniz used double subscripts as fictitious numbers: The first figure denotes the equation, the second the power the coefficient belongs to. In order to reduce these two cubic equations to two quadratic equations he needed two steps:

1. The first equation is multiplied by 20, the second by 10, the multiplied first equation is subtracted from the multiplied second equation.
2. The first equation is multiplied by 23, the second by 13, the multiplied second equation is subtracted from the multiplied first equation.

Thus we get:

$$
\begin{array}{lll}
10.21\ x^2 & +10.22\ x & +10.23 \qquad \text{or } (20)x^2 + (21)x + (22) = 0\\
-11.20 & -12.20 & -13.20 = 0\\
10.23\ x^2 & +11.23\ x & +12.23 \qquad \text{or } (10)x^2 + (11)x + (12) = 0\\
-13.20 & -13.21 & -13.20 = 0
\end{array}
$$

The original coefficients – now put between parentheses – are 'explicated' as Leibniz called this procedure: $(20) = 10.21 - 11.20, (21) = 10.22 - 12.20, (22) = 10.23 - 13.20, (10) = 10.23 - 13.20$ etc. It has to be repeated until one gets the resultant sought.

Leibniz elaborated a 'tabula dichotomica', a dichotomic table for these repeated explications of the double subscripts with 2 as first figure:

etc.

If one continuously substitutes the explications for the foregoing double subscripts 22, 21, 20, one gets Leibniz's dichotomic table (Knobloch, 1974, pp. 162f.):

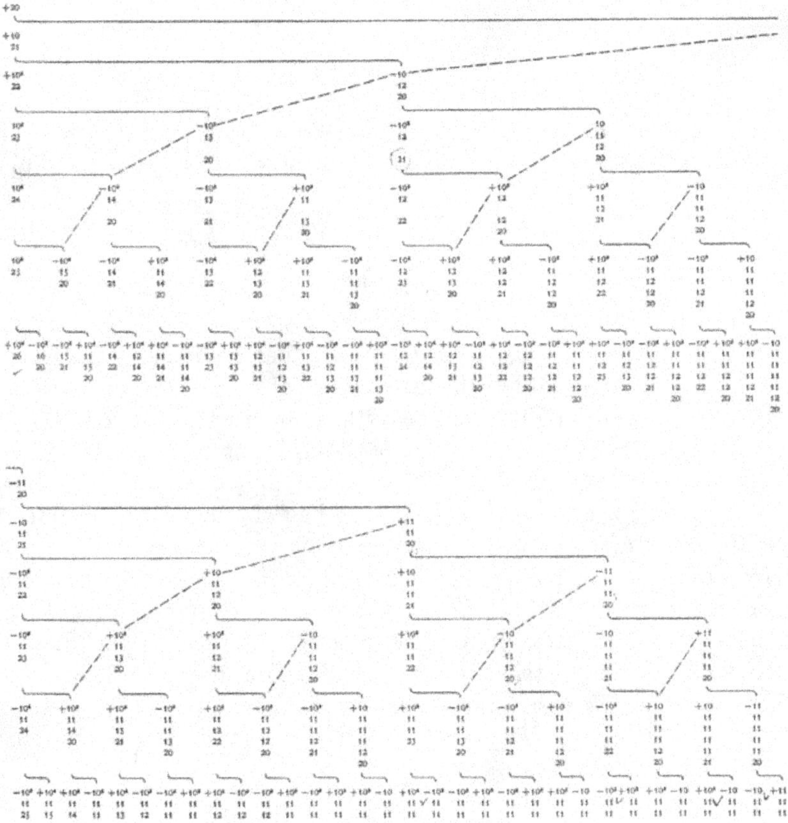

Fig. 9.2: Leibniz's dichotomic table regarding elimination theory (Knobloch, 1974, 162f.)

Eventually the series of terms of a certain line of this table directly provides the resultant looked for. It might be instructive to compare Euler's calculation and Leibniz's notation in this respect. To that end let us take a linear equation and an equation of fourth degree. Euler wrote (Euler, 1748 vol. II, p. 261):

$$P + Qy = 0$$
$$P + qy + ryy + sy^3 + ty^4 = 0$$
$$Q^4 p - PQ^3 q + PPQQr - P^3 Qs + P^4 t = 0$$

Leibniz wrote:

$$10x + 11 = 0$$
$$20x^4 + 21x^3 + 22x^2 + 23x + 24 = 0$$

The fourth line of explication provides the resultant ($12 = 13 = 14 = 0$):

$$10^4.24 - 10^3.11.23 + 10^2.11^2.22 - 10.11^3.21 + 11^4.20 = 0$$

What is the hidden structure of this table? Leibniz described it by means of a special notation:

- In every line the capitals A, B, C, D, etc. denote domains that are the second half, the second fourth, the second eighth, the second sixteenth of the whole set of terms. The capitals within parentheses (A), (B), (C), (D), etc. denote series based on a further bisection ('subbisectio') of the domains A, B, C, D, etc.:
- The series (A) is contained in the second half of a line. There we take the first term of the second half, of the second fourth, of the second eighth, etc. The series (B) is contained in the second fourth. There we take the first term of the second half, of the second fourth, of the second eighth, etc.
- Combinations of capitals have to be interpreted from the right to the left. The series $(A)A$ is contained in the second half of the second half of a line. There we take the first term of the second half, of the second fourth, of the second eighth, etc. The series $(B)AB$ is contained in the second fourth of the second half of the second fourth of a line. There we take the first term of the second half, of the second fourth, of the second eighth, etc. Let us take examples from the seventh line that begins with $+10^6.26$. The following 'series' consist of only one term:

$$(D)A = -10^3.11.11.14.20$$
$$(C)AA = +10^2.11.11.11.13.20$$
$$(B)AAA = -10.11.11.11.11.12.20$$

34 corollaries describe peculiarities of this table. Yet, Leibniz neither mentioned nor proved the following most interesting theorem:
Any permutation of the capitals (with or without parentheses) of such a combination of capitals represents the same series.
Without Leibniz's notation this theorem cannot be even formulated.

9.6 Epilogue

In October 1674 Leibniz wrote his life maxim:

Malo enim bis idem facere, quam semel nihil (LSB VII, 3, p. 539)
For I prefer to do the same twice instead of doing nothing once.

References

Abbreviations: LH = Leibniz-Handschriften, cited according to Bodemann's catalogue (Bodemann, 1966), LSB = Gottfried Wilhelm Leibniz, *Sämtliche Schriften und Briefe*, herausgegeben von der Preußischen Akademie der Wissenschaften, now von der Berlin-Brandenburgischen Akademie der Wissenschaften und der Akademie der Wissenschaften in Göttingen, since 1923. LSB VI, 4B means: Series VI, volume 4B.

1. Bodemann, Eduard, 1966. *Die Leibniz-Handschriften der Königlichen öffentlichen Bibliothek zu Hannover*. Mit Ergänzungen und Register von Gisela Krönert und Heinrich Lackmann, sowie einem Vorwort von Karl-Heinz Weimann. Hildesheim: Olms (Reprografischer Nachdruck der Ausgabe Hannover 1889).
2. Cajori, Florian, 1925. "Leibniz, The master-builder of mathematical notation". *Isis* 7: pp. 412-429.
3. Euler, Leonhard, 1748. *Introductio in Analysin infinitorum*, 2 vols. Lausanne: Bousquet (= *Opera omnia* vols. I, 8 and I,9). [I cite the edition of the Opera omnia]
4. Euler, Leonhard, 1750. "Démonstration sur le nombre des points où deux lignes des ordres quelconques peuvent se couper. Mémoires de l'Académie des sciences de Berlin" 4 (1748), pp. 234-248 (= *Opera omnia* vol. I, 26, pp. 46-59).
5. Euler, Leonhard, 1766. "Nouvelle méthode d'éliminer les quantités inconnues des équations" 20 (1764), pp. 91-104 (= *Opera omnia* vol. I, 6, pp. 197-211).
6. Knobloch, Eberhard, 1974. "Unbekannte Studien von Leibniz zur Eliminations- und Explikationstheorie". *Archive for History of Exact Sciences* 12: pp. 142-173.
7. Knobloch, Eberhard, (ed.) 1976. "Die mathematischen Studien von G. W. Leibniz zur Kombinatorik". Textband. *Studia Leibnitiana Supplementa* XVI. Wiesbaden: Steiner.
8. Knobloch, Eberhard, (ed.) 1980. "Der Beginn der Determinantentheorie, Leibnizens nachgelassene Studien zum Determinantenkalkül". *Arbor scientiarum* Reihe B, Band II. Hildesheim: Gerstenberg.
9. Knobloch, Eberhard, 1982. "Zur Vorgeschichte der Determinantentheorie" in *Theoria cum praxi, Zum Verhältnis von Theorie und Praxis im 17. und 18. Jahrhundert, Akten des III. Internationalen Leibnizkongresses Hannover, 12. bis 17. November 1977*, Bd. IV Naturwissenschaft, Technik, Medizin, Mathematik. *Studia Leibnitiana Supplementa* XXII. Wiesbaden: Steiner, pp. 96-118.
10. Krämer, Sybille, 1992. "Symbolische Erkenntnis bei Leibniz". *Zeitschrift für philosophische Forschung* 46, pp. 224-237.
11. Leibniz, Gottfried Wilhelm, 1676a. *De natura numerorum primorum et in genere multiplorum*. LSB VII,1, 594-598.
12. Leibniz, Gottfried Wilhelm, 1676b. *De characteribus et compendiis*. LSB VI,3,433-434.
13. Leibniz, Gottfried Wilhelm, 1686. *Discours de métaphysique*. LSB VI, 4B, 1529-1588.
14. Leibniz, Gottfried Wilhelm, 1688. *De characteribus et de arte characteristica*. LSB VI, 4A, 916.
15. Poser, Hans, 1979. "Signum, notio und idea – Elemente der Leibniz'schen Zeichentheorie". *Zeitschrift für Semiotik* 1, pp. 309-324.

16. Poser, Hans, 1979. "Erfahrung und Essenz – Zur Stellung der kontingenten Wahrheiten in Leibniz' ars characteristica". In A. Heinekamp, Fr. Schupp (eds.): *Die intensionale Logik bei Leibniz und in der Gegenwart*, Symposium der Leibniz-Gesellschaft Hannover, 10. und 11. November 1978. Studia Leibnitiana Sonderheft 8. Wiesbaden: Steiner, pp. 67-81.

17. Waring, Edward, 1762. *Miscellanea analytica de æquationibus algebraicis, et curvarum proprietatibus*. Cambridge: J. Bentham.

18. Waring, Edward, 1770. *Meditationes algebraicæ*. Cambridge: J. Bentham.

www.ingramcontent.com/pod-product-compliance
Lightning Source LLC
Chambersburg PA
CBHW061135220326
41599CB00025B/4241